AF749419

3 0104 00924658 2

ECOLOGICAL PRINCIPLES OF
NATURE CONSERVATION

CONSERVATION ECOLOGY SERIES: Principles, Practices and Management

Series Editor:

Dr E. Duffey
Cergne House, Church Street, Wadenhoe, Peterborough PE8 5ST, UK

Editorial Board

Professor L. Hansson (*Sweden*)
Professor W. Erz (*Germany*)
Professor P. Blondin (*France*)
Dr S. Hopper (*Australia*)

Titles published in the Series:

Ecological Principles of Nature Conservation
L. Hansson

Further titles are in preparation.

ECOLOGICAL PRINCIPLES OF NATURE CONSERVATION

Applications in Temperate and Boreal Environments

Edited by

LENNART HANSSON

Department of Wildlife Ecology, Swedish University of Agricultural Sciences, Uppsala, Sweden

UNIVERSITY COLLEGE STOCKTON LIBRARY
BOOK No. 42044944
PREFIX
CLASS 639.9
CODE Nos. HAN
FORM
DEPT 63
TF 4/94
HEAD 1
CAT S.Y.
CHECK TW

ELSEVIER APPLIED SCIENCE

LONDON and NEW YORK

ELSEVIER SCIENCE PUBLISHERS LTD
Crown House, Linton Road, Barking, Essex IG11 8JU, England

Sole distributor in the USA and Canada
ELSEVIER SCIENCE PUBLISHING CO., INC.
655 Avenue of the Americas, New York, NY 10010, USA

WITH 32 TABLES AND 72 ILLUSTRATIONS

© 1992 ELSEVIER SCIENCE PUBLISHERS LTD

British Library Cataloguing in Publication Data

The ecological principles of nature conservation.
I. Hansson, Lennart
639.9

ISBN 1851667180

Library of Congress Cataloging-in-Publication Data

The Ecological principles of nature conservation/edited by Lennart Hansson.
p. cm.—(Conservation ecology series)
Includes bibliographical references and index.
ISBN 1-85166-718-0
1. Nature conservation. 2. Ecology. I. Hansson, Lennart, 1938–
II. Series.
QH75.E27 1992
333.95'16—dc20

91-26208
CIP

639.9
HAN

No responsibility is assumed by the Publisher for any injury and/or damage to persons or property as a matter of products liability, negligence or otherwise, or from any use or operation of any methods, products, instructions or ideas contained in the material herein.

Special regulations for readers in the USA
This publication has been registered with the Copyright Clearance Center Inc. (CCC), Salem, Massachusetts. Information can be obtained from the CCC about conditions under which photocopies of parts of this publication may be made in the USA. All other copyright questions, including photocopying outside the USA, should be referred to the publisher.

All rights reserved. No part of this publication may be reproduced, stored in a retrieval system, or transmitted in any form or by any means, electronic, mechanical, photocopying, recording, or otherwise, without the prior written permission of the publisher.

Photoset by Interprint Ltd. Malta.
Printed in Great Britain at the University Press, Cambridge.

Preface

This volume is the first in a series entitled *Conservation Ecology: Principles, Practices and Management*, a theme which Elsevier's pioneering journal *Biological Conservation* has promoted since its foundation thirty-three years ago. The science of conservation ecology is now widely acknowledged as an essential component in the planning and development of activities which change or modify our natural environment. Nevertheless in spite of much research and publicity, there is still a wide gap between theory and practice. Today it is especially important to try to bridge this gap by interpreting the results of ecological research so that they are understandable and relevant to a wide range of land managers, agriculturalists, foresters, and those working in the many categories of protected areas. The volumes in this series are designed to fulfil this purpose, and also to play an important educational role for students of the environmental sciences in schools, universities and other institutions.

In most industrial countries the public is becoming acutely aware of the demands made on our rural landscape by intensive agriculture and forestry, new roads, urban expansion and industrial development with the consequent loss of habitat diversity, wildlife, hedgerows, wetlands, ancient woodland and meadows. Ecologists are now considering how these trends can be reversed and the research required to recreate a varied and attractive landscape including the restoration of native woodlands, meadows, heaths and other wildlife communities. The *Conservation Ecology* series will review the available information and show how it can be applied to current problems in nature conservation.

When the proposed series was first discussed with the Editorial Board we were delighted when Professor L. Hansson (through a research team of the Swedish Environmental Protection Agency of the Agricultural University of Uppsala) offered us an introductory volume on the general principles of conservation ecology. The twelve chapters cover a wide range of environmental disciplines and present much new scientific data in the examination of ecological concepts relevant to wildlife and landscape management. Together they form an excellent basis for the succeeding volumes in the series, which will discuss conservation priorities and management of major ecosystems and landforms.

The principles discussed in this first volume are widely applicable in many parts of the world, especially temperate and boreal regions. On the other hand, the Scandinavian viewpoint is of particular interest as it provides new ideas on well-known concepts based on a long-established tradition of ecological research applicable to nature conservation. Many of the case studies described will be new to readers in other countries—for example the preservation needs of the faunas and floras of the extensive Scandinavian archipelagoes and boreal forests. Professor Hansson and his colleagues are to be congratulated on the high standard of science achieved.

The *Conservation Ecology: Principles, Practices and Management* series has five other titles in preparation dealing with marine conservation, recreational ecology, management of freshwater ecosystems, and the problems of wetland conservation and coastal formations. The ultimate aim of the series is to cover a comprehensive range of ecosystems throughout the world where sufficient ecological knowledge is available to discuss the scientific basis for conservation management.

Eric Duffey

Acknowledgements

This book is largely a result of a fairly comprehensive research programme at the Swedish Environmental Protection Agency on 'Remnant Habitats in Production Landscapes'. Most of the present authors have been conducting research projects with grants from this Agency. We are very grateful for the continuing support from the Agency and its research officers, and especially thank Tor-Björn Larsson for initiatives and seminars leading towards this synthesis.

We thank Elsevier Science Publishers, and especially Eric Duffey for their great interest in our project and their rapid and exhaustive replies to our queries. We are impressed by the swift and careful subediting of our chapters by Eric and Rita Duffey. As English is a necessary yet foreign language for all of us, they have performed a considerable task in amending incorrect wording or plain abuse of their language. We admire the patient discussions of terms and concepts that took place during this subediting. In addition, we also received many hints about unknown papers and alternative interpretations.

Lennart Hansson
and all authors

List of Contributors

P. ANGELSTAM
Swedish Environmental Protection Agency, Grimsö Wildlife Research Station, S-77031 Riddarhyttan, Sweden

S. ÅS
Department of Zoology, Uppsala University, P.O. Box 561, S-751 22 Uppsala, Sweden

J. BENGTSSON
Department of Ecology and Environmental Research, Swedish University of Agricultural Sciences, P.O. Box 7072, S-750 07 Uppsala, Sweden

T. EBENHARD
Department of Zoology, Uppsala University, P.O. Box 561, S-751 22 Uppsala, Sweden

B. EHNSTRÖM
Department of Plant and Forest Protection, Swedish University of Agricultural Sciences, S-750 07 Uppsala, Sweden

L. ERICSON
Department of Ecological Botany, University of Umeå, S-901 87 Umeå, Sweden

P.-A. Esseen
Department of Ecological Botany, University of Umeå, S-901 87 Umeå, Sweden

L. HANSSON
Department of Wildlife Ecology, Swedish University of Agricultural Sciences, P.O. Box 7002, S-750 07 Uppsala, Sweden

O. JENNERSTEN
Department of Zoology, Uppsala University, P.O. Box 561, S-751 22 Uppsala, Sweden

J. Loman
Department of Ecology, Lund University, Ecology Building, S-223 62 Lund, Sweden

A.P. Møller
Department of Zoology, Uppsala University, P.O. Box 561, S-751 22 Uppsala, Sweden

C. Nilsson
Department of Ecological Botany, University of Umeå, S-901 87 Umeå, Sweden

S.G. Nilsson
Department of Ecology, Ecology Building, S-223 62 Lund, Sweden

J. Robertson
Department of Wildlife Ecology, Swedish University of Agricultural Sciences, P.O. Box 7002, S-750 07 Uppsala, Sweden

K. Sjöberg
Department of Wildlife Ecology, Swedish University of Agricultural Sciences, S-901 87 Umeå, Sweden

L. Söderström
Department of Ecological Botany, University of Umeå, S-90183 Umeå, Sweden

C. Solbreck
Department of Plant and Forest Protection, Swedish University of Agricultural Sciences, S-750 07 Uppsala, Sweden

L. Svensson
Department of Systematic Botany, University of Lund, Ö. Vallgatan 18–20, S-223 61 Lund, Sweden

B. Widén
Department of Systematic Botany, University of Lund, Ö. Vallgatan 18–20, S-223 61 Lund, Sweden

Contents

Chapter 1

Introduction: Applied Ecological Principles and their Implementation in Nature Conservation

LENNART HANSSON

Department of Wildlife Ecology, Swedish University of Agricultural Sciences, S-750 07 Uppsala, Sweden

Ecology is a young science. It was defined by Haeckel (1869) as the total relationship between animals (or organisms) and their environment, including other organisms. However, it was not generally accepted as a separate discipline until well into the 20th century and did not reach its present status until the 1960s and 1970s. Professionals within ecology have largely emerged from the 18th–19th century tradition of natural history, both internationally (McIntosh, 1985) and in Scandinavia (Söderqvist, 1986). They also have often had a deep feeling for nature and nature protection. During recent decades the nature conservation movement has attracted another type of ecologist, more interested in theoretical ideas and especially in their mathematical aspects, and who may be pursuing such studies more as an intellectual challenge.

It is those people with a genuine interest in the natural history of plants and animals who above all have argued for the protection of species and habitats, and who have joined voluntary and official organizations for nature conservation. The tradition of natural history embraces an accumulation of data on distribution, numbers, habits and habitats of plant and animal species, while less interest is devoted to the relationships between isolated observations. Causes and effects are, if considered at all, often just guessed at. In this tradition it is not customary to make integrated generalizations, but to 'let the data speak for themselves'. The new, theoretically inclined, ecologists have attempted to contribute in the form of very general ideas derived from evolution, genetics, population dynamics and abstract mathematical reasoning. The resulting predictions are therefore very generalized, and usually not applicable to individual species or communities. In addition they often have a considerable degree of uncertainty due to variability in the relevance of their basic assumptions, and so are difficult to apply in the local context, frequently subject to unique conditions. Certain recent international handbooks on nature conservation (e.g. Schonewald-Cox *et al.*, 1983; Soulé, 1986, 1987) are largely based on such general ideas. They do, however, provide hypotheses, and sometimes more reliable principles, on causal effects which make data gathered by naturalists more useful.

Evaluation of these theoretical deductions should be performed with care. This can be readily illustrated from another area of applied ecological research, i.e. game harvesting, where such general principles were suggested earlier than for nature conservation. Certain deductions from the theoretical logistic equation for population growth were used to define sustainable and maximum sustainable yield for hunted populations (general description in Caughley, 1977). However, these suggestions were hardly ever implemented in practical game management, being either little known or rejected by game managers and sportsmen. Later evaluations (Beddington & May, 1977; Caughley, 1982) have also claimed that they would have caused population disasters and possible extinctions if strictly practised. Similarly, general theoretical ideas for conservation, useful as they may be, have to be evaluated in regional (e.g. Scandinavian) and local contexts (e.g. for separate species or nature reserves).

A well-known example of applying ecological ideas to habitat conservation comes from the island biogeographic theory developed by MacArthur and Wilson (1967). One principal deduction shows, from a presumed equilibrium between immigration and extinction rates on islands with lognormal species distributions, that the number of species on an island increases as a power function of island area, with the exponent around 0·3. This relationship was in the first instance derived for real islands but has been used in nature conservation discussions for devising size and configuration of habitat islands in terrestrial systems, in order to preserve a maximum number of species. It has also been a basis for the SLOSS controversy (Diamond, 1975 and May, 1975 vs Simberloff & Abele, 1976 and Higgs, 1981, and Scandinavian examples in Lahti &

Ranta, 1985) with regard to whether a Single Large Or Several Small reserves will best protect most species. However, later studies and evaluations have demonstrated that the suitability of the surrounding habitat and the types of species considered are very important for the final outcome (Soulé & Simberloff, 1986; Boecklen & Bell, 1987). For instance, certain communities and habitats (springs, caves) only occur as small patches, and species from such areas are better considered in relation to dynamics within and between archipelagoes (metapopulation dynamics) than to island size. In this book we place emphasis on general principles but many additional or locally relevant factors will be taken into consideration when applying the principles to particular species or communities.

General principles applicable to nature conservation have thus been derived from population genetics, population dynamics and community ecology. They are applicable, respectively, to the genetical integrity of local populations, preservation of viable populations of individual species and requirements with regard to habitat and landscape composition for natural communities. The scientific ideas and principles which emerge will be considered, together with regionally and locally modifying factors, and representative examples will be given from recent research in Scandinavia.

Nature conservation has strong support in Scandinavia. Usually it is connected with a desire for clean air and pure water and expressed in a context of preserving 'ecological relationships'. However, such proposals, even if often favoured politically, are poorly defined and difficult to satisfy as regards nature reserves and fauna and flora. Nevertheless, conservation managers turn to ecologists for research and advice regarding the 'preservation' of nature. This book is thus concerned with the types of advice that scientific ecologists can provide.

The maintenance of functioning communities is considered to be a top priority in conservation (cf. Janzen, 1983) and our survey will start with mechanisms which threaten community integrity, often originating from the surroundings of valuable or protected areas (chapter 2). We recognize also the importance of conserving individual taxa, especially when functioning as 'keystone species' for subcommunities, or when representative of particular environments ('indicator species'), or with a large part of their world populations in regional, e.g. Scandinavian, habitats. Thus, we shall try to determine what kind of species are especially vulnerable and under what conditions (chapter 3). The importance of inbreeding and genetic drift in small populations has been stressed in recent handbooks on nature conservation, but mainly for animals. Here it is examined particularly for plants (chapter 4), but certain conclusions may be generally applicable. Dispersal of both plants and animals is very important in fragmented habitats and this topic, and its implications for conservation, are discussed (chapter 5). Principles derived from studies of island biogeography and metapopulation dynamics are examined in the context of habitat islands in the final chapter, on ecological mechanisms.

Handbooks on nature management of regional biomes are starting to appear (e.g. Saunders *et al.*, 1987 on Australian habitats). This book is the result of work by Swedish ecologists engaged for many years in conservation research, especially on remnants of communities and ecosystems which occurred more commonly in the past. However, research has embraced both general ecological ideas related to species and community persistence and practical application in various environments, and may also be germane to many areas outside Sweden—the ecological principles in a general context, and the suggestions for habitat management more particularly relevant to most other boreal and temperate regions. The northern holarctic region is globally important for conservation, still providing large expanses of fairly natural habitats with a sparse human population (McCloskey & Spalding, 1989).

In order to provide a setting for the work on which the ecological principles are based, and their present application (chapters 7–11), we give below a short description of the geography and habitats of Scandinavia.

The climate of Scandinavia is comparatively humid and mild, with large amounts of snow in winter due to the ameliorating effects of the Gulf Stream, which influences most Scandinavian environments (Sjörs, 1965). Both forests and humans occur more extensively at higher latitudes than anywhere else in the world. This may be the most pronounced feature to consider when making comparisons with other boreal and temperate areas. As in other mainly boreal areas, nature here is very young (e.g. Nilsson, 1972). The Pleistocene ice did not start to withdraw from Scandinavia until some 15 000 years BP and the last remnants had not disappeared until 8000–9000 BP. During several thousand years Scadinavia was covered by tundra environments, where humans mainly hunted reindeer for a living. The boreal taiga forest as we now know it, with its dominance of Norway spruce, has not existed in Scadinavia for more than three or four thousand years (e.g. Sjörs, 1956)—less in many western and southern areas due to the recent immigration by the spruce. Similarly, beech and hornbeam forests are as young, or even younger, in south Scandinavia. Thus, man has a much longer history in Scandinavia than many of today's boreal environments, including the dominant forests. This means that human influence has been profound, at least in the more heavily populated parts.

Yet the human population has been very unevenly distributed in Scandinavia. Since neolithic times, at least 5000 years BP, agriculture has been fairly common in Denmark and southern Sweden, along the Norwegian coast beyond the Polar Circle, and also along the coast of the Baltic Sea. It has also been fairly widespread in Finland (slash-and-burn agriculture) but rare in Sweden and Norway above the Highest Coast Line with its hard-tilled moraine soils. Even so, some immigrant Finns were modifying these areas as early as the 16th to 17th centuries. Sediment soils below the HCL were used at an early stage for cultivation, while the interior parts of Scandinavia were visited for several thousand years only by hunters and fishermen, e.g. by the Sami people. However, even hunters may have affected the forest and its fauna and flora by their activities, including the use of fire for attracting or driving game (Komarek, 1971). Later, the Sami turned to reindeer herding, further influencing both forests and mountain tundra. In the 17th century coastal agriculturalists began to move into the vast northern forests. There they mainly used mires and other wetlands for pastures and cultivation but also strongly influenced forests by timber cutting, cattle grazing and hunting. In keeping with the very rapid rise in the human population in the 19th century most exploitable (and many impossible) parts of the southern forest and heath areas, earlier grazed, were used for agriculture, and the pressure on land also increased in northern Scandinavia.

Due to large-scale emigration to North America, technical advances, the common use of fertilizers and general urbanization, distant agricultural sites were abandoned in the 20th century and agriculture became concentrated on the most productive areas in southern and south-central Scandinavia. The human population of the countryside showed a pronounced decline in 1950–60 and new changes in the same direction are now close at hand. As a consequence, the area of forest has at one time been retracting and then increasing, and now makes up 63% of the land area of Sweden (lakes constituting 8% of the total geographic area). Around 12% of the land area is mire and wetlands, 12% mountain heathland, 9% agricultural land and 4% urban and related environments (Skogsdata 88, 1988). Further changes may appear in the not-too-distant future: forest damage by acidification or pollution has been reported from southwestern Scandinavia for some years and now seems also to occur in northern Sweden, and people are starting to discuss the landscape effects of global heating... However, pollution problems will be considered only marginally in this book.

These human-induced changes raise a particularly important question concerning conservation: What should the natural state of Scandinavia be? Seemingly unmodified environments may only be found in inaccessible cliffs but even there the fauna of large predators, both carnivores and raptors, has been changed, so that original (or 'indigenous') communities do not occur. There should thus be a time horizon for nature conservation: what stage in the historical development of nature should be restored? Many conservationists see the time before extensive agriculture as a point of departure. However, such a point of time hardly existed for southern Scandinavia if the present common spruce, beech and hornbeam forests are considered! Some people argue that fairly unaffected forests should be left to develop spontaneously, but it is unclear what they will finally represent in such a region.

The main terrestrial habitats in Sweden are described and analysed with regard to conservation requirements and implementation in chapters 7 to 11. The main emphasis is placed on the boreal forests, which makes up the dominant part of the land area (chapter 7). However, the boreal conifer forests are by no means homogeneous: earlier they consisted of intermingled seres of secondary successional stages and climax forests, with new successions usually being started by extensive forest fires. Now modern forestry is creating a chequerboard pattern of various age classes intermingled with mires which still cover very large parts of Finland and Sweden (chapter 8). Forested wetlands, and the numerous lakes, rivers and streams, are surrounded by especially productive and species-rich riparian shrublands and forests (chapter 9). The multitude of forested islands in the Baltic may be a unique feature of Finland and Sweden and have attracted an intensive biogeographic research effort with obvious applications in conservation (chapter 6). Southernmost Sweden and Denmark belong to the temperate zone with nemoral forests but there is also a transition zone with mixed deciduous and coniferous stands. The southern conditions are described in chapter 10 with analyses of the particularly difficult conservation problems of the area. Finally, comparisons of the traditional and modern agricultural areas are presented in chapter 11 together with research on the possibilities of securing old agricultural habitats and their natural inhabitants.

Thus we present an evaluation of the concepts and principles used in conservation ecology under separate habitat chapters, and add considerable natural history information which may be used in deriving sound management practices. The proposed management will cover both the selection and protection of representative and species-sustainable areas. Although we are convinced of the need to understand causes and effects rather than merely accumulating data, we perceive the limitation and uncertainty of present ecological management. Hence, experimental management would be preferable for many reserves or common landscapes, leaving large parts open to free development or whatever is the traditional management. New important ideas and findings in conservation biology might thus easily be incorporated in the future preservation of species and environments.

REFERENCES

Beddington, J.R. & May, R.M. (1977). Harvesting natural populations in a randomly fluctuating environment. *Science, N.Y.*, **197**, 463–5.

Boecklen, W.J. & Bell, G.W. (1987). Consequences of faunal collapse and genetic drift to the design of nature reserves. In *Nature Conservation*, ed. D.A. Saunders, G.W. Arnold, A.A. Burbridge & A.J.M. Hopkins. Surrey Beatty & Sons, Chipping Norton, pp. 141–9.

Caughley, G. (1977). *Analysis of Vertebrate Populations*. Wiley, Chichester.

Caughley, G. (1982). Harvesting of wildlife: past, present, and future. In *Game Harvest Management*, ed. S. Beasom & S. Robertson. Kleberg Wildlife Research Institute, Texas, pp. 3–14.

Diamond, J.M. (1975). The island dilemma: lessons of modern biogeographic studies for the design of natural reserves. *Biol. Conserv.*, **7**, 129–46.

Haeckel, E. (1869). Ueber Entwicklungsgang und Aufgabe der zoologie, *Jen. Z. Med. Naturwiss.*, **5**, 353–70.

Higgs, A.J. (1981). Island biogeography theory and nature reserve design. *J. Biogeogr.*, **8**, 7–24.

Janzen, D. (1983). No park is an island: increase in interference from outside as park size decreases. *Oikos*, **41**, 402–10.

Komarek, E.V. (1971). Principles of fire ecology and fire management in relation to the Alaskan environment. In *Fire in the Northern Environment*, ed. C.W. Slaughter, R.J. Barney & G.M. Hansen. US Forest Service, Portland, Oregon, pp. 3–22.

Lahti, T. & Ranta, E. (1985). The SLOSS principle and conservation practice: an example. *Oikos*, **44**, 369–70.

MacArthur, R.H. & Wilson, E.O. (1967). *The Theory of Island Biogeography*. Princeton University Press, Princeton, New Jersey.

May, R.M. (1975). Island biogeography and the design of wildlife preserves. *Nature, Lond.*, **254**, 177–8.

McClosky, J.M. & Spalding, H. (1989). A reconnaissance-level inventory of the amount of wilderness remaining in the world. *Ambio*, **18**, 221–7.

McIntosh, R.P. (1985). *The Background of Ecology. Concept and Theory.* Cambridge University Press, Cambridge.

Nillsson, T. (1972). *Pleistocene.* Esselte, Lund.

Saunders, D.A., Arnold, G.W., Burbridge, A.A. & Hopkins, A.J.M. (eds) (1987). *Nature Conservation. The Role of Remnants of Native Vegetation.* Surrey Beatty & Sons, Chipping Norton.

Schonewald-Cox, C.M., Chambers, S.M., MacBryde, B. & Thomas, L. (ed.) (1983). *Genetics and Conservation. A Reference for Managing Wild Animal and Plant Populations.* Benjamin/Cummings, Menlo Park, California.

Simberloff, D. & Abele, L.G. (1976). Island biogeography theory and conservation practice. *Science, N.Y.*, **191**, 285–6.

Sjörs, H. (1956). *Nordisk växtgeografi.* Bonniers, Stockholm.

Sjörs, H. (1965). Features of land and climate. *Acta Phytogeogr. Suec.*, **50**, 1–12.

Skogsdata 88 (1988). *Aktuella Uppgifter om de Svenska Skogarna Från Riksskogstaxeringen.* Institutionen för Skogstaxering, Umeå.

Söderqvist, T. (1986). *The Ecologists. From Merry Naturalists to Saviours of the Nation.* Almquist & Wiksell, Uppsala.

Soulé, M.E. (ed.) (1986). *Conservation Biology. The Science of Scarcity and Diversity.* Sinauer Associates, Sunderland, Massachusetts.

Soulé, M.E. (ed.) (1987). *Viable Populations for Conservation.* Cambridge University Press, New York.

Soulé, M.E. & Simberloff, D. (1986). What do genetics and ecology tell us about the design of nature reserves? *Biol. Conserv.*, **35**, 19–40.

Chapter 2

Conservation of Communities —The Importance of Edges, Surroundings and Landscape Mosaic Structure

PER ANGELSTAM

Swedish Environmental Protection Agency, Grimsö Wildlife Research Station, S-73091 Riddarhyttan, Sweden and Department of Zoology, Uppsala University, Sweden

2.1 INTRODUCTION

The alteration, fragmentation, and finally loss, of natural habitats are the major causes of the increasingly rapid decline in overall biotic diversity on Earth (Burgess & Sharpe, 1981; Harris, 1984; Saunders *et al.*, 1987). Usually the effects are measured as the local, regional or irreversible loss of species. For communities and important processes, however, conditions may be altered long before any of the individual species actually disappear. From a conservation perspective this often means that even if all species survive locally, their densities may become reduced and their ecological situation changed due to a different pattern of interaction with surrounding communities or environmental conditions. It therefore becomes meaningful to distinguish among several types of 'extinctions' such as the extinction of functioning processes in a natural community, of a wild population that is able to maintain itself without man's aid, of a managed or subsidized population, of captive individuals, and finally of the genes characteristic of a given species. To solve such problems one must consider not only the dynamics of the target species or process, but also the changes in the biotic and abiotic surroundings. This usually requires studies covering larger geographical areas than is usual in population and community studies, i.e. on a metapopulation or landscape scale.

Consequently, because of the need to study how changing spatial patterns affect the dynamics of fauna and flora, new interdisciplinary paradigms such as patch dynamics (Merriam, 1984, 1988; Pickett & White, 1985), western-style landscape ecology (Forman & Godron, 1981, 1986; Risser *et al.*, 1984) and conservation biology (e.g. Soulé, 1986; Ginsberg, 1987) have all developed, and are currently trying to develop, guidelines for the preservation of biological diversity at different geographical and temporal levels. Because of the large scales and interactions with society, these kinds of paradigms also involve a closer link between ecologists and the administrators and legislators acting on the knowledge gathered by ecologists. The urgent need for research based on these paradigms is to explain, predict and hopefully repair the effects of habitat alteration, fragmentation and loss on the structure and function of (natural) ecological communities, including all taxa and processes.

It should not be forgotten, however, that conservation biology or landscape ecology issues *per se* are not brand new. On the contrary, they have developed from roots in wildlife management, which was formerly the only scientific branch that dealt with conservation problems, albeit only with larger mammals and birds. Those who lack the wildlife tradition should consult Leopold (1933), Grange (1946), Dasmann (1964) or read the recent excellent review by Simberloff (1988).

At first glance, fragmented or patchy habitats show a great deal of resemblance with insular habitats. For that reason, many studies dealing with conservation issues have applied the theory of island biogeography (MacArthur & Wilson, 1967) to explain variations in the number of species in relation to patch or fragment size in many kinds of insular habitat (see chapter 6). The theory of island biogeography has often been misused, to a greater or lesser extent. However it was originally not intended for studies of habitat islands. As explicitly stated by MacArthur and Wilson (1967, p. 114), habitat patches are different from oceanic islands, the setting for which the theory was originally developed. A habitat island is usually not surrounded by a habitat type which is as hostile as the ocean is to most terrestrial organisms. The surroundings of a habitat island contain both potential competitors and predators that may enter the island and interact with species living inside it (Simberloff & Abele, 1982; Janzen, 1983; Wilcox & Murphy, 1985; Andrén &

Angelstam, 1988). By contrast, the theory assumes that islands in an archipelago are homogeneous habitats, and does not usually consider how habitat patchiness (Kobayashi, 1985), beta-diversity (Hubbell & Foster, 1983) and total habitat diversity (Simberloff, 1974) influence the distribution and abundance of species, without taking the species' identity into account. Moreover, the island theory does not give any explanation for the mechanisms that create area-dependent patterns such as loss of species and individuals per unit area as patch size declines. Apart from colonization and extinction processes, mechanisms causing such patterns may be edge-effects of different kinds, affecting large patches less than small ones, since sufficiently large patches will have an unaffected core-area which is lacking in small patches (Levenson, 1981).

As recently as the middle of the 1980s, this important difference between true islands and habitat patches became more widely recognized (Ambuel & Temple, 1983; Wilcove, 1985; Verner *et al.*, 1986). Depending on the type of matrix or ecological neighbourhood (Antonovics & Levin, 1980; Addicott *et al.*, 1987), two patches that are identical with respect to size, quality, shape and isolation may be subject to different processes or conditions depending on the type of matrix that surrounds them. To study how a habitat island is affected by influences from its surrounding matrix (in other words to place edge-effects in the focus of attention) has recently become a most fruitful approach (di Castri *et al.*, 1988; Harris, 1988*a*).

When landscape mosaics change, certain key elements may affect community structure and processes in several steps. For example, ripple-effect extinctions may occur as a consequence of the disappearance of keystone species in coevolved complexes (Gilbert, 1980) or by elevated populations of predators, parasites or competitors (Karr, 1982; Janzen, 1986).

Very often a species requires more than one type of patch in order to succeed in completing its annual or life cycle (see chapter 5). But even that knowledge is not enough for the complete understanding of how populations and communities are affected by habitat alteration, fragmentation and loss. Another complicating factor is the grain-size (MacArthur & Levins, 1964; MacArthur & Wilson, 1967; Levins, 1968; Southwood, 1977) of landscape mosaics as well as the juxtaposition of different habitat types. Many examples for insects (Webb, 1989), mammals (Soulé *et al.*, 1988), birds (Ambuel & Temple, 1983), lichens (Esseen, 1983) and vascular plants (Ranney *et al.*, 1981) clearly show that effects from the matrix to the patch, and sometimes vice versa, and other changes in landscape structure are vital problems that must be dealt with in nature conservation. In this chapter, I will present some examples of the importance of considering what happens in the edges and surroundings of habitat patches, as well as in different landscape mosaics in man-dominated landscapes in order to understand the life in the patch itself. Although this book deals mainly with boreal forest landscapes, the examples used here will also be taken from other biota, since patterns and processes are often universal.

2.2 DEFINITIONS

2.2.1 Community ecology

The community concept has its origin in the observation that plants and animals are not homogeneously distributed in the landscape. The history of the science of community ecology was at an early stage divided into two different views. On the one hand, Clements (1916) argued that communities were almost as closely integrated as individual organisms, while on the other hand Gleason (1917) considered communities only as a set of species that happen to exist at the same place at the same time and where each species responds to its specific requirements. The Clementsian view dominated community ecology for several decades until around 1950, when many plant ecologists started to consider the Gleasonian alternative, since they were able to show that there was often little interdependence in species' distributions.

However, the number of species with strong biological interactions is limited within any kind of habitat (Christiansen & Fenchel, 1977). Thus, it is reasonable to suggest that any community consists of small groups of species with strong interactions (predator–prey interactions, competition, mutualism, etc.) that coexist within assemblages of a large number of species with which there are weak or no interactions. This also implies that there are keystone species or processes (*sensu* Paine, 1966; Laws, 1970; Simenstad *et al.*, 1978; Estes *et al.*, 1982; Dayton *et al.*, 1984; Paine *et al.*, 1985; Hansson & Angelstam, 1988); that is, if some species or processes are lost due to alteration, fragmentation and finally loss of natural habitats, this has more important overall community effects than if other species or processes are lost.

Regarding the effects of landscape alterations on community structure, two problems should be noted: (1) the importance of an orignal biotope mosaic—including keystone species or properties—for preserving ecosystems with natural interactions; and (2) the importance of area in the preservation of biotope value, including absence of influence from the surrounding matrix. For conservation purposes, however, we need a better resolution than the species level, as provided by metapopulation dynamics and landscape ecology.

2.2.2 Metapopulation dynamics

A very important tool for studies on conservation at a landscape scale is metapopulation dynamics (see also chapter 6). While community ecology involves studies at the species level, the theory of metapopulation dynamics increases the magnification and deals with the extinction and recolonization of populations (Hanski, 1989, 1991). As many other 'new' fields in ecology, metapopulation dynamics also has its predecessors. For example, Andrewartha and Birch (1954, p. 335) describe the concept 'underpopulation' as having two aspects: 'One is the final extinction of a population which has been well established in an area but which has been brought down to low numbers by natural vicissitudes or the deliberate destructiveness of man. The other is the failure of a small colony of immigrants to become established in a new area which is favourable in all respects except for the sparseness of the colonizing population.'

Andrewartha and Birch (1954, p. 559) also explicitly discuss the fact that the 'places where each sort of animal may live are usually distributed unevenly and this imparts a fundamental patchiness to the distribution of animals' and that 'the prevailing scarcity of most species results in many suitable places being quite empty, purely by chance' (see also Andrewartha & Birch, 1984).

2.2.3 Landscape ecology

Broadly speaking, conservation biology (CB) and landscape ecology (LE) have the same aims as interdisciplinary research fields—i.e. to maintain Earth's biodiversity. However, to some extent CB and LE are like the two sides of a coin. An analysis of the published articles within the two fields shows that while CB is more oriented towards preserving rare species *per se*, LE is more concerned with maintaining biodiversity in the overall everyday landscape by studies of flows and processes (Forman & Godron, 1986; Golley, 1989; Hansson & Angelstam, 1991).

According to the central European tradition, Naveh and Lieberman (1984, p. 3) define LE as 'a young branch of ecology that deals with the interrelationship between man and his open and built-up landscapes'. In North America, however, the emphasis is more directly on management in order to conserve biodiversity. From this perspective Risser *et al.* (1984) stated that 'landscape ecology considers the development and dynamics of spatial heterogeneity, spatial and temporal interactions and exchanges across heterogeneous landscapes, influences of spatial heterogeneity on biotic and abiotic processes, and management of spatial heterogeneity'.

A patchy landscape can be characterized by the size and type of patches as well as by how these patches are arranged in space and time. Specifically, the structure of such a landscape can be described using the following properties: (1) the internal quality of a patch; (2) patch size, and (3) patch context (i.e. how patches are situated in relation to each other, and in which type of matrix); the last two together describe the grain and juxtaposition of patches in the landscape. The landscape ecology literature contains a whole set of ways of describing landscape structure and configuration. Those interested in this more spatial aspect of landscape may consult Forman and Godron (1986) and Turner (1987).

2.2.4 Scale problems

Elton (1927) stated that '... the term animal communities is really a very elastic one, since we can use it to describe on one hand the fauna of the equatorial forest, and on the other the fauna of a mouse's caecum'.

A soil particle, a leaf, a tree, a characteristic habitat type as well as a landscape type all have different kinds of patchiness. In other words all 'environments' are patchy, and whether we consider their patchiness important or not depends on our prejudices and scale of resolution. Therefore, in order to make sense, all kinds of patchiness must be given a scale. Depending on the size and/or area requirements of the animal, the plant or the process studied, the hierarchical structure of territories, populations, metapopulations and demes may be shifted up or down the geographical scale (Fig. 2.1). In other words, what is large to one type of organism, may be small to another type.

Because of these aspects of scale there is often confusion among scientists working with different taxa about what is large and what is small (Grieg-Smith, 1979; Wiens, 1989). What is a large geographical area to plants and insects is usually small to vertebrates. Järvinen and Ranta (1987) illustrate this nicely by comparing how the number of species of different taxa increases in relation to island area in the archipelago at the south-western corner of Finland (Fig. 2.2, see also Nilsson *et al.* (1988)). The patch structure of a given environment also changes through time. Thus, the responses of individuals and populations to patchiness is

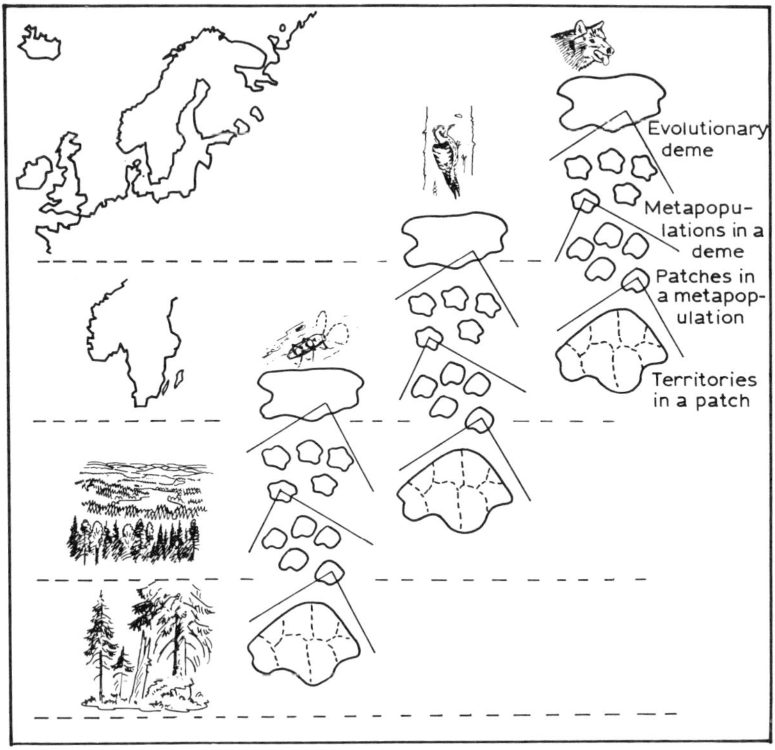

Fig. 2.1. Communities are structured hierarchically and the area requirements increase with increasing level of organization. Also, different taxa and types of species have different area requirements (see also Fig. 2.2). Therefore the different taxa and species have to be matched to the scale according to how humans perceive or manage nature for different species. The figure presents three different types of species, one invertebrate, one specialized bird, and one large top predator, with respect to area requirements.

dynamic. The consequences for community organization and processes must therefore be defined both with respect to space and time.

The scale of the examples and ideas presented in this chapter will all be related to the degree of resolution one would find on, for example, a forestry map showing the different stand types or a map of agricultural fields as revealed from an aerial photograph. In other words, this chapter will mirror the scale at which land-use management is performed, i.e. at a landscape ecological level in the sense that landscape changes originate from man's impact.

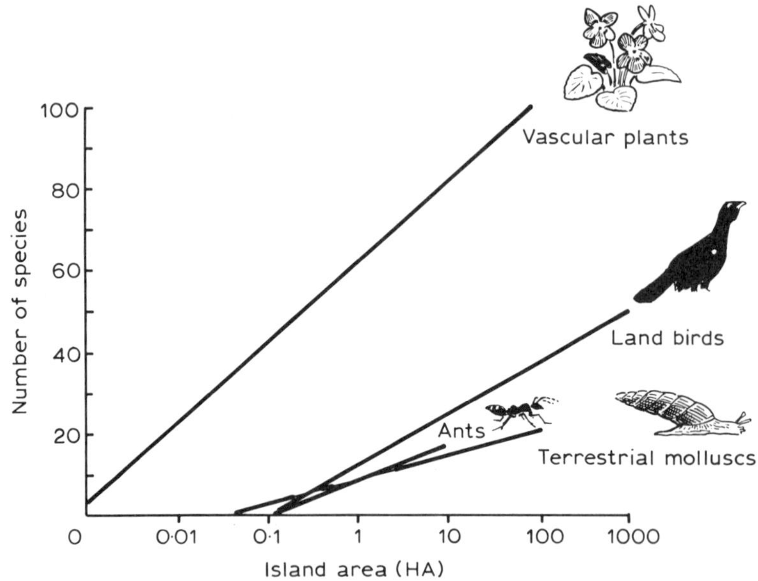

Fig. 2.2. Different groups of species have species–area curves with different slopes and intercepts. (Redrawn after Järvinen & Ranta, 1987.)

2.3 SWEDISH LANDSCAPE TYPES

Based on geomorphology and topography, Sweden has been divided into several distinct landscape types (Anon., 1984; Fig. 2.3(a)). Starting from the north and leaving out the mountain range, there is a narrow and often broken strip of northern boreal forests in a natural or near-natural state (in Swedish: 'fjällnära skogar'). In northern Sweden, loss and fragmentation of pristine forest has otherwise been almost complete, and the landscape has been altered into a managed mosaic with even-aged and homogeneous forest stands of different sizes. In the far north there are two principal successional stages—recently (< 30 year) cut forest and not-yet-cut old forest (these stands have, however, been logged by selecting the large and valuable trees). It is only in this part of the country that near natural forest communities still exist (for details, see chapter 7). Farther south there is a wide belt of managed, mainly coniferous forest with few, small and scattered patches of natural (usually wet—see chapter 8) vegetation. Today most of the forested landscape in Fennoscandia is boreal, even if other types of forest have been dominant during previous

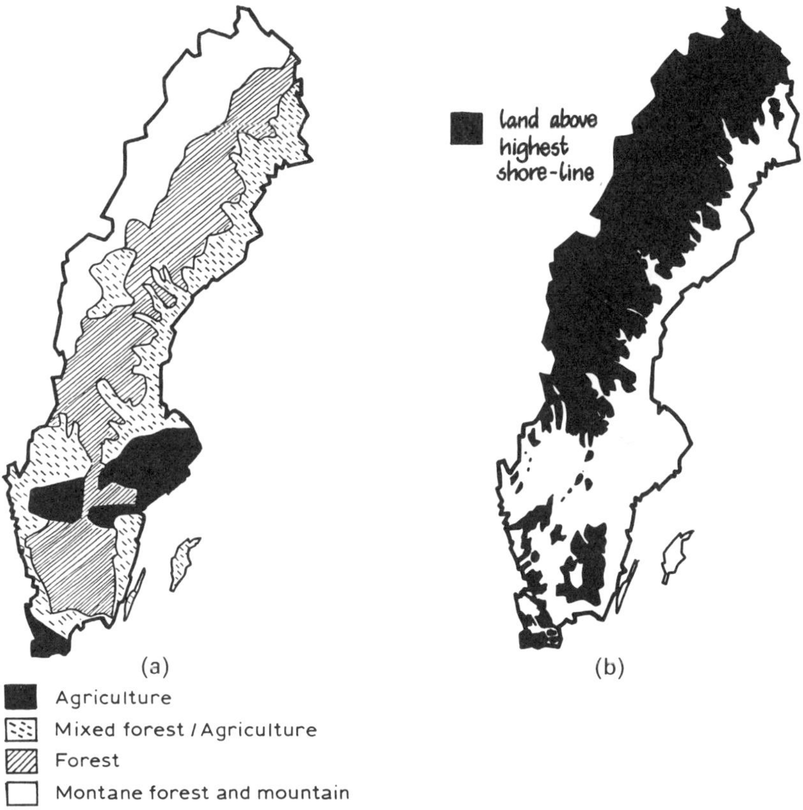

Fig. 2.3. (a) Sweden is divided into four major regions from south to north. The division between regions with farmland and those with forest is largely related to altitude above sea level. (b) Land situated above the highest post-glacial shoreline is covered by glacial till, while richer clayey soils developed as fine particle sediment at the bottom of the sea and of large ice-dammed lakes (as in west-central Sweden, Fig. 2.3(a)). Mainly the latter soil types have been developed for agricultural purposes.

periods after the latest glaciation (Sjörs, 1965). The forests in the southern part of the boreal zone in south-central Sweden have been used for commercial purposes (mining and iron industry) for up to 400 years (Mattson & Stridsberg, 1980). The country-wide habitat alterations started during the latter part of the 19th century, with a massive fragmentation of the northern boreal forest since around 1960 (Gamlin, 1988).

Along river valleys, and coasts in the north and in the south, the forest gives way to agricultural land that becomes continuous in the far south. The distribution of agricultural land in Sweden is largely limited by soil conditions. On the whole, these areas are largely associated with post-glacial deposits when ice-lakes and the sea still covered the landscape (Fig. 2.3(b) shows how the marine limit coincides with the agricultural regions (Fig. 2.3(a))).

In a broad sense, therefore, Fennoscandia today contain three main kinds of landscapes—virgin forest, managed forest, and agriculture. Among these there are two types of transition that form two major edges—between pristine and managed forest and between managed forest and agricultural land (Fig. 2.4). Due to ongoing as well as future changes in our use of the landscape, shifts in patch-matrix ratios occur—the former transformation is irreversible within a reasonable time-span, whereas the latter is reversible at least when viewed over several decades.

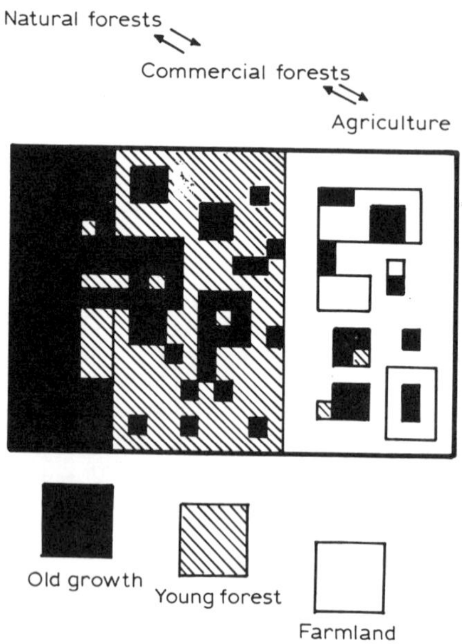

Fig. 2.4. At a landscape scale there are two major transition zones in Sweden: one between unexploited forests in the northwest, and the other between the forest-dominated landscape and the agricultural landscapes in the south. Both these boundaries are reversible, but at different time scales, i.e. less than 50 years vs hundreds of years.

In developed countries, former agricultural land is often brought back to forest as agricultural production becomes more efficient. From settlement in prehistoric times to the 1930s in Sweden the border between farmland and forest moved northwards. Since then, however, it has moved southward again (Anon., 1989). For the United States, Williams (1989) recently published a detailed account of the clearing of forests for various purposes, showing how this development has turned around during the last century. The forests that once covered large portions of the USA were relentlessly destroyed by settlement, lumbering and industry as soon as the white man entered the continent. The impact was somewhat tempered around 1870, but then changed into a rebirth of forest, especially so in New England where for several reasons forests are building up rather than declining: (1) the new harvest of forest on land where the virgin forest was harvested earlier is coming into a harvestable state; (2) as early as 1840, farmland was being abandoned in New England and gradually returned to forest; (3) regeneration by plantation; (4) intermediate stand treatment and insect and disease control; (5) the supression and control of fires; (6) *per capita* consumption of timber production has been reduced.

For Great Britain, Wilcove *et al.* (1986) describe a similar but even more pronounced example of reversion of open landscapes to forest.

As this development proceeds, important changes in patch-matrix ratios will occur, and it is therefore necessary to understand the consequences for nature conservation.

In view of the development described above, it is important to discuss the aims of nature conservation and landscape planning in a historic perspective (Brusewitz & Emmelin, 1985). Ever since man colonized what is now Sweden, the proportions of old and young forest and agricultural land have changed drastically (Figs 2.5, 2.6, 2.7). These two examples are valid for many parts of central and southern Sweden and show that it is often difficult to ascertain which are the most representative landscapes.

2.4 LOCAL AND REGIONAL MANIFESTATIONS OF EDGE-EFFECTS

Contrary to earlier views, it is now widely accepted that ecological systems are usually heterogeneous, being composed of patches and components of differing structure and dynamics (Forman & Godron, 1981, 1986; Pickett & White, 1985; Wiens, 1989). This has led to the con-

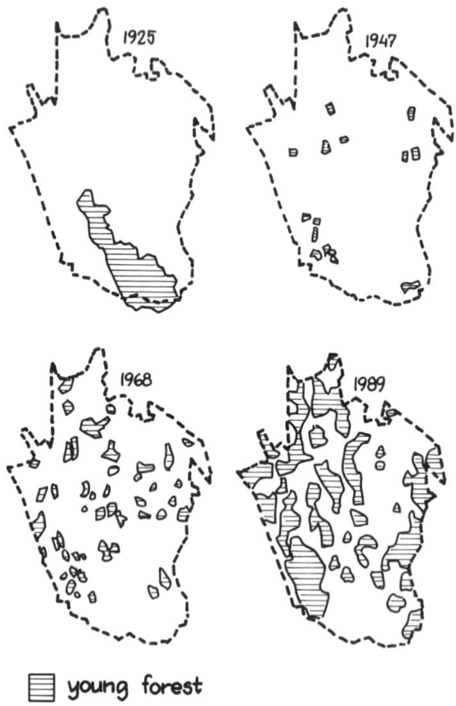

Fig. 2.5. As shown from the distribution of young (approx. 5–20 years old) and old (approx. 80–120 years) forest within the Grimsö Wildlife Research Area in south-central Sweden, the proportions of these two habitats have changed considerably. In 1925 all forest was logged by selective cutting and thinning and the whole landscape had the structure of older forest. The large patch of young forest in 1925 was due to a fire started by a railway steam engine in 1914. In 1947 only small clear-cuts occurred. In 1968 large-scale clear-cutting had started and by 1989 the proportion of young forest had increased to 34%.

cept of patch dynamics, emphasizing that patches are often ephemeral, maintaining a specific pattern only for some finite period. In a landscape, there is also often a hierarchical system of patches, with different spatial and temporal extensions that are superimposed on each other (O'Neill *et al.*, 1986; Urban *et al.*, 1987). This of course produces structural and functional edges of different kinds.

Fragmentation of natural environments, whether boreal forests, natural moors, mires or meadows, into habitat islands means that the new

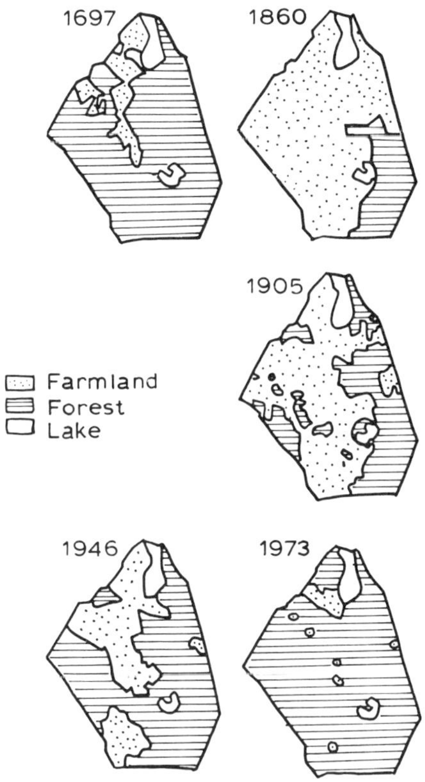

Fig. 2.6. The changes in the amount and distribution of agricultural land and forest in one parish in the Swedish county of Östergötland during the past three centuries. (Redrawn after Kardell, 1973.)

habitats inserted between the fragments are bound to increase in importance. Graphical models also show that islands of different size have shape-specific ratios of perimeter to area (Levenson, 1981; Forman & Godron, 1986, p. 110). At the landscape level this may be expressed as changes in patch-matrix ratios (Forman & Godron, 1981; Wilcove *et al.*, 1986) whereby the local edge effects may be translated to the whole landscape.

The term 'edge-effect' is an expression that summarizes a diverse group of phenomena that occur at ecotones. During recent years a growing body of evidence shows that the properties or processes that occur at edges (or ecotones, ecoclines, landscape or fragment boundaries, transi-

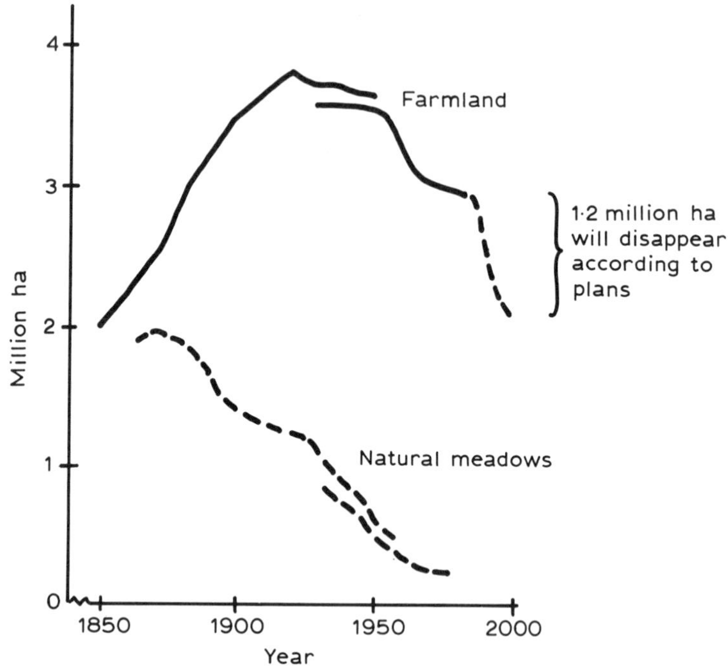

Fig. 2.7. Changes in the amount of agricultural land from 1850 to 1990, and the expected changes after reduced amounts of agricultural subsidies in 1990.

tion zones or any other term that has been used) alter the ecologies of many kinds of habitat islands and in several scales. Hansen *et al.* (1988) list three different reasons for this: (1) patch boundaries may influence ecological flows between patches: (2) unique patterns of biodiversity may occur in ecotones; (3) humankind is substantially altering landscape boundary patterns without knowledge of the consequences.

Originally, the term 'edge-effect' was used to describe the increase in species diversity associated with the edge between two adjacent habitats (spanning from Shelford (1913) to Odum (1971)), and became strongly associated with wildlife management (Leopold, 1933; Grange, 1946; Allee *et al.*, 1949; Wing, 1951; Dasmann, 1964; Burger, 1973; Yahner, 1988). The high diversity of plants and animals associated with edges and ecotones even became the edge-effect principle, a basic concept in wildlife management and sometimes in ecology (Odum, 1971, pp. 157ff.; Wiens, 1976).

By this definition an edge-effect was a positive effect. A commonly cited example is the increase in browse for deer. Specific management rules to maximize the edge-effect were even postulated: create openings with a diameter similar to the tree height (Wing, 1951, p. 132). Most evidence for the generally positive effects of edges has been circumstantial, and the benefit to wildlife in general has been made into a rule by intensive use rather than by scientific studies (Reese & Ratti, 1988; Yahner, 1988). However, the increasing emphasis on plant and non-game wildlife conservation in recent years has shown that edge-effects are not at all necessarily positive (Harris, 1988*a*). As a consequence of this more complex view of edge-effects, land managers in North America are now debating whether one should prescribe the creation of edges or not. The current use of the term 'edge-effects' therefore implies no value; the effect may be either positive or negative. In this section I will focus on the importance of ecotones and edges between some different types of matrices and patches that are typical for Nordic conditions, even if the examples are sometimes from other environments.

2.4.1 Types and definitions of edges

Needless to say, a patchy landscape has edges that separate the different patch types (often vegetation). In the absence of disturbance, vegetation types are based on local differences in moisture, soil type, topography and climate (Thomas, 1979; Forman & Godron, 1986; di Castri *et al.*, 1988). Pioneering studies in plant ecology such as those by Clements (1897), Livingston (1903) and Griggs (1914) were made on the transition zones between plant communities, i.e. ecotones. The word 'ecotone' is derived from the greek words 'oikos' (household) and 'tonos' (tension), thereby emphasizing the tension zone between species from adjacent communities.

In the following I use the word 'edge' both for natural edges, such as the ecotone between natural plant communities, as well as for induced (i.e. caused by man, or some other kind of disturbance), such as the junction between successional stages in a forest, or between forest and farmland. In this context we are mainly dealing with the latter, induced, type.

The massive literature on habitat fragmentation is dominated by studies from areas where a forest is a forest, i.e. stands with large trees. In both tropical and temperate areas, the forest is usually regenerated in small gaps in the forest as a single tree or a group of trees falls down (Falinski, 1986). Even with a great turnover of individual trees, such a forest landscape does not look fragmented on an aerial photograph. In boreal forest, however, the natural regeneration occurs in large areas simultaneously, mainly due to forest fire. Therefore the boreal forest landscape also naturally includes other successional stages, and forest land is not always covered with stands containing large trees.

2.4.2 Direct and indirect abiotic effects

Effects from the surrounding to the interior

Forest edges generate microclimatic gradients which produce an environment that is different from both the open habitat outside the forest and its interior (Gysel, 1951; Hagar, 1960; Trimble & Tryon, 1966; Wales, 1967; Jakucs, 1972; Dieschke, 1974; Ranney, 1977; Ranney *et al.*, 1981; Chasko & Gates, 1982; Willmanns & Bruun-Hool, 1982; De Walle, 1983; Lovejoy *et al.*, 1986). Edges facing the equator are typically wider than those facing the poles (Carpenter, 1935; Wales, 1972; Forman & Godron, 1981; Ranney *et al.*, 1981; Whitney & Runkle, 1981). This may be due either to better plant growth and reproduction or to the deeper penetration of climatic effects, affecting some species negatively (e.g. drought- or wind-sensitive species). In newly created edges, increased light and evapotranspiration affect shade-tolerant and intolerant species differently. This microclimatic effect is often said to extend about three times the canopy height into the forest in boreal areas (Harris, 1984). As an edge matures, the structural heterogeneity may increase, as well as humidity and soil moisture. Therefore, edges are often wider and more of a distinct transition habitat type when a patch is older (Wales, 1972; Yi, 1976).

Typically, a cross-section of the edge between forests and fields or pastures has a mantle of smaller trees or shrubs that borders, or is under, the canopy trees. Then follows a herbaceous layer called the 'saum' or 'perennial herb border'.

There is, however, a strong difference between the managed boreal forest, where edges are much sharper, and agricultural landscapes where old edges often are a specific type of a habitat with long continuity (Peterken, 1981; Ranney *et al.*, 1981; Harris, 1984). The edge between a forest and a clear-cut may be very sharp and is usually not a habitat in itself. Here the effects of wind and temperature easily penetrate into the forest. Consequently, wind damage is more frequent near edges and becomes serious when only small stands are present (Franklin & Forman, 1987).

Plant responses in experimentally isolated boreal forest fragments—a case study P.A. Esseen (pers. comm.) carried out an experiment in a north-boreal spruce forest located on the mountain of Norra Gardfjället, Västerbotten, north Sweden, to study the population and community response of plants to fragmentation, with the aim of evaluating mechanisms behind edge-effects. The experiment included studies on tree dynamics, understorey plant composition, demography of forest herbs and the impact of keystone herbivores in order to evaluate links between different organism groups. The square forest remnants (1/16 ha, 1/8 ha, 1/4 ha, 1/2 ha and 1 ha on a 40-ha clear-felled area, no replication) were isolated in February 1986 and followed during four years, along with an adjacent 100-ha control area.

The results show that altered abiotic conditions are more important determinants of plant community composition in forest edges than biotic factors in the first years after fragmentation. A major impact was extensive storm-felling of spruce trees and other damage to the forest stand, mainly caused by increased wind speeds (see also Franklin & Forman, 1987). Tree mortality ranged from 25% in the 1-ha stand to 95% in the 1/16-ha stand, and was concentrated in the western parts that were more exposed to prevailing winds. Of all groups studied, arboreal lichens were most severely affected by fragmentation and by tree-falls, increased wind speed and decreased air humidity when compared with the control area of undisturbed forest.

The phenological development of vascular plants in the understorey started more than one week earlier on the clear-felled area and in the smallest remnants, but was delayed several days in quadrants with much tree litter when compared with the control. Species richness of vascular plants was little affected by fragmentation. In contrast, marked changes in species abundance patterns occurred in both remnants and on clear-felled ground. The changes included a marked increase in cover of grasses such as *Deschampsia flexuosa* and of many herbs. Many species showed an increased flowering and increased recruitment by seedlings or by vegetative propagation. This well-known response of early successional plants to clear-felling (Sjörs, 1965) successively spread from edges of remnants inwards, with a delay of one to two years compared to the clear-felled area. Annuals, for example *Melampyrum pratense* and *M. sylvaticum*, showed a very rapid response to fragmentation with an increase in population size, plant size, numbers of flowers and seed production. Clonal herbs showed both positive and negative population responses to fragmentation. Shoot numbers in species such as *Geranium sylvaticum*, *Orthilia secunda* and *Moneses uniflora* increased up to two-fold in remnants and up to four-fold in the clear-cut area from 1986 to 1989. Other species, such as the orchid *Listera cordata*, showed a marked reduction in shoot numbers with decreasing stand area. Mortality of clonal forest herbs was higher in remnants than in the clear-felled area, mainly because of litter-fall and soil disturbance following storm damage. Marked desiccation damage occurred mainly during the first year after fragmentation.

It was predicted that the increased plant growth on the clear-felled area should increase population densities of, or use by, generalist herbivores such as microtine rodents (especially *Microtus agrestis*), moose *Alces alces* and some insects. This in turn could result in higher levels of herbivory on understorey plants in forest remnants (see section 2.5.1). Compared with the control area, however, both lower and higher levels of herbivory were obtained in different plant species.

This experiment demonstrates that a multitude of factors is involved in the process of forest fragmentation and that different plant species may show very contrasting responses. Initially, changes in abiotic conditions through increased wind speeds, increased light availability and decreased air humidity in remnants evidently contributed more to changes in plant abundance than did biotic factors. Indirect abiotic influence by storm-felling of trees and soil disturbance was the single most important factor affecting plant mortality.

The conclusion was that isolated forest fragments sized 1 ha or less consist entirely of edge habitat and lack a central undisturbed core area. Consequently, even after such a short period as four years, significant edge-effects reach more than 56 m into remnants of old-growth spruce forests in boreal areas.

Effects from the interior to the surrounding

However, a forest also affects its surroundings. At a distance that is three times the tree height out into a clear-cut wind speed is reduced to 50%, and by a distance of 30 times the canopy height there is no reduction in wind speed (Söderström, 1978). Similarly, hedgerows in agricultural landscapes modify evaporation for a distance corresponding to 16 times the hedgerow height and 28 times for the wind speed (Forman & Baudry, 1984).

Edges and wildlife

Because edges are different from both the habitats outside the forest and from the forest interior they sometimes produce distinctive effects on animals. Results on birds regarding the effects on species richness in edges may at first seem somewhat contradictory. Gates and Gysel (1978) and

Chasko and Gates (1982) both found that bird species richness as well as nest density increased from the forest interior to the forest edge. Kroodsma (1982), however, found such patterns only during one out of three seasons studied. Finally, Helle and Helle (1982) and Hansson (1983) found that bird density peaked in forest/clear-cut edges but species richness was higher away from the forest edge. Bird counts usually measure the location of singing males only, and since high perches such as the trees in a forest edge are good singing posts, the edge-effect may simply be an artefact.

However, if the edges are divided into those that are more permanent and more of a specific type of habitat, such as those between a forest stand and farmland in landscapes where the edge has existed for a long time, certain bird species actually find their specific habitat just along the edge (Wiens, 1989, vol. 2, p. 208). As an example, Szaro and Jakle (1985) found a strong increase in bird density along the edge between the wash (seasonally flooded land) and the upland habitats in a riparian woodland in Arizona. The study of Morgan and Gates (1982) clearly demonstrated that the more structurally complex the edge, the higher the species richness, abundance and diversity of birds. Finally, Helle (1983) concluded that, based on data collected from boreal forest in north Finland, the amount of bush growth at forest/open ground edges best explains the variation in breeding bird density. In his study edges could be divided into two categories: (1) those with a poor shrub layer (found in edges bordering open mires, recently abandoned meadows and young clear-cuts: and (2) those with a well-developed shrub layer (in edges bordering riverside meadows, pastures and fields, clear-cuts older than 25 years). The average breeding density in the latter type was twice as high (143 pairs/km^2) as in the former type (77 pairs/km^2). The average density deeper into the forest was 79 pairs/km^2.

There are other ways, however, in which edges may actually be dangerous. The combination of the attractiveness of nesting in edges and the actual decline in production of young in such habitats has been labelled an 'ecological trap' (Gates & Gysel, 1978, and section 2.4.3.) Also, patch-interior species that cannot tolerate edges will become prone to extinction in small patches (Wilcove *et al.*, 1986). Therefore, attempts to maximize local diversity by promoting edge habitats may lead to a reduction in regional diversity due to loss of forest-interior species (Noss, 1983, and see also section 2.5.2.). Harris (1984) argued that one way to buffer the old-growth stands in northwestern US forests from many microclimatic effects is to create adjacent younger stands. However, as will be seen from the many other kinds of edge-effects, such a strategy may in some situations actually threaten the biodiversity of the habitat patch we want to save, as, for example, by the invasion of successional plants and animals that interfere with species restricted to the older communities (Janzen, 1983, 1986).

Edges may sometimes serve as corridors or pathways and facilitate movements along habitat borders (see also chapter 9). According to Knopf (1986), some woodland birds may have been able to spread across the North American prairie by moving along riparian zones. Similarly, riparian zones in the coniferous forests of the Pacific northwest and many other kinds of corridors are thought to be used as movement corridors for vertebrates (see Harris, 1984; Harris & Gallagher, 1989; and Saunders *et al.*, 1987; and Saunders & Hobbs, 1991 for detailed reviews on the role of habitat corridors).

2.4.3 Predation

The loss of bird species from forest or other vegetation types as fragmentation progresses is well-documented (Svensson, 1978; Whitcomb *et al.*, 1981; Ambuel & Temple, 1983; Wilcove & Whitcomb, 1983; Helle, 1984; McLellan *et al.*, 1986; Saunders, 1989). Apart from the obvious fact that area requirements are not satisfied (see Haila *et al.*, 1979), a number of explanations have been suggested for this decline in the number of species (see Whitcomb *et al.*, 1981; Temple & Cary, 1988). In North America much of the evidence shows that poor fecundity is a major component of this reduction (Gates & Gysel, 1978; Gates & Mosher, 1981; Ambuel & Temple, 1983; Brittingham & Temple, 1983; Johnson & Temple, 1986; Temple, 1986). This poor fecundity is apparently related to interspecific interactions from outside the habitat islands.

One factor is increased brood parasitism (Brittingham & Temple, 1983). The brown-headed cowbird *Molothrus ater* is the only obligate nest parasite in continental North America and inhabits open landscapes and edges. Like the European cuckoo *Cuculus canorus* it lays its eggs in the nests of other bird species. Many passerine species have avoided heavy levels of nest parasitism by breeding in large tracts of forest but when forest patch size declines some species can no longer breed (Ambuel & Temple, 1983).

Another factor is an increased nest predation (Robbins, 1980; Ambuel & Temple, 1983; Wilcove, 1985) in small forest fragments as compared to large ones.

Predation in the forest–farmland edge

In North America Wilcove (1985) tested the idea that nest predation rates differ between large and small forest fragments and found higher predation rates in small as compared to large fragments. The same pattern in predation has been found in the Old World. Andrén and Angelstam (1988) found a consistently negative trend in the predation pressure on dummy nests from the more productive habitat (agricultural land on post-glacial clay) into a less productive habitat (coniferous old-growth on glacial till). The effect could be seen up to about 200 m into the forest stands. Similarly, but for real nests, M. Björklund (pers. comm.) found that predation rates on scarlet rosefinch *Carpodacus erythrinus* nests were significantly higher within 25 m from the forest/farmland edge (62%) than farther into the forest (21%).

Other studies of loss rates of nests in forest edges contradict the effect of edge (Yahner & Wright, 1985; Angelstam, 1986; P.C. Mankin & R.E. Warner, pers. comm.). However, Yahner and Wright's study was carried out within forest fragments so small that dummy nests placed in the middle were still within the distance (*c.* 50 m) where the edge-effect is large. The same argument applies to P.C. Mankin and R.E. Warner (pers. comm.), who studied predation on artificial nests in patches of forest in an agricultural matrix in Illinois and could not find any predation patterns in edges. However, as I understand it, in many highly fragmented regions such as theirs, there are no areas left that are sufficiently large to include a core area at all; thus, such patches appear to be free from edge-effects such as predation simply because everything is edge. Finally, Angelstam (1986) explained the absence of an effect as a consequence of farmland patches being small and sparse and incapable of supporting a high density of generalist predators.

Temple and Cary (1988) used a computer model to examine the effects of varying degrees of habitat fragmentation on forest-interior birds. They found that the dynamics of the populations was most sensitive to variations in fecundity, which varied as a function of how far the bird's territory was from the forest edge. Hence, without immigration from other regions where reproduction is better, habitat-interior populations in a landscape with only small fragments can go extinct. However, this is actually not much more than a complicated way of expressing the idea that small patches are dominated by edge (see Levenson, 1981).

Thus, predation is an important edge-effect and may produce the patterns in population densities found in North America for bird species requiring forest-interior characteristics. There are, however, alternative untested explanations that could work in the same direction, such as the rapid disappearance of forest in the tropical winter-quarters of these forest-interior species. In Europe there are no recent comparable patterns of population declines of predation-prone species, but these extinction processes quite possibly occurred several hundred years ago (Wilcove *et al.*, 1986), i.e. when fragmentation really occurred. At least one possible example (capercaillie *Tetrao urogallus*) is implicitly provided by Yapp (1962, p. 290).

Predation in forest and farmland landscapes

In northern Europe, however, predation as an edge-effect may alter the dynamics of populations regionally. It appears to be a general rule that population cycles of birds and mammals in the holarctic zone disappear towards the southern part of the taiga (Hansson & Henttonen, 1988; Sutherland, 1988). In parts of Fennoscandia, for example, all grouse and hare populations undergo regular population fluctuations every three to four years (Fig. 2.8; Hörnfeldt, 1978; Hörnfeldt *et al.*, 1986; Angelstam *et al.*, 1984, 1985; Lindén, 1988). By contrast, grouse populations farther south on the European continent are non-cyclic (De Franceschi, 1982; Ellison *et al.*, 1982; Ruwet, 1982; Angelstam *et al.*, 1985; Ellison & Magnani, 1985).

Several hypotheses have been suggested as explanations for the cyclic fluctuations in Fennoscandian grouse species (for reviews, see Watson & Moss, 1979; Angelstam *et al.*, 1985). Apart from explaining the cyclicity itself, any hypothesis about the cycles among tetraonids in northern Fennoscandia must also explain: (a) the synchrony between grouse and small rodent cycles; (b) the time lag of one year between spring densities of small rodents and grouse; and (c) the general disappearance of cyclicity towards southern Fennoscandia. Angelstam *et al.* (1985) found that hypotheses involving intrinsic factors or classic predator–prey oscillations failed to explain the system of regular short-term population fluctuations among boreal grouse species in Fennoscandia. Instead, temporal and spatial variation in predation pressure may explain all the above characteristics of Fennoscandian grouse cycles. The main explanation, the alternative prey hypothesis, is that predation pressure from a generalist predator community upon grouse (an alternative prey with low abundance) varies in relation to the abundance of small rodents (the main prey of generalist predators) (Angelstam *et al.*, 1984; Lindström *et al.*, 1988).

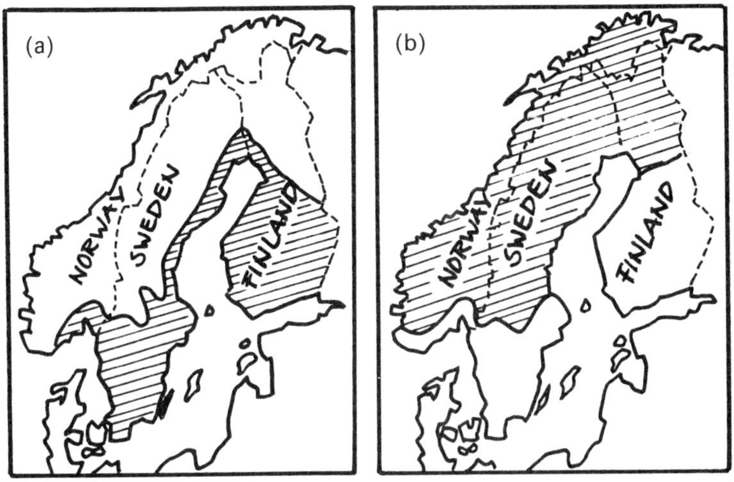

Fig. 2.8. Distribution of synchronous population cycles of small mammals, their predators and the alternative prey of predators in Fennoscandia (a) and of landscapes with agricultural land (b). (Redrawn after Angelstam *et al.* 1986.)

Regarding the disappearance of population cycles towards the south, overall predation seems to decrease from south to north (Bergerud & Mossop, 1984). In parallel, grouse habitats become increasingly surrounded by man-altered habitats to the south (cf. Wolff, 1980; Sievert & Keith, 1985). A similar explanation probably applied to the distribution of grouse cyclicity in Fennoscandia (Angelstam *et al.*, 1985). Where cycles exist, i.e. in the boreal forest north of the forest farmland edge, predators have few alternative food items among which to choose. Towards the south, human population density increases, as does the proportion of the landscape that is used for agricultural purposes (cf. Fig. 2.4). This leads to a more complex set of food items available to predators. Where interspecific synchrony does not exist, the diet of a main predator (red fox *Vulpes vulpes*) was more diverse than where cycles were found. The disappearance of population cycles as well as dietary change was most pronounced at the southern border of the boreal forest zone. According to Angelstam *et al.* (1985), these changes may swamp out the effects of small-rodent cycles (and, as it turned out, also the small-mammal cycles) on predator populations and hence also produce a high and more or less constant predation pressure on grouse.

The idea that fragmentation leads to an increase in prey species diversity and hence in the number of generalist predators and, as a result, an increased predation pressure, was tested by Andrén *et al.* (1985) for predation on nests. As predicted, higher predation rates were found in the south than in the north. The rate of predation on experimental dummy nests was correlated with the abundance of corvid birds (the major group of nest predators), which in turn was positively related to the proportion of agricultural land, human density and the degree of fragmentation of boreal forests. Similarly, Hansson (1988) studied the grazing pressure of small rodents and Angelstam *et al.* (unpublished data) vole population dynamics across a gradient of increasing degree of fragmentation of boreal habitats into a matrix of agricultural land. The same gradient in the degree of fluctuation could also be seen in a gradient from large to small boreal forest patches to farmland within the agricultural landscape.

The disappearance of the vole cycles with increasing fragmentation and with decreasing size of patches of boreal forest is consistent with the idea that predation from generalist predators forms an edge-effect. Hence, the consequences of predation as an edge-effect can also be seen at a landscape level near the edge of the continuous boreal forest (Fig. 2.4).

Predation as an edge-effect in forest/clearcut edges within the taiga

There also appear to be differences in predation pressure among forested landscapes with different degrees of fragmentation at the northern edge of the managed boreal forest even without the great differences in productivity found in forest/farmland edges at the southern end of the managed boreal forest. In fact, studies on nest predation by mustelids on nests situated at different distances from forest/clear-cut edge show that predation may also be an edge-effect in this type of edge (Angelstam, unpublished data). Kuitunen and Helle (1989) measured nesting success of a hole nester, the common treecreeper *Certhia familiaris* breeding in nest boxes and found that a greater proportion (37%) of nests was robbed at forest edges (>20 m) than in the interior parts of stands (26%). Similarly, Angelstam (unpublished data) found that the survival rate of nests in nest-boxes depended on the distance from the habitat edge. The predation rate on nests at or within 10 m of the clear-cut was high (up to 90%) and decreased into the forest.

Extending the edge-effect concept to a landscape level in the boreal forest

It has been argued that in a forest landscape that is fragmented by clear-cuts, the overall carrying capacity for predators is higher than in a natural, unfragmented forest landscape because there is more prey in the former than in the latter (Hansson, 1979). Therefore, predation on ground-nests in mature forest patches may be matrix-dependent (Christianssen, 1979; Rolstad & Wegge, 1989). Support for this idea is provided both from the Old and the New Worlds. In Pennsylvania, USA, Yahner and Scott (1988) measured predation on dummy nests in mature forest stands with zero, 25 and 50% of the surrounding matrix fragmented by clear-cutting. Nest predation was highest in the 50% situation and lowest at 0%.

In Sweden, the argument for this idea in Hansson (1979) is: (1) Increased proportion of clear-cut areas and young forest in the landscape leads to (2) an increase in the amount of food for herbivores, which leads to (3) increased densities of moose and small rodents, which leads to (4) more food and thus higher densities of small and medium-sized predators, which finally leads to (5) an increased predation pressure on ground-nesting birds.

The super-normal densities of moose in Sweden (see section 2.5.1) may provide indirectly generalist predators with substantial amounts of food. Quantitative estimates of the amount of moose carrion and calculations of the food requirements of the major nest predators suggest that carcasses alone will cover large proportions of the food demands of red fox and raven *Corvus corax* populations (10–120% and 160–1900% depending on the time of the year, respectively) (Angelstam, unpublished data), and may hence explain the long-term positive population trends of these species.

A testable prediction from this hypothesis is that the predation rate of nests and breeding birds in old forest stands should increase from a landscape with few or no clear-cuts to a landscape containing a high proportion of clear-cuts. In a test of this idea performed in Swedish Lapland in 1989, Angelstam and Welander (unpublished data) found that in large ($>100\,km^2$) tracts without clear-cuts the predator community was more diverse, including brown bear *Ursus arctos*, wolverine *Gulo gulo* and lynx *Lynx lynx*, while in fragmented landscapes it was dominated by medium-sized generalist predators, raven, hooded crow *Corvus cornix* and red fox. Predation rates on artificial nests in the very large mature forest was lower than in medium-sized (*c.* $1\,km^2$) and small (*c.* $0{\cdot}1\,km^2$) stands. Together, these results suggest that fragmentation of surrounding forests may negatively affect the breeding success of ground-nesting birds, and that a base line for what is the unaffected state requires a landscape approach to studies (see also section 2.7).

Other types of edges

Other types of edges representing natural and altered types show similar patterns. In SW Skåne in Sweden, the dunlin *Calidris alpina schintzii* breeds on small remnant patches of coastal meadows, which can be considered as islands in a matrix that is able to support high densities of generalist predators. In his study on the population dynamics of that species, Jönsson (1988) found that this population had the lowest hatching success of any dunlin population studied so far, the main reason being very high rates of nest predation from crows and foxes. This population is therefore no longer self-sustaining and is declining to an expected extinction within the next decade, in spite of an estimated annual immigration rate of 5%.

In a study of bird assemblages on mires, Boström and Nilsson (1983) found that mires with shapes that deviated from the circular had fewer wader bird species. They suggested predation from the matrix as a viable hypothesis.

Predation in edges with varying productivity gradients

In the American tropics, the species composition in forest fragments is extremely sensitive to influences from the surrounding habitats. However, the influences from the surrounding habitat are greater if they are secondary forest than if they are farmland (Janzen, 1983). The magnitude of the edge-effect has been suggested to be inversely related to the habitat quality of adjacent habitats (Angelstam, 1986; Harris, 1988*b*). Indeed, the survival rate of nests and breeding birds in old-growth stands should gradually increase from a landscape with few or no clear-cuts to one containing a high proportion of clear-cuts to one where forest patches are surrounded by agricultural land. Angelstam *et al.* (unpublished data) confirmed by experiment the hypothesis that predation as an edge-effect increases with the difference in productivity/predator carrying capacity between the habitat patches, as suggested by Angelstam (1986). Dummy nests, designed to reveal which species had been the predator by track sampling, were put out in forest–clear-cut, forest–farmland and forest–lake ecotones in central Sweden. Predation rate was significantly higher in forest–farmland ecotones than in forest–clear-cut ecotones. Predation rate decreased from farmland to forest but not from clear-cuts into the forest. Predation rates were also higher in the former type of edge than in the latter. In the forest–lake ecotone, predation rate was highest near the edge. Predators were determined to species in 85% of the cases. There were more predator species in forest–farmland and forest–lake ecotones than in forest–clear-cut ecotones. The increased predation rate in forest–farmland and forest–lake ecotones was mainly due to increased numbers of generalist bird and mammal predators.

2.4.4 Other interactions

In North America several studies have been directed at determining how large browsers use food resources on clear-cuts at different distances from the edge between a forest and a clear-cut. Lyon and Jensen (1980) showed for elk *Cervus canadensis* and black-tailed deer *Odocoileus hemionus* that use of larger, open clear-cuts was hampered if adequate plant cover was not present. For white-tailed deer *Odocoileus virginianus* in eastern Canada, Drolet (1978, Fig. 3) found the same pattern but was also able to quantify the effects: clear-cuts over 70 ha were used less than expected compared with smaller clear-cuts. Finally, Hamilton *et al.* (1980) showed that the moose (*Alces alces*) underexploits large clear-cuts in winter but that predation pressure, moose density, and snow type all influence the use of clear-cuts. The evolutionary background behind these patterns is probably related to the fact that large herbivores need protection from severe microclimatic conditions (Moen, 1973) as well as against predators (Hamilton *et al.*, 1980). Many species can exploit a given food resource only if adequate plant cover is present in order to decrease predation risks. In an experiment, Thompson (1982) was able to increase small mammal densities by increasing cover. Similarly, hazel grouse *Banasa bonasia* avoid eating away from plant cover and will consume almost all food (*Alnus* catkins) where cover is adequate and none if cover is insufficient (J. Swenson, pers. comm.).

Janzen's (1983) idiom 'No park is an island' was based on examples of seed dispersal and predation causing intrusion of new plant species to apparently pristine forest. Seed predation may also be a factor that affects the edge and the interior of a stand differently. For beech *Fagus sylvatica*, Nilsson and Wästljung (1987) found that vertebrate predation on seeds was increased in stands below 1 ha, suggesting that seed predation was an edge effect extending about 50 m into a stand.

Jennersten (1988) analysed the effects of habitat fragmentation on the pollination success of a perennial herb *Dianthus deltoides*. The fragmented area has a lower diversity and abundance of both flowering plants and insects visiting the flowers. *Dianthus* flowers had fewer visits, and produced fewer seed, in the fragmented area than in the unfragmented area. Experiments with hand-pollination clearly demonstrated that the difference in natural seed set between the two kinds of habitats could be explained as a consequence of depauperization of the polinator fauna following habitat fragmentation.

2.4.5 Conclusions about edge-effects

While edges are defined by the structural components in landscapes, edge-effects must also be defined from the point of view of the processes. Edge-effects may therefore penetrate into the interior or the surroundings for considerable distances. 1. Forest-interior birds avoid the 50 m nearest to the edge (Lovejoy *et al.*, 1986). 2. Forest micro-climate is affected up to 100 m into the forest stand (Lovejoy *et al.*, 1986). 3. Predation on nests by corvid birds extends up to 200 m form the surrounding agricultural landscape (Wilcove, 1985; Andren and Angelstam, 1988). 4. Light-loving butterflies can be found up to 300 m into a forest stand (Lovejoy *et al.*, 1986). 5. Secondary or non-native plants may colonize new sites several km into the forest (Hubbel and Foster, 1986; Janzen, 1986). 6. Generalist predators and large herbivores affect whole landscapes (Alverson *et al.*, 1988). 7. Pollution from industries and urban centers cast their effects from one country to another.

A number of questions, however, remain to be answered.

(1) How constant are edge functions within the same fragment and among fragments in the same fragment type?
(2) Are some edge-effects specific for certain organism types or habitats?
(3) How do edge functions change during secondary succession?
(4) How do edge-effects interact with the effects of area? (They must be sampled across edges using matrix-edge–interior transects in a range of patch sizes.)
(5) Are edge-effects multiplicative?

Most seems to be known about predation as a kind of edge-effect. The intensity of this type of edge-effects depends, however, on several factors:

(1) The extent of penetration of individual predators into a patch including
(a) the distance from a habitat edge
(b) the strength of the penetration of individual predators, which depends on the type of habitat that surrounds a patch—i.e. the quality gradient
(2) The size of the patch
(3) The shape of the patch

(4) The type of prey and alternative prey
(5) The type of predator; what is the 'sea' to one species may be the 'island' to another.

Summarizing, predation as an edge-effect is a very important mechanism that may lead to:

(1) Impoverishment of the bird fauna in small habitat patches (Ambuel & Temple, 1983, and subsequent papers along those lines).
(2) Alteration of population fluctuation patterns. In northern Fennoscandia many bird and mammal species fluctuate in synchrony with voles. In southern Fennoscandia there is no such connection.
(3) Altered predator–prey relationships leading to limitation of the geographical distribution of species (Sievert & Keith, 1985).

2.5 LANDSCAPE MANAGEMENT

In the previous section I reviewed how the effects we may observe in edges of different kinds may translate into the whole landscape. However, we cannot always conclude that edge-effects are necessarily positive nor always negative because their consequences may be viewed on at least two scales. We must first understand the consequences of spatial heterogeneity at the edge itself (vegetation, sharpness, etc. and the consequences thereof) and, second, as a function of the spatial heterogeneity that this creates in the landscape as a whole.

The changes we see on a local scale contribute to the change in the distribution and abundance of wildlife species on a broader geographical scale. Whereas a more complex vegetation structure at the edge may increase wildlife diversity (Willson, 1974; Morgan & Gates, 1982), the overall landscape effects on other organisms may involve a loss of biological diversity (Alverson *et al.*, 1988). The losses are also often predictable. Species with large home ranges and/or specialists are the first to disappear from a system that is fragmented or altered (e.g. Harris & Gallagher, 1989, p. 21). If keystone species or parts in coevolved complexes (Gilbert, 1980) are lost or the populations of predators, parasites or competitors are elevated (Karr, 1982; Janzen, 1986), we may get ripple effects cascading down the trophic ladder or even altering the ecosystem. This often leads to a two-staged change in the community. First, the few keystone species slowly disappear; then, more rapidly, the secondary effects show up.

2.5.1 Ripple effects due to loss of keystone species

Hairston *et al.* (1960) proposed that populations at an intermediate trophic level may severely affect the lower trophic level, but are usually unable to do that because of higher trophic level control of the intermediate level. These three levels may under certain circumstances be interpreted as plants, herbivores and carnivores, respectively.

In aquatic ecosystems, the effects of removal of predators are clear and often well-documented. Paine (1966) clearly demonstrated that the loss of top predators in a rocky marine intertidal ecosystem can lead to drastic simplification of the remaining community. Since then several aquatic examples have been reported (Simenstad *et al.*, 1978; Estes *et al.*, 1982; Dayton *et al.*, 1984; Paine *et al.*, 1985).

For terrestrial environments Pacala and Roughgarden (1984) and Schoener and Spiller (1987) observed similar explosions in spider populations in the absence of lizards on tropical islands. Introductions of herbivores on islands often lead to disturbance of the local vegetation (Klein, 1968; Caughley, 1970; Dilks & Wilson, 1978; Meurk, 1982; Rodriguez Luengo & Rodriguez Pinero, 1990) and may even cause extinctions in the local flora (Coblentz, 1978). However, in insular situations effects may be exaggerated since plants may not have evolved defences against herbivory (Carlquist, 1974).

Subsequently it has become a common viewpoint that animals at intermediate trophic levels under certain circumstances can have stable levels of density (endemics) when controlled by the higher trophic level. If control is removed, however, populations may fluctuate widely in abundance and occasionally reach far higher levels of density (epidemics) (e.g. Sutherland, 1974; Simenstad *et al.*, 1978; Oksanen *et al.*, 1981).

Yet we know very little about the consequences for fauna and flora of the loss of keystone species like top predators in terrestrial non-island systems involving larger vertebrates.

Effects of changing predator densities on prey

In North America, the ranges of large carnivores such as mountain lions *Felis concolor* and grizzly bears *Ursus arctos* have been constricted due to the reduction in the extent of large undisturbed tracts (Wilcove & May, 1986, Wilcove *et al.*, 1986). Matthiae and Stearns (1981) describe the disappearance of large mammals of the north-central states in the US and Harris *et al.* (1982) the differential loss of wide-ranging carnivores as the habitats of the Cascade Mountains became fragmented. Maehr (1984) and Pelton (1986) describe in detail how the loss of connective habitats isolate habitat fragments and thereby restrict the distribution of black bears in the eastern United States. When large top predators have become locally extinct, in tropical locations, smaller omnivores and predators have undergone population explosions, in some cases becoming 4–10 times more abundant than usual (Terborgh & Winter, 1980). Based on observations from the Barro Colorado Island, a former hilltop that became an island when the Panama Canal was built in 1891–1914, Terborgh and Winter (1980) suggested that the presence of top predators may affect the composition of tropical forest bird communities. Since the canal was built, medium-sized and large generalist mammals have become unwary and remarkably abundant. These species usually constitute the bulk of prey taken by three species of top predators, all of which had bred on Barro Colorado for decades but are now locally extinct. It is inferred that these species have increased in numbers because they live in a top predator-free environment. Since its isolation from the mainland the island has lost bird species. Among 18 forest-dwelling species, the 15 that share the habit of nesting on or near the ground, i.e. being sensitive to predation from generalists, have gone extinct. Similar consequences for temperate locations are poorly documented. Soulé *et al.* (1988), however, provide some inferential evidence that coyotes control the populations of smaller predators such as foxes and domestic cats. Based on this they proposed that the absence of coyotes may lead to higher levels of predation on birds by a process they name 'mesopredator release' (Fig. 5 in Mackintosh, 1989). Finally, Harris (1989) mentions the effects of racoon predation on the nests of turtles, tortoises and alligators.

As mentioned earlier, the strong reduction of large predators and released competition from cattle as well as clear-cutting (Strandgaard, 1982) have allowed a marked increase in the moose population, which seems to have increased predator densities (especially corvid birds; cf. Tomialoj'c, 1990) and raised predation pressure in the boreal forest (Storaas & Wegge, 1985).

Effects of changing herbivore densities on their food plants

Release of large herbivores because of a lack of large mammal predators may also yield radically raised browsing pressure. Because large herbivorous animals can often invade habitats that are unfavourable, this edge-effect may extend over the whole landscape.

Several examples clearly show that super-normal levels of herbivory can profoundly alter plant community composition and/or plant forms. One famous example presented by Rasmussen (1941) describes the development of the deer herd on the Kaibab plateau on the north side of the Grand Canyon in Arizona. Prior to 1907, the deer herd numbered about 400. In 1907 a bounty was placed on large predators and within 15–20 years they were almost extirpated. This was followed by an immediate eruption of deer, which peaked at 100 000 in 1924. By then the vegetation was so severely damaged that 60% of the herd died in two winters. As discussed by Caughley (1970), factors other than predation explained better the eruption of deer. There are, however, other suggestive examples.

Inspired by Hough (1965), Alverson *et al.* (1988) studied the effects on vegetation caused by browsing by white-tailed deer *Odocoileus virginianus*. Enclosure studies and population studies showed that deer densities as low as 4 deer/km^2 may prevent successful recruitment of once common tree species such as Canada yew *Taxus canadensis*, eastern hemlock *Tsuga canadensis* and white cedar *Thuja occidentalis*, as well as several herbs. Recently, deer densities have often reached levels of 5–12 deer/km^2, and the stated goals of the Forest Service are for still higher numbers, calling for sufficient habitat to support super-abundant deer populations. The starting point for this increase was extensive timber cutting beginning in the late 19th century and the decline in large predators. Active management for wildlife by creating clearings and ecotones to improve deer food resources (i.e. using the traditional meaning of the edge-effect concept) has allowed a further increase in deer numbers. Because of the large home ranges of deer, they can penetrate deeply into the remaining stands of old forests and modify the plant community far away from the areas that actually support these high deer densities.

Altogether, this series of events poses serious threats to the natural plant communities in all of the north-central United States. There are three ways to create habitats for deer-sensitive plant species: enclosures, increased hunting, or habitat management to reduce deer densities. The first two are either expensive (enclosures) or probably not feasible for political reasons (hunting), while the third one in fact has been made into a proposed recommendation to create 'Diversity Maintenance Areas' (Alverson *et al.*, 1988). Such areas must be very large. Because the normal travel distances of deer between summer ranges and winter yards average 8 km, only the core area of *c.* 200 km^2 would be unaffected by deer browsing. Of course, historical patterns of movement as well as deer behaviour also affect the size required for securing a core area without deer browsing. Areas of this size are much larger than any of the remaining old-growth forest blocks in northern Wisconsin (around 2000 ha, i.e. comparable to the size of Hough's (1965) study area of 1650 ha.

To conclude, Alverson *et al.* (1988) stress the need to set aside large (200–400 km^2) continuous areas of old forest, with increasing hunting in such areas to retain plant species that are sensitive to the high browsing pressure by white-tailed deer. This proposal is politically very controversial, although it is currently being considered by the US Forest Service.

The same consequences on plants my be seen in Sweden where moose densities are supernormal. Van Ballanberghe's (1987) review of the role of predation on moose shows that wolves and bears are capable of keeping moose densities below the carrying capacity of food abundance. In Sweden these large predators disappeared as ecological factors at the end of the 19th century (Ekman, 1910). Later, the large-scale clear-cutting of Swedish forests produced a drastic increase in the amount of food for moose, which allowed an exponential growth of the moose bag from the late 1920s to the mid 1980s, when a further increase in moose densities was halted by hunting (Strandgaard, 1982). Nevertheless, the annual bag of moose in Sweden today is equal to that of the boreal zone of the entire North American continent (Kelsall, 1987). The very high moose densities cause Swedish forestry considerable damage by heavy browsing on one of the only two conifer tree species, *Pinus silvestris*. However, preferred food species like aspen *Populus tremula*, rowan *Sorbus aucuparia* and goat willow *Salix caprea* are so heavily browsed that they never grow into the natural tree form (Angelstam & Andrén, unpublished data). This is a serious long-term problem, especially for hole-nesting birds, since aspen is the tree species in which the majority of nest holes are found in boreal forest (Angelstam, 1990). More subtle secondary effects of browsing were reported by Danell and Huss-Danell (1985), who found that heavily browsed birches had a richer insect fauna than those that were ungrazed.

Animal influences on the physical environment

It is well-known that predation or herbivory may cause behavioural changes in the food organisms, induce evolutionary changes such as physical or chemical defence, or change the composition of communities. However, somc examples even suggest that animal populations may exert influences at the ecosystem, as well as at the population and community levels. For example, elephants cause major changes in habitats and landscapes in Africa that affect fire frequencies as well as the food supply and habitats for other wildlife species (Laws, 1970). On short-grass prairies in the USA, mima mounds—earthmounds up to 2 m high and sometimes 100 m across—may occur at very high densities (100/ha). Until recently they were thought to have been created by erosion or other geomorphological processes. Cox and Gakahu (1986), however, showed that they are caused by pocket gopher *Thomomys* spp. activity over several centuries.

In boreal forests, two species are especially interesting from this point of view and reveal implications that are complex but more substantial and long-lasting than we might expect. The first example is a study on the moose by Pastor *et al.* (1988) in Minnesota. If released by lack of predation, intensive browsing by moose on deciduous plants may shift the forest community to conifers *Picea glauca* and *Picea mariana*, both of which produce low-quality litter. Over longer times this change in the proportion of conifers to deciduous plants affects soil formation and nitrogen cycling, ultimately lowering the productivity of spruce as well as moose populations.

Likewise, Naiman *et al.* (1988) argue that beaver *Castor canadensis* change the structure and dynamics of streams and thus literally alter the landscape. By cutting wood and building dams, they alter stream morphology and hydrology. These activities retain sediment and organic matter, create and maintain wetlands and shallow lakes, and ultimately influence the composition and diversity of plant and animal communities (Naiman *et al.*, 1986). In a detailed analysis of the Kabetogama Peninsula in northern Minnesota Naiman *et al.* (1988) found that following the creation of a large supply of aspen by fire and logging, beaver density (measured as the number of dams) increased from 71 dams in 1940 to 835 dams in 1986. Less than 1% of the area was impounded by beaver in 1940 as compared with 13% in 1986. In addition, 12–15% of the uplands near streams and rivers were altered during the same period by beaver browsing.

2.5.2 Landscape grain and interdispersion of habitat types

Habitat fragmentation as a process involves both reduction and loss of a given habitat type and a corresponding increase of other habitat types. Equally important, however, is the graininess of the landscape mosaic as well as the juxtaposition of different habitat types. This aspect has rarely been studied at the landscape scale. The grain concept (MacArthur & Levins, 1964; MacArthur & Wilson, 1967; Levins, 1968; Southwood, 1977) applies both to the habitat as well as to the response by individuals. A landscape can be fragmented in a coarse-grained or in a fine-grained

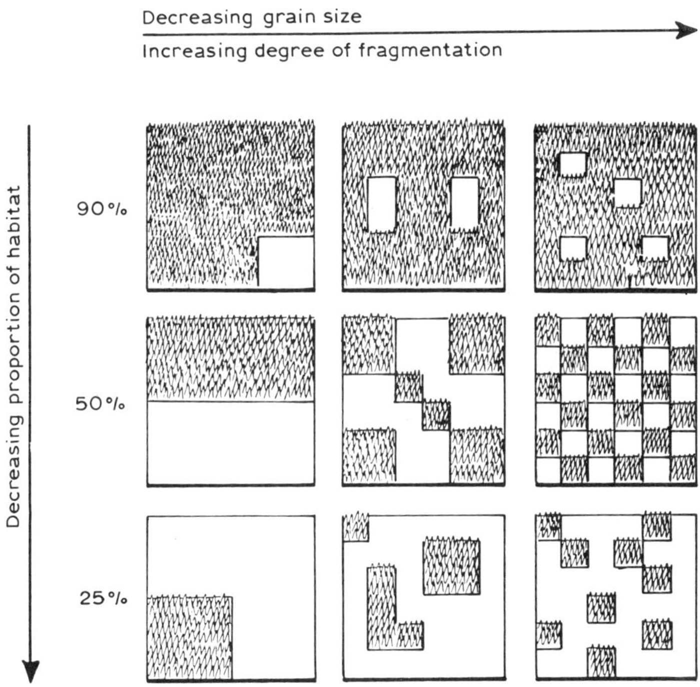

Fig. 2.9. The effects of exploitation have two components in the spatial configuration of two patch types such as old forest (dark): habitat fragmentation and habitat loss. Therefore landscapes may have, or be managed for, a coarse-grained (left) or fine-grained (right) structure.

pattern (Fig. 2.9). Correspondingly, we may characterize the 'grain response' of a given species as fine-grained if individuals utilize the habitat mosaic patches according to availability, and coarse-grained if individuals prefer certain habitat types (MacArthur & Levins, 1964). In general, the grain response is expected to become fine-grained with increasing degree of fine-grainedness of the habitat mosaic (Wiens, 1976).

The capercaillie—a case study

For birds, several population studies show that species depending on or benefiting from the old forest have shown strongly negative population trends (e.g. Väisänen *et al.*, 1986; Järvinen & Miettinen, 1988). For the capercaillie, the population has been declining continuously in Finland and is now at about 45% of the density in the 1940s.

The main factor explaining variations in capercaillie density is the proportion of the landscape covered by old forest (defined as forests which have a good field layer with *Vaccinium myrtillus* and pines of sufficient size in order to carry the weight of browsing capercaillie) (Rolstad, 1989). As the boreal forest is gradually removed, the size of old forest stands, and the proportion of old forest in the landscape, decline. This means that the territories of individual cocks are lost, leading to a reduction in the number of cocks per lek and, finally, to the disappearance of leks.

There are three different kinds of critical area requirements for the amount of old forest that the capercaillie needs in order to maintain a stable population level:

(1) 20–50 ha. The territory size of one individual capercaillie cock at the lek;
(2) 200–500 ha. The size of all the cock territories, i.e. a local lek population;
(3) >10 000 ha. The size of an area required in order to have a core area with a local 'viable population' (3–4 adjacent leks covering 2000 ha) that is unaffected by intruding generalist predators (see above).

Regarding the grain size of the forest landscape, the general question is: given a certain amount of old forest, how does the population size vary in relation to the grain of the landscape mosaic? If the size of clear-cuts is sufficiently small, the capercaillie may perceive the forest landscape as one single old forest area instead of as a set of forest stands. In other words, it may be able to tolerate the removal of a larger proportion of the old forest if clear-cuts are made sufficiently small (Rolstad & Wegge, 1987, 1989; Fig. 2.10).

Community effects of varying grain size

Spatial heterogeneity or horizontal patchiness often leads to increased local wildlife diversity (Roth, 1976; Hunter, 1990). Maintaining a high regional diversity, however, requires large forests for the species having large area requirements (Forman *et al.*, 1976).

Using bird census data Helle (pers. comm. based on Helle 1984, 1985*a*,*b*) modelled the effects of variations in landscape grain size on north taiga forest bird communities from the start of exploitaton by clear-cutting of continuous forest. Three grain sizes were used—5, 20 and 50 ha—and the rate of exploitation was 75% logged in 100 years. He then

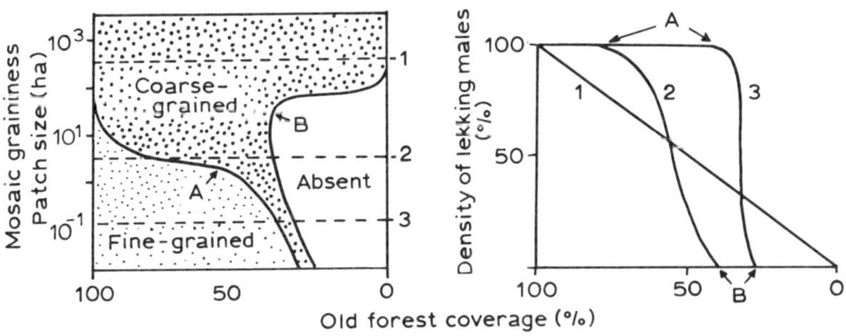

Fig. 2.10. A predictive model of the grain response of capercaillie in relation to the percent cover of old forest and the grain-size of different landscapes. The left-hand portion of the figure indicates the transition zone between a fine-grained (cf. Fig. 2.9, lower right)—male density independent of the proportion of old forest—and coarse-grained (cf. Fig. 2.9, lower left)—male density tracks the amount of old forest—habitat use. The right-hand portion of the figure denotes the critical values at which local populations are expected to go extinct due to an insufficient area of old forest remnants. 1, 2 and 3 depict male densities at three levels of landscape grain, i.e. coarse-, intermediate, and fine-grained, respectively. (Redrawn after Rolstad & Wegge, 1987.)

simulated the effects on total bird density, biomass, species richness and number of sedentary species. The simulation predicts that bird density will be highest when the smallest clear-cut area is used, and the adult biomass will be highest with the largest grain size. Bird species diversity shows the greatest values when medium-sized clear-cuts were used. However, the species suffering from forest management in north Finland are those which are sedentary (Väisänen *et al.*, 1986; Virkkala, 1990); hence, if the desired option is to reduce their decline a coarse-grained landscape is to be preferred provided that the age distribution is even. For migratory birds, the smallest cut was the optimal option.

In conclusion, there is no consensus of what is good and what is bad regarding either the amount of edges in a landscape or its grain-size (Van Horne, 1983). This topic forms a major gap in community and landscape ecology. A provisional recommendation is to create a diversity of landscape grains (Hunter, 1990).

The interdispersion of habitat patch types

While many species may be termed single-patch users and will be satisfied by one type of habitat to fulfil their daily, seasonal or life-time demands, others are multi-patch users and require certain combinations of habitat types as well as certain densities of patches in the landscape.

Leopold (1933) presented these ideas as the 'law of interdispersion', i.e. the spatial arrangement of habitats. At first, the interdispersion concept seems synonymous with graininess. However, this is true in an environment with two path types only. In a landscape with three or more patch types, there may be different types of interdispersion in mosaics of the same graininess. Hence, while the grain concept applies to the size of the habitat patches, interdispersion applies to the spatial relationship of different patch types. The juxtaposition of habitats is an important characteristic of a given environment, at any scale.

The red grouse *Lagopus lagopus scoticus* is perhaps the best example. By managing heather moors in Scotland using fire, a patchwork of different successional stages of heather is maintained so that each territory has all the components needed (Hudson, 1986).

In Minnesota and Pennsylvania, the chequerboard pattern created by logging is aimed at maintaining high densities of ruffed grouse *Bonasa umbellus* and has resulted in an increased local diversity of breeding birds (Gullion, 1984; Yahner, 1984). Franklin and Forman (1987) modelled the differences of certain landscape characteristics along a continuum from old-growth only to clear-cut only. The border length was at a maximum when the landscape consisted of equal parts of the two habitat types, and they proposed that species diversity should be highest at this ratio. In Sweden, Andrén (1989) confirmed this expectation for corvid birds over a gradient from all boreal forest to all agriculture in landscape segments of $4\,km^2$.

2.5.3 Key or indicator species, properties and processes

Contrary to the keystone species concept, the key (or perhaps better, the indicator) species concept is often questioned (e.g. Landres *et al.*, 1988). Keystone species may often be indicator species (like top predators), but indicator species do not have to be keystone species (for example, *Usnea longissima*; see chapter 7).

Linked to the key or indicator species concept is the current debate on whether conservation of communities should be achieved by applying general rules without studying individual species (Gilpin, 1987) or whether each system or even populations of a species should be studied on a case-by-case basis (Simberloff, 1988). Zimmerman and Bierregard (1986) feel that the general rules are speculative enough to be overruled even by an informed 'naturalist's feel'. As pointed out by Simberloff (1988), the fear that detailed specific studies will take too long is alleviated by the possibility of using certain keystone species (*sensu* Paine, 1966) whose interactions with other members of the community are crucial as indicator species. If we understand the mechanisms related to such an indicator species, it is likely that we will be able to conserve a large fraction of the entire community (Gilbert, 1980; Terborgh, 1986; but see Landres *et al.*, 1988).

If we attempt to design a conservation strategy for an environment about which we know nothing, there is one rule that all can agree upon—large is better than small. However, provided that our knowledge is reasonably detailed and that we know what the indicator species or topic represents, management or considerations to conserve this single species or feature may be a very good way to secure a large number of other species. However, we should not only try to conserve species but, perhaps more importantly, we must conserve mechanisms. In order to do so, some indicator features for boreal communities should be distinguished.

Compared with the virgin landscape, where forest fire was the main disturbance, today's boreal forests have changed in two important ways that affect boreal communities. First, the age distribution in a fire-dominated forest versus a managed forest are different. In the latter, every site is disturbed equally often, whereas in the former, some sites are disturbed often while others are disturbed very seldom and therefore they will have a long continuity (Van Wagner, 1978). Second, the diversity (especially of a keystone feature like tree species diversity) of today's forest has been lowered due to management for conifer timber and pulp (for details see also chapter 7).

For the preservation of boreal communities, I propose that the following indicator features be considered.

Forests with long continuity

Several plant species and invertebrates are found only in certain forests with a long continuity that has not been broken by fire. Apart from the old spruce forest itself, some lichens such as *Usnea longissima*, *Evernia divaricata* and *Lobaria* spp. are good indicators of such communities (for details see chapter 7).

Old deciduous trees and birds

In most forest sites with frequent disturbance by fire, deciduous trees form a major component that may be considered as an indicator phenomenon. Bird communities (Angelstam & Sand, unpublished data) and wood-boring insects (S. Ås, pers. comm.) also provide good examples of how the fauna is affected by the changes in forest structure and diversity that occur from natural forest with a high degree of disturbance to managed boreal forest stands. As described in chapter 7, silviculture produces stands that have a lower structural diversity, fewer mature deciduous trees and less dead wood. For birds, tree species diversity (i.e. the proportion of hardwood species (*Betula* spp. and *Populus tremula*)) was the main factor explaining variation in species number, total abundance and diversity. More than one-third of the resident bird species occurred only in forest stands containing higher diversity of deciduous tree species than found in commercially managed forests (Fig. 2.11). In Fennoscandia, for birds in boreal forest, three species may be proposed as key or indicator species which indicate—if all three are present—that the assemblage of bird species is complete. The presence of capercaillie indicates a sufficient patch size, hazel grouse that the proportion of deciduous tree species is sufficient, and three-toed woodpecker *Picoides tridactylus* or white-backed woodpecker *Dendrocopus leucotos* that there is a sufficient amount of dead wood.

Preservation of processes

However, many of the natural processes and interactions within the environment may already have been severely altered even if all the above features are preserved. One such process is predation on large herbivores, which, if it is too low, negatively affects other members of the boreal community. It is likely that viable populations of large predators like the brown bear and the wolf *Canis lupus* are indicator features for a more normal predation pressure on large prey. The black woodpecker *Dryocopus martius*, on the other hand, may be considered as an indicator species for the continuous production of large nest-holes. The need for forest fire is developed in depth in chapter 7.

To conclude, in the future much greater care must be taken to secure the long-term preservation of natural processes than has previously been the case (Fig. 12).

2.5.4 Effects on communities in terms of pollution on a regional scale

Heavy pollution from large industries and cities is now affecting considerable areas, even extending to global warming (Dobson *et al.*, 1989). Today pollution effects have begun to create problems in Swedish forests that may in just a few decades become much more severe than habitat fragmentation.

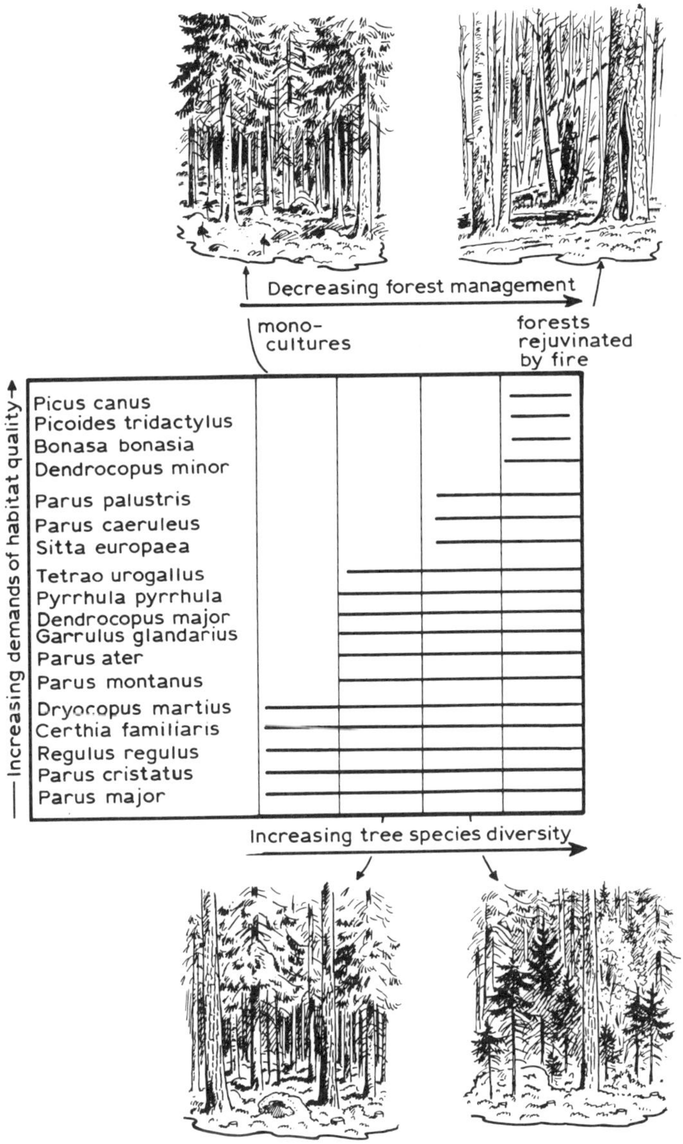

Fig. 2.11. The occurrence of resident boreal forest bird species in relation to different levels of tree-species diversity. From left to right the columns represent: single tree species coniferous stand, mixed pine and spruce stands, not intensively managed, structurally multi-layered stands with some old deciduous trees, and a forest stand that has grown up without management after a forest fire or after grazing has ceased (Angelstam & Sand, unpublished).

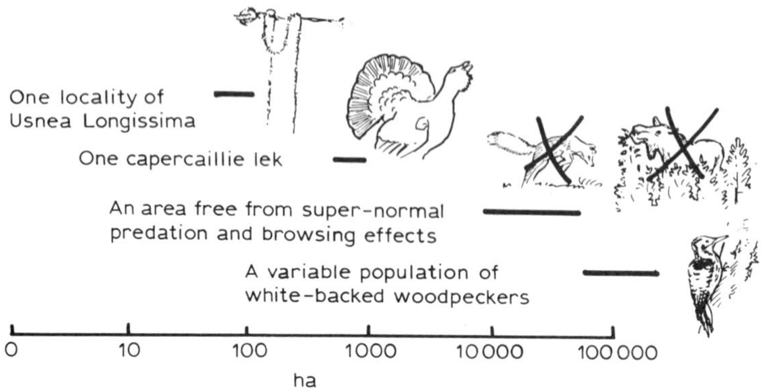

Fig. 2.12. The areas required for maintaining natural processes and viable populations are very large compared with those required for the short-term conservation of a local population. Note that the scale is logarithmic.

Nitrogen

In West Germay eutrophication is now considered the main threat to the preservation of biological diversity (a very comprehensive review is Ellenberg *et al.*, 1989). In The Netherlands the main threat to natural communities is no longer habitat fragmentation but rather the gradual change of plant communities due to a superabundant supply of nitrogen from airborne pollution (Roelofs *et al.*, 1989). Here, fast growing nitrophilous grasses are outcompeting more slowly growing plant species, thus changing heaths and peatlands into grasslands. This has led to specific types of habitat management in order to maintain certain habitats. To prevent natural coastal heathlands dominated by *Calluna* and *Erica* from turning into *Molinia* meadows, the soil is scraped and organic matter is removed to reduce nutrient loads, thereby maintaining a natural disturbance regime (Niewold & Nijland, 1987). In these areas up to 150 kg N/ha/year is coming into the terrestrial system by airborne pollution.

In other areas, already productive ecosystems such as estuaries on the North Sea coast have been provided with superabundant nitrogen supplies that lead to strong eutrophication. This is suggested to have caused the exponential growth in the numbers of seabirds (Vauk *et al.*, 1989).

Things are not yet so serious in Sweden (Nilsson & Grennfelt, 1988), although the development is alarming. In northern Sweden, the deposition of N is between 0 and 5 kg/ha/year and the fertilizing effect is considered positive from a wood fibre production point of view, while there are not yet any negative effects on plant community composition. In the south-central part of Sweden the depositions of N amount to approximately 10 kg/ha/year, which is already above the limit where effects on plant communities begin to appear. In south-western Sweden the depositions of N amount to 15–30 kg/ha/year, which may be up to ten times the dose producing a neutral response. With these high doses the ground has become saturated with N. Draining of forest land, soil scarification and the massive release of nutrients when forests are clear-cut all increase the leaking of N into the aquatic systems. In terrestrial systems we may expect changes mainly in the ground-layer vegetation.

Acidification

The precipitation of acid compounds is today two to ten times that which the environment can stand without changes. Compared with N, this problem is more of a national than an international problem, which is also most serious in the south-western part of the country, where pH has reached such low levels that important base minerals as well as heavy metals and aluminium are leaking from the soil. Recent unpublished investigation in northern Sweden show that acidification is also becoming a problem in the far north (J.-E. Lundmark, pers. comm.).

This kind of pollution produces cascading effects in aquatic as well as terrestrial communities; these will not be considered in detail in this context.

Altogether, such regional pollution effects along with locally distributed fertilizers have caused declines in many fungi in central Europe (Jansen, 1990). In Sweden botanists consider 500 of the 3000 species of fungi as extinction-prone.

2.6 MANAGEMENT IMPLICATIONS

2.6.1 Be careful when analysing data on communities and assemblages

The results presented in this book clearly show that the alteration of the boreal forest landscape has already created a loss in the diversity of habitats and hence in the diversity of plant and animal life. Unless adequate measures are taken, these changes will continue.

The loss in biological diversity involves mainly more or less sedentary species that use the boreal environment throughout their lives, such as plant species, invertebrates and resident birds. These species require disturbance regimes that are different from those in today's landscape. It is therefore misleading to analyse a whole group of species without considering the conditions that such sensitive species require when generalizing about long-term faunal changes (Järvinen & Ulfstrand, 1980). Birds are a good example. Because of the positive effects of structurally diverse edges in altered temperate environments, edge species are added to forest bird communities in patches of forest. This produces total population densities in forest patches that are generally negatively correlated with patch size (Martin, 1981). In tropical forest islands the opposite pattern may appear (Wright, 1981). However, temperate-zone bird faunas comprise both residents (usually forest interior) and migrants (usually edge and young forest species) whereas tropical species are mainly resident. Results from boreal forests in Finland (Väisänen *et al.*, 1986) show that habitat changes affect these two groups of species differently. While residents decline, migrants increase in numbers (see also Virkkala, 1990). Because migrants greatly outnumber residents, patterns among the former are likely to mask patterns among the latter. Hence, the positive relationship between density and size of a habitat island found among some resident birds (Angelstam, 1983; Rolstad, 1989; P. Helle, pers. comm.; see also Oksanen, 1987) does not show up when one considers an entire bird fauna. Thus, it is possible that the differences observed between bird faunas in temperate areas may primarily reflect differences in the proportions of migrant and resident species.

2.6.2 Conservation by creating reserves or by management

In the literature on biological conservation there has been a continued debate regarding how we should go about conserving biodiversity. Usually, this has focused only on the formation of nature reserves or national parks. However, the role of reserves versus management of man-used landscapes differs with respect to the conservation of different taxa, trophic and organizational levels. The size of an area required to maintain a viable population of *Usnea longissima*, for example, is certainly different from the area we need to maintain viable populations of the white-backed woodpecker; buying land to preserve the former is feasible but not for the preservation of the white-backed woodpecker.

It is not unfair to propose that the Swedish nature reserves and national parks have been set aside mainly for aesthetic reasons and not to secure the long-term conservation of communities and processes.

According to the criteria and laws regulating the formation of national parks in Sweden, they should meet the following standards:

(1) consist of areas with representative or unique types of Swedish landscape in a system covering the whole country;
(2) consist of untouched natural landscape or landscape that is nearly natural;
(3) contain landscape formations, features or natural environments that are magnificent or highly unusual and that have a high scientific value;
(4) cover a large area, normally at least 10 km^2;
(5) can be used within reasonable limits for outdoor recreational purposes and research without natural values being threatened.

According to the recent (1989) plan for the establishment of national parks in Sweden, the park system should increase from a total of 20 parks covering 618 300 ha to a total of 33 parks covering 2 257 100 ha, the latter corresponding to approximately 5% of the size of Sweden. With respect to the boreal forest, however, the vast majority of that area lies along the mountain chain in the northwestern part of the country and no large parks are present in the southern part of the country.

Even if we had had a good network of national parks it would presumably be difficult to preserve species, let alone processes. Several studies predict faunal collapses in nature reserves in temperate and tropical areas (Willis, 1974; Western & Ssemakula, 1981; Soulé *et al.*, 1979, 1988). However, only one of them is from the northern temperate zone. Glenn and Nudds (1989) used species–area curves derived from range maps depicting mammal distributions in Canada prior to European settlement to predict the number of species to have been present in areas with the size equalling Canadian national parks during the pre-settlement era. Comparisons with the present number of species showed that in densely populated areas parks had fewer species than expected and that the difference was larger for small than for large parks. Compared to the situation in other parts of the world, however, the threat of a faunal collapse in Canadian parks is by no means acute—except for the densely populated south eastern part of the country. Note, however, the large size of Canadian parks and that this type of calculation puts a great deal of confidence in the species–area curves, which are usually rather crude.

Maybe the Swedish national park and nature reserve system will be able to preserve at least some plant species and some of the invertebrates. For forest vertebrates and organisms demanding natural frequent-disturbance forest regimes, however, the proposed system is not sufficient for two reasons. First, for rare species the system is far too concentrated in areas of poor productivity and in the northern part of Sweden (C. Nilsson and F. Götmark, pers. comm.). Second, many species are preserved for aesthetic reasons or because they are hunted and should therefore be present throughout the landscape.

The long-term conservation of boreal forest communities must therefore be dependent on maintaining viable populations of the various species within the framework of a managed landscape in which consideration of fauna and flora is a necessary part. Parks and reserves may be important core areas in this system, as proposed by Harris (1984) (see also Wilcove, 1989). To preserve natural processes (such as low levels of predation and herbivory) in such core areas, they must be very large.

We must also accept that it is dangerous to be too orthodox when setting aside reserves. It is better not to be a purist or idealist and instead accept an assemblage of species that may not be complete but that nevertheless allows the main interactions to operate. Therefore, a reserve that is near-natural is better than no reserve at all (Kardell, 1982). In southern Sweden we may have to accept larger deviations from the ideal than in northern Sweden. Above all, however, we must manage our productive landscapes so that they become as similar as possible to the natural processes of communities and ecosystems. Management must also be adjusted so that our man-dominated landscapes are at least somewhat prepared to meet the threat of airborne pollution. To restore communities may sometimes be the only solution (cf. Niewold & Nijland, 1987).

2.6.3 Conservation of resources versus nature conservation

As correctly pointed out by Peterken (1981), it is important to distinguish between conservation in the sense of the conservation of a renewable biological resource that may be used on a sustained basis and nature conservation. In Fig. 2.13 I have transformed his ideas to the managed Swedish forest systems.

Today, foresters and conservation biologists agree that we should stay in the right part of the graph shown in Fig. 2.13. The discussion is merely about the vertical position. The development in today's society is from 2 to 3 after a successful step from 4 and 1 to 2 during the last century.

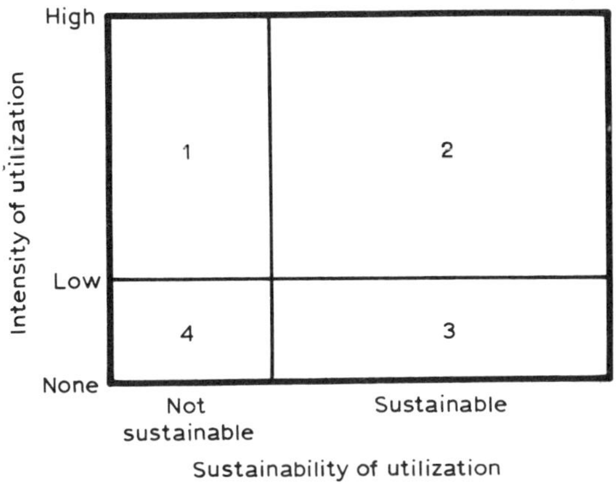

Fig. 2.13. Conservation of resources vs nature conservation (inspired by Peterken, 1981). 1, Exploitation of forest resources without ensuring that new forest regeneration is initiated within a reasonable time-span (the only recent Swedish example is the exploitation of some of the forests near the mountain ranges in the north-western part of the country). 2, The intensive forestry throughout most of Sweden. 3, A less intensive forestry where management is directed towards producing not only timber but also other values, such as in the objectives for national forests in the USA (Bean, 1983; Wilcove, 1989; see also Saastamoinen *et al.*, 1984). 4, Transitions from forest to open ground due, for example, to intensive grazing. Today this type is common in tropical and subtropical areas and has been suggested to be one of the main reasons behind the global change in climate. However, this type also occurred earlier in Sweden, when domestic cattle grazed in forests. It also occurs regionally where large herbivores browse on highly preferred tree species (e.g. *Populus tremula, Sorbus aucuparia, Salix* spp.).

2.6.4 To manage or to be passive—that is the question

So far, the literature on conservation biology has concentrated on populations with negative trends in density. It has therefore become acceptable to advocate measures to halt population declines, extending from a ban on hunting, habitat management and the creation of reserves, to capturing the wild population for breeding in captivity (e.g. the Californian condor). I have argued here, however, that the faunal changes following intensive exploitation of the environment by man also allow certain species to attain supernormal densities. It is important that it becomes equally acceptable to break the abnormally positive trends of, for example, generalist predators such as racoons and opossums in North America (Soulé *et al.*, 1988), gulls in many parts of the world (Krunnit island—Helle *et al.*, 1988) and corvid birds in many temperate man-dominated landscapes. Likewise, we must deal with the direct effects of supernormal densities of large herbivores, such as the white-tailed deer in North America and the moose in parts of Fennoscandia, on plants and plant communities, as well as the indirect effects on generalist predators.

2.7 A PLEA FOR THE FUTURE

One of the most urgent ecological problems of the present and near future is the increasing rate of species extinctions that will lead to an irretrievable loss of biological diversity on the earth. It is mainly caused by destruction of natural habitats. The threat is greatest in luxuriant tropical forests but is also acute in the Fennoscandian boreal forests. Provided that extensive forests are not fragmented and management is not too intensive, widespread losses of species in such forests are not inevitable. It is a responsibility of each nation to secure the survival of their native flora and fauna. How should this be done? See also chapter 7.

2.7.1 Preserve the remaining natural forests

Natural forests larger than a few square kilometres have entirely vanished from Fennoscandia except in the northernmost parts. It is absolutely necessary to have natural reference areas to understand ecological changes occurring in forests fragmented by forestry. Consequently, it is of the utmost priority to preserve the few remaining extensive areas of natural forest.

As shown by Alverson *et al.* (1988), and Angelstam and Welander (unpublished data), the size of areas that must be set aside to create forest with natural processes is unrealistically large. Therefore, populations must usually be managed. However, such really large areas are important as base-line areas. Assuming that we would need a study area of 2500 ha (5 × 5 km) to avoid the effect of intruding predators or herbivores, we must add to this study area a border wide enough to prevent invasions we suspect to be important. Assuming the width of this buffer zone to be 5 km (approximate maximum diameter of red fox and raven home ranges), we must find areas without managed forests of about $(5+5+5)\times(5+5+5)\,\text{km}=225\,\text{km}^2$, or 22 500 ha. There are just a handful of such areas in the whole of Fennoscandia, however, and all of them are in the north (Järvinen & Miettinen, 1988).

2.7.2 Create buffer zones around reserves

Most existing Fennoscandian reserves lack adequate buffer zones needed to protect the most valuable core areas. No reserve is isolated from its surrounding environment. Therefore, habitat changes outside reserves inevitably lead to changes in ecological processes within reserves, this effect being more acute the smaller the reserve, it is therefore highly desirable to establish such buffer zones around natural forest reserves.

To buy clear-cuts adjacent to reserves will thus in the long run be a good strategy.

2.7.3 Develop strategic land-use planning

When the amount of natural forest declines in the landscape, the numbers of vulnerable species may follow this decline only to a critical level beyond which a sudden population crash may follow. To avoid such unexpected catastrophes, we recommend regional planning to ensure a good balance of different habitat types and forest age classes.

Several bird species requiring forest-interior habitats show long-term negative trends both in North America (Whitcomb *et al.*, 1981) and Northern Europe (Väisänen *et al.*, 1986). It is therefore important to plan the exploitation of a forest resource so that sufficient amounts of old forest are left in each region. It should also be made clear that changes in the ecological conditions are non-linear as a landscape changes from all forest to all clear-cut (Franklin & Forman, 1987). Local populations may suddenly disappear even if there appears to be a great deal of habitat left. For capercaillie, for example, the social system seems to break up when 70% of the old growth has been removed (Rolstad, 1989).

2.7.4 Enhance the natural variation in forests

A major ecological characteristic of natural forests is small-scale diversity in environmental conditions such as topography, microclimate, soil conditions and plant cover. Such variation is destroyed by uniform and extensive silviculture. There is an urgent need to develop silvicultural practices that are sympathetic to and protect the natural diversity of forest environments.

Arguments for this point are mainly found in the chapters about wet forests and boreal forests, which also nicely illustrates the tyranny of small decisions. Many locally unimportant decisions such as draining after logging, precommercial thinnings that selectively remove certain tree species, removing snags etc., show up as profound landscape changes in the long run.

2.7.5 Increase numbers of (old) broad-leaved trees

The biological diversity of managed forests can be restored by using knowledge obtained from natural forests. An important, realistic measure is to increase the proportion of deciduous trees in mature forest stands. For example, the proportion of old broad-leaved trees in forest stands in Swedish coniferous forests today is only 4%, but it should be increased to at least 10%.

2.7.6 Plant native trees

The introduction of exotic tree species as *Pinus contorta* and *Picea sitchensis* may have unexpected, disastrous consequences for native ecological systems. We urge great caution and far more explicit regulation and legislation concerning introductions.

The observation that *Pinus contorta*, contrary to common belief, can produce offspring without the aid of forest fire to break serotinous cones may produce unexpected chain effects if the species spreads into new habitats.

ACKNOWLEDGEMENTS

This chapter summarizes my personal view on landscape ecology and conservation as it has developed during the last five years of research. Per-Anders Esseen kindly provided a summary of the forest fragmentation experiment at Gardfjället, and Pekka Helle data from his models on the effects of grain-size on bird communities. The plea was discussed in great detail at an international workshop among forestry and wildlife scientists, 'Effects of forest fragmentation on the bird fauna: theoretical aspects and practical implications' at Grimsö Wildlife Research Station, Sweden, on 21–25 November, 1988.

I thank especially Henrik Andrén, Per-Anders Esseen, Lennart Hansson, Pekka Helle, Gray Merriam, Jørund Rolstad and John Wiens for detailed comments on the manuscript.

REFERENCES

Addicott, J.F., Aho, J.M., Antolin, M.F., Padilla, D.K., Richardson, J.S. & Soluk, D.A. (1987). Ecological neighbourhoods: scaling environmental patterns. *Oikos*, **49**, 340–6.

Allee, W.C., Emerson, A.E., Parc, O., Parc, T. & Schmidt, K.P. (1949). *The Principles of Animal Ecology*. Saunders, Philadelphia.

Alverson, W.S., Waller, D.M. & Solheim, S.L. (1988). Forest too deer: edge effects in northern Wisconsin. *Conserv. Biol.*, **2**, 348–58.

Ambuel, B. & Temple, S.A. (1983). Area-dependent changes in the bird communities and vegetation of southern Wisconsin forests. *Ecology*, **64**, 178–86.

Andrén, H. (1989). Predation processes in fragmented boreal forest landscapes. PhD thesis, *Acta Universitatis Uppsaliensis*, **201**.

Andrén, H. & Angelstam, P. (1988). Elevated predation rates as an edge effect in habitat islands: experimental evidence. *Ecology*, **69**, 544–7.

Andrén, H., Angelstam, P., Lindström, E. & Widén, P. (1985). Differences in predation pressure in relation to habitat fragmentation: an expriment. *Oikos*, **45**, 273–7.

Andrewartha, H.G. & Birch, L.C. (1954). *The Distribution and Abundance of Animals*. University of Chicago Press, Chicago, Illinois.

Andrewartha, H.G. & Birch, L.C. (1984). *The Ecological Web*. University of Chicago Press, Chicago.

Angelstam, P. (1983). Population dynamics of tetraonids, especially the black grouse *Tetrao tetrix* L., in boreal forests. PhD thesis, Uppsala University.

Angelstam, P., Lindström, E. & Widén, P. (1985). Synchronous short-term population fluctuations of some birds and mammals in Fennoscandia—occurrence and distribution. *Holarctic Ecology*, **8**, 285–98.

Angelstam, P. (1986). Predation in ground-nesting birds' nests in relation to predator densities and habitat edge. *Oikos*, **47**, 365–73.

Angelstam, P. (1990). Factors determining the composition and persistence of local woodpecker assemblages in taiga forest in Sweden. In *Conservation of Woodpeckers*, ed. A. Carlsson & G. Aulén. SLU Info/repro Uppsala, pp. 147–64.

Angelstam, P., Lindström, E. & Widén, P. (1984). Role of predation in short-term fluctuations of some birds and mammals in Fennoscandia. *Oecologia* (*Berl.*), **62**, 199–208.

Anon. (1984). *Naturgeografisk Regionindelning av Norden*. Nordiska Ministerrådet, Berlings, Arlöv, Sweden.

Anon. (1989). *Det Föränderliga Odlingslandskapet*. Arbetsmaterial, Riksantikvarieämbetet, Stockholm.

Antonovics, J. & Levin, D.A. (1980). The ecological and genetic consequences of density-dependent regulation in plants. *Ann. Rev. Ecol. Syst.*, **11**, 411–52.

Bean, M.J. (1983). *The Evolution of the National Wildlife Law*. Praeger, New York.

Bergerud, A.T. & Mossop, D.H. (1984). The pair bond in ptarmigan. *Can. J. Zool.*, **62**, 2129–41.

Boström, U. & Nilsson, S.G. (1983). Latitudinal gradients and local variations in species richness and structure of bird communities on raised peat-bogs in Sweden. *Ornis Scand.*, **14**, 213–26.

Brittingham, M.C. & Temple, S.A. (1983). Have cowbirds caused forest songbirds to decline? *BioScience*, **33**, 31–35.

Brusewitz, G. & Emmelin, L. (1985). *Det Föränderliga Landskapet*. Esselte Herzogs, Uppsala.

Burger, G.V. (1973). *Practical Wildlife Management*. Winchester Press, New York.

Burgess, R.L. & Sharpe, D.M. (1981). *Forest Island Dynamics in Man-dominated Landscapes*. Springer Verlag, New York.

Carlquist, S. (1974). *Island Biology*. Columbia University Press, New York.
Carpenter, J.R. (1935). Forest edge birds and exposures of their habitats. *Wilson Bull.*, **47**, 106–8.
Caughley, G. (1970). Eruption of ungulate populations, with emphasis on Himalayan thar in New Zealand. *Ecology*, **51**, 54–72.
Chasko, G.G. & Gates, J.E. (1982). Avian habitat suitability along a transmission-line corridor in an oak-hickory forest region. *Wildl. Monogr.*, **82**, 1–4.
Christianssen, E. (1979). Skog-och jordbruk, smågnagere og rev. *Norw. J. For.*, **3**, 115–19.
Christiansen, F.B. & Fenchel, T.M. (1977). *Theories of Populations in Biological Communities*. Springer-Verlag, New York.
Clements, F.C. (1897). Peculiar zonal formations of the Great Plains. *Amer. Nat.*, **3**, 968.
Clements, F.C. (1916). *Plant Succession*. Publs. *Carnegie Inst. Wash.*, No. 242.
Coblentz, B.E. (1978). The effects of feral goats *Capra hircus* on island ecosystems. *Biol. Conserv.*, **13**, 279–86.
Cox, G.W. & Gakahu, C.G. (1986). A latitudinal test of the fossorial rodent hypothesis of mima mound origin. *Ann. Geomorphol.*, **30**, 485–501.
Danell, K. & Huss-Danell, K. (1985). Feeding by insects and hares on birches earlier affected by moose browsing. *Oikos*, **44**, 75–81.
Dasmann, R.F. (1964). *Wildlife Biology*. Wiley, New York.
Dayton, P.K., Currie, V., Gerodette T., Keller, B.D., Rosenthal, R. & Ven Tresca, D. (1984). Patch dynamics and stability of some California kelp communities. *Ecol. Monogr.*, **54**, 253–89.
De Franceschi, P. (1982). Fluttuazioni della popolazioni di tetraonidi sulli Alpi Carniche. *Estratto da Dendronatura*, **3**, 19–38.
De Walle, D.R. (1983). Wind damage around clearcuts in the ridge and valley province of Pennsylvania. *J. For.*, **81**, 158–9.
di Castri, F., Hansen, A.J. & Holland, M.M. (eds.) (1988). A new look at ecotones: emerging international projects on landscape boundaries. *Biology Int.*, **17**. International Union of Biological Sciences, Paris.
Dierschke, H. (1974). *Saumgesellschaften im Vegetations- und Standortsgefälle an Waldrändern*. Verlag Erich Goltze KG, Göttingen.
Dilks, P.J. & Wilson, P.R. (1978). Feral sheep and cattle and royal albatrosses on Campbell Island; populations trends and habitat changes. *N.Z.J. Zool.*, **6**, 127–39.
Dobson, A., Jolly, A. & Rubenstein, D. (1989). The greenhouse effect and biological diversity. *Trends Ecol. Evol.*, **4**, 64–8.
Drolet, C.A. (1978). Use of forest clear-cuts by white-tailed deer in southern New Brunswick and Central Nova Scotia. *Can. Field-Nat.*, **92**, 275–82.
Ekman, S. (1910). Om människans andel i fiskfaunans spridning till det inre Norrlands vatten. *Ymer*, **30**, 133–40.
Ellenberg, H., Rüger, A. & Vauk, G. (eds.) (1989). Eutrophierung—das gravierendste Problem im Naturschutz. *Nordd. Naturschutzakad., Ber.*, **2** (1).
Ellison, L. & Magnani, Y. (1985). Eléments de dynamique de population du tétras lyre (*Tetrao tetrix*) dans les Alpes Francaises. *Gibier Faune Sauvage*, **4**, 63–84.
Ellison, L., Magnani, Y. & Corti, R. (1982). Comparison of hunted and three protected black grouse populations in French Alps. *Proc. Int. Symp. on Grouse, 2nd*, 1981, ed. T. Lovel, 175–87.

Elton, C. (1927). *Animal Ecology*. Sidgwick & Jackson, London.
Esseen, P.-A. (1983). Ecology of lichens in boreal coniferous forests with reference to spatial and temporal patterns. Ph.D. thesis, University of Umeå, Sweden.
Estes, J.A., Jameson, R.J. & Rhode, E.B. (1982). Activity and prey selection in the sea otter: influence of population status on community structure. *Amer. Nat.*, **120**, 242–58.
Falinski, J.B. (1986). Vegetation Dynamics in Temperate Lowland Primeval Forest. *Geobotany*, **8**, 1–537.
Forman, R.T.T. & Baudry, J. (1984). Hedgerows and hedgerow networks in landscape ecology. *Environ. Manage.*, **8**, 495–510.
Forman, R.T.T. & Godron, M. (1981). Patches and structural components for a landscape ecology. *BioScience*, **31**, 733–40.
Forman, R.T.T. & Godron, M. (1986). *Landscape Ecology*, Wiley, New York.
Forman, R.T.T., Galli, A.E. & Leck, C.F. (1976). Forest size and avian diversity in New Jersey woodlots with some land-use implications. *Oecologia* (*Berl.*), **26**, 1–8.
Franklin, J.F. & Forman, R.T.T. (1987). Creating landscape patterns by forest cutting: ecological consequences and principles. *Landscape Ecol.*, **1**, 5–18.
Gamlin, L. (1988). Sweden's factory forest. *New Scient.*, **117**, 41–7.
Gates, J.E. & Gysel, L.W. (1978). Avian nest dispersion and fledgling success in field-forest ecotones. *Ecology*, **59**, 871–83.
Gates, J.E. & Mosher, J.A. (1981). A functional approach to estimating habitat edge width for birds. *Amer. Midl. Nat.*, **105**, 189–92.
Gilbert, L.E. (1980). Food web organization and conservation of neotropical diversity. In *Conservation Biology: An Evolutionary-Ecological Perspective*, ed. M.E. Soulé & B.E. Wilcox. Sinauer Associates, Sunderland, Massachusetts, pp. 11–34.
Gilpin, M.E. (1987). Theory vs practice. *Trends Ecol. Evol.*, **2**, 16.
Ginsberg, J.R. (1987). What is conservation biology? *Trends Ecol. Evol.*, **2**, 262–4.
Gleason, H.A. (1917). The structure and development of the plant association. *Bull. Torrey Bot. Club*, **43**, 463–81.
Glenn, S.M. & Nudds, T.M. (1989). Insular biogeography of mammals in Canadian parks. *J. Biogeogr.*, **16**, 261–8.
Golley, F.B. (1989). Landscape ecology and biological conservation. *Landscape Ecol.*, **2**, 201–2.
Grange, W.B. (1946). *The Way to Game Abundance.* Charles Scribner's Sons, New York.
Grieg-Smith, P. (1979). Pattern in vegetation. *J. Ecol.*, **67**, 755–79.
Griggs, R.F. (1914). Observations on the behavior of some species on the edge of their ranges. *Bull. Torrey Bot. Club*, **41**, 25–49.
Gullion, G.W. (1984). *Managing Northern Forests for Wildlife*. The Ruffed Grouse Society, Miscellaneous Journal Series, Minnesota Agricultural Station, St Paul.
Gysel, L.W. (1951). Borders and openings of beech–maple woodlands in southern Michigan. *J. For.*, **49**, 13–19.
Hagar, C.C. (1960). The interrelationship of timber, birds, and timber regeneration in the Douglas fir region of northwestern California. *Ecology*, **41**, 116–25.

Haila, Y., Järvinen, O. & Väisänen, R. (1979). Effect of mainland population changes on the terrestrial bird fauna of a northern island. *Ornis Scand.*, **10**, 48–55.

Hairston, N., Smith, F. & Slobodkin, L. (1960). Community structure, population control and competition. *Amer. Nat.*, **94**, 421–5.

Hamilton, G.D., Drysdale, P.D. & Euler, D.L. (1980). Moose winter browsing patterns on clear-cuttings in northern Ontario. *Can. J. Zool.*, **58**, 1412–6.

Hansen, J., Fung, I., Lacis, A., Lebedeff, S., Rind, D., Ruedy, R., Russel, G. & Stone, P. (1988). Prediction of near-term climate evolution: What can we tell decision-makers now? *Preparing for the Climate Change: Proc. N. Amer. Conf. Preparing for Climate Change*. Climate Institute, Rockville, Maryland.

Hanski, I. (1989). Metapopulation dynamics: does it help to have more of the same? *Trends Ecol. Evolut.*, **4**, 113–14.

Hanski, I. (1991). Single-species metapopulation dynamics: concepts, models and observations. *Biol. J. Linn. Soc.*, **42**, (1/2), pp. 17–38.

Hansson, L. (1979). On the importance of landscape heterogeneity in northern regions for the breeding population densities of homeoterms: a general hypothesis. *Oikos*, **33**, 182–9.

Hansson, L. (1983). Bird numbers across edges between mature conifer forest and clearcuts in central Sweden. *Ornis Scand.*, **14**, 97–103.

Hansson, L. (1988). Grazing impact by small rodents in a steep cyclicity gradient. *Oikos*, **51**, 31–42.

Hansson, L. & Angelstam, P. (1988). New ecological processes in small reserves. In *Nature Conservation–Symposium in Budapest, 27 March 1987*, ed. L. Svensson. National Swedish Environmental Protection Board, Report 3496, pp. 25–32.

Hansson, L. & Angelstam, P. (1991). Landscape ecology as a theoretical basis for nature conservation. *Landscape Ecol.*, **5**(4), 191–201.

Hansson, L. & Henttonen, H. (1988). Rodent dynamics as community processes. *Trends Ecol. Evol.*, **3**, 195–200.

Harris, L.D. (1984). *The Fragmented Forest. Island Biogeography Theory and the Preservation of Biotic Diversity*. University of Chicago Press, Chicago.

Harris, L.D. (1988*a*). Edge effects and conservation of biotic diversity. *Conserv. Biol.*, **2**, 330–2.

Harris, L.D. (1988*b*). The nature of cumulative impacts on the biotic diversity of wetland vertebrates. *Environ. Manage.*, **12**, 675–93.

Harris, L.D. (1989). A rationale for wildlife habitat superiority of southern bottomland hardwood communities. In *Bottomland Hardwood Wetlands; Ecological and Regulatory Perspectives*, ed. J. Gosselink, L. Lee & T. Muir.

Harris, L.D. & Gallagher, P.B. (1989). New initiatives for wildlife conservation. The need for movement corridors. In *In Defense of Wildlife: Preserving Communities and Corridors*, ed. G. Mackintosh. Defenders of Wildlife, Washington DC, pp. 11–34.

Harris, L.D., Maser, C. & Mckee, A. (1982). Patterns of old growth harvest and implications for Cascades wildlife. *Trans. N. Amer. Wildl. Nat. Res. Conf.*, **47**, 374–92.

Helle, P. (1983). Bird communities in open ground-climax forest edges in northeastern Finland. *Oulanka Reports*, **3**, 47–9.

Helle, P. (1984). Effect of habitat area on breeding bird communities in Northeastern Finland. *Ann. Zool. Fennici*, **21**, 421–5.

Helle, P. (1985*a*). Habitat selection of breeding birds in relation to forest succession in Northeastern Finland. *Ornis Fennica*, **62**, 113–23.

Helle, P. (1985*b*). Effects of forest regeneration on the structure of bird communities in northern Finland. *Holarct. Ecol.*, **8**, 120–32.

Helle, E. & Helle, P. (1982). Edge effect on forest bird communities on offshore islands in the northern Gulf of Bothnia. *Ann. Zool. Fennici*, **19**, 165–9.

Helle, E., Helle, P. & Väisänen, R.A. (1988). Population trends among archipelago birds in the Krunnit sanctuary, northern gulf of Bothnia, in 1939–85. *Ornis Fennica*, **65**, 1–12.

Hough, A.F. (1965). A twenty-year record of understory vegetational change in a virgin Pennsylvania forest. *Ecology*, **46**, 370–3.

Hubbell, S.P. & Foster, R.B. (1983). Diversity of canopy trees in a neotropical forest and implications for conservation. In *Tropical Rain Forest: Ecology and Management*, ed. S.L. Sutton, T.C. Whitmore & A.C. Chadwick. Blackwell Scientific Publications, Oxford, pp. 25–42.

Hubbell, S.P. & Foster, R.B. (1986). Commonness and rarity in a Neotropical forest: complications for tropical tree conservation. In *Conservation Biology: The Science of Scarcity and Diversity*, ed. M.E. Soulé. Sinauer Associates, Sunderland, Massachusetts, pp. 205–31.

Hudson, P. (1986). *Red Grouse—the Biology and Management of a Wild Gamebird.* The Game Conservancy Trust, Bourne Press, Bournemouth.

Hunter, M.L. (1990) *Wildlife, Forests and Forestry. Principles of Managing Forests for Biological Diversity.* Prentice-Hall, New Jersey.

Hörnfeldt, B. (1978). Synchronous population fluctuations in voles, small game, owls and tularemia in northern Sweden. *Oecologia* (*Berl.*), **32**, 141–52.

Hörnfeldt, B., Löfgren, O. & Carlsson, B.-G. (1986). Cycles in voles and small game in relation to variations in plant production indices in Northern Sweden. *Oecologia*, (*Berl.*), **68**, 496–502.

Jakucs, P. (1972). *Dynamische Verbindung der Wälder und Rasen.* Akademiai Kiado, Verlag Ungarischen Akademie, Budapest.

Jansen, A.E. (1990) Conservation of fungi in Europe. *Mycologist*, **4**(2), 83–5.

Janzen, D.H. (1983). No park is an island: increase in interference from outside as park size decreases. *Oikos*, **41**, 402–10.

Janzen, D.H. (1986). The eternal external threat. In *Conservation Biology: The Science of Scarcity and Diversity*, ed. M.E. Soulé. Sinauer Associates, Sunderland, Massachusetts, pp. 286–303.

Järvinen, O. & Miettinen, K. (1988). *Sista Paret ut. Om Naturvårdens Biologi.* Naturskyddsföreningen, Miljöförlaget, Helsinki, Finland.

Järvinen, O. & Ranta, E. (1987). Patterns and processes in species assemblages on northern Baltic islands. *Ann. Zool. Fennici*, **24**, 249–66.

Järvinen, O. & Ulfstrand, S. (1980). Species turnover of a continental bird fauna: Northern Europe, 1850–1970. *Oecologia* (*Berl.*), **46**, 186–95.

Jennersten, O. (1988). Pollination in *Dianthus deltoides* Caryophyllaceae: effects of habitat fragmentation on visitation and seed set. *Conserv. Biol.*, **2**, 359–66.

Johnson, R.G. & Temple, S.A. (1986). Assessing habitat quality for nesting in fragmented tallgrass prairies. In *Wildlife 2000: Modeling Habitat Relation-*

ships of Terrestrial Vertebrates, ed. J. Verner, M.L. Morrison & C.J. Ralph. University of Wisconsin Press, Madison, Wisconsin, pp. 245–50.

Jönsson, P.E. (1988). Ecology of the southern dunlin *Calidris alpina schintzii*. PhD thesis, Department of Animal Ecology, Lund University.

Kardell, L. (1973). Skogbrok och friluftsliv—en olöslig konflikt. *Sveriges Naturs Årsbok*, **64**, 199–208.

Kardell, L. (1982). *Tivedens Nationalpark—en Skogshistorisk Betraktelse*. Swedish University of Agricultural Sciences, Section of Environmental Forestry, Report No. 22.

Karr, J. (1982). Avian extinction on Barro Colorado Island: a reassessment. *Amer. Nat.*, **119**, 220–39.

Kelsall, J.P. (1987). The distribution and status of moose *Alces alces* in North America. *Swed. Wildl. Res.*, Suppl. **1**, 1–10.

Klein D.R. (1968). The introduction, increase, and crash of reindeer on St Matthew island. *J. Wildl. Manage.*, **32**, 350–67.

Knopf, F.L. (1986). Changing landscapes and the cosmopolitism of the eastern Colorado avifauna. *Wildl. Soc. Bull.*, **14**, 132–42.

Kobayashi, S. (1985). Species diversity preserved in different numbers of nature reserves of the same total area. *Res. Popul. Ecol.*, **27**, 137–43.

Kroodsma, R.L. (1982). Bird community ecology on power-line corridors in East Tennessee. *Biol. Conserv.*, **23**, 79–94.

Kuitunen, M. & Helle, P. (1988). Relationship of the common treecreeper *Certhia familiaris* to edge effect and forest fragmentation. *Ornis Fennica*, **65**, 150–5.

Landres, P.B., Verner, J. & Thomas, J.W. (1988). Ecological uses of vertebrate indicator species: A critique. *Conserv. Biol.*, **2**, 316–28.

Laws, R.M. (1970). Elephants as agents of habitat and landscape change in East Africa. *Oikos*, **2**, 1–15.

Lay, D. (1938). How valuable are woodland clearings to birdlife? *Wilson Bull.*, **50**, 254–6.

Leopold, A. (1933). *Game Management*. Charles Scribner's & Sons, New York.

Levenson, J.B. (1981). Woodlots as biogeographic islands in southeastern Wisconsin. In *Forest Island Dynamics in Man-dominated Landscapes*, ed. R.L. Burgess & D.M. Sharpe. Springer-Verlag, New York, pp. 13–39.

Levins, R. (1968). *Evolution in Changing Environments*. Princeton University Press, Princeton, New Jersey.

Lindén, H. (1988). Latitudinal gradients in predator–prey interactions, cyclicity and synchronism in voles and small game populations in Finland. *Oikos*, **52**, 341–9.

Lindström, E., Angelstam, P., Widen, P. & Andrén, H. (1988). Do predators synchronize vole and grouse fluctuations? An experiment. *Oikos*, **48**, 121–4.

Livingston, B.E. (1903). The distribution of the upland societies of Kent County. *Mich. Bot. Gaz.*, **35**, 36–55.

Lovejoy, T.E., Bierregard, A.B., Rylands, A.B., Malcolm, J.R., Brown, K.R., Quintela, C.E., Harper, L.H., Powell, A.H., Powell, G.V.N., Schubart, H.O.R. & Hays, M.B. (1986). Edge and other effects of isolation on Amazon forest fragments. In *Conservation Biology. The Science of Scarcity and Diversity*, ed. M.L. Soulé. Sinauer Associates, Sunderland, Massachusetts, pp. 257–85.

Lyon, L.J. & Jensen, C.E. (1980). Management implications of elk and deer use of clear-cuts in Montana. *J. Wildl. Manage.*, **44**, 352–62.
MacArthur, R.H. & Levins, R. (1964). Competition, habitat selection, and character displacement in a patchy environment. *Proc. Natn. Acad. Sci. U.S.*, **51**, 1207–10.
MacArthur, R.H. & Wilson, E.O. (1967). *The Theory of Island Biogeography.* Princeton University Press, Princeton, New Jersey.
Mackintosh, G. (ed.) (1989). *In Defence of Wildlife: Preserving Communities and Corridors*, Defenders of Wildlife, Washington, DC.
Maehr, D. (1984). Distribution of black bears in eastern North America. In *7th Proc. Eastern Workshop on Black Bear Research and Management*, ed., D. Maehr & J. Brady. Florida Game and Freshwater Fish Commission, Gainesville, Florida, pp. 74.
Martin, T.E. (1981). Limitation in small habitat islands: chance or competition? *Auk*, **98**, 715–34.
Martin, T.E. (1988). Habitat and area effects on forest bird assemblages: is nest predation an influence? *Ecology*, **69**, 74–84.
Matthiae, P. & Stearns, F. (1981). Mammals in forest islands in southeastern Wisconsin. In *Forest Island Dynamics in Man-dominated Landscapes*, ed. R.L. Burgess & D.M. Sharpe. Springer Verlag, New York, pp. 55–66.
Mattson, L. & Stridsberg, E. (1980). *Skogens Roll i Svensk Markanvändning.* Department of Forest Economics, Swedish Institute of Agricultural Sciences, Umeå.
McLellan, C.H., Dobson, A.P., Wilcove, D.S. & Lynch, J.F. (1986). Effects of forest fragmentation on New and Old-World bird communities: empirical observations and theoretical implications. In *Wildlife 2000: Modeling Habitat Relationships of Terrestrial Vertebrates*, ed. J. Verner, M.L. Morrison & C.J. Ralph. University of Wisconsin Press, Madison, Wisconsin, pp. 305–13.
Merriam, G. (1984). Connectivity: A fundamental ecological characteristic of landscape pattern. In *Proc. First. Int. Sem. Meth. Landscape Ecol. Res. Plann.* Theme 1, ed. J. Brandt & P. Agger. Roskilde. Denmark, pp. 5–15.
Merriam, G. (1988). Landscape dynamics in farmland. *Trends Ecol. Evol.*, **19**, 16–20.
Meurk, C.D. (1982). Regeneration of subantarctic plants on Campbell Island following exclusion of sheep. *N.Z.J. Ecol.*, **5**, 51–8.
Moen, A.N. (1973). *Wildlife Ecology: An Analytical Approach.* W.H. Freeman, San Fransisco, California.
Mooney, H.A. & Gordon, M. (eds) (1983). *Disturbance and Ecosystems. Components of Response.* Springer-Verlag, Berlin, Heidelberg.
Morgan, K.A. & Gates, J.E. (1982). Bird population patterns in forest edge and strip vegetation at Remington Farms, Maryland. *J. Wildl. Manage.*, **46**, 933–44.
Naiman, R.J., Johnston, C.A. & Kelley, J.C. (1988). Alteration of North American streams by beaver. *BioScience*, **38**, 753–62.
Naiman, R.J., Melillo, J.M. & Hobbie, J.E. (1986). Ecosystem alteration of boreal forest streams by beaver (*Castor canadensis*). *Ecology*, **67**, 1254–69.
Naveh, Z. & Lieberman, A.S. (1984). *Landscape Ecology–Theory and Application.* Springer Verlag, New York.

Niewold, F.J.J. & Nijland, H. (1987). The outlook for the West European moor- and heathland blackcock. *Zeit. Jagdwiss.*, **33**, 227–41.

Nilsson, J. & Grennfelt, P. (1988). Critical loads for sulphur and nitrogen. *Nordic Council of Ministers*, **15**.

Nilsson, S. G., Bengtsson, J. & Ås, S. (1988). Habitat diversity or area *per se*? Species richness of woody plants, carabid beetles and land snails on islands. *J. Anim. Ecol.*, **57**, 685–704.

Nilsson, S.G. & Wästljung, U. (1987). Seed predation and cross-pollination in mast-seeding beech *Fagus sylvatica* patches. *Ecology*, **68**, 260–5.

Noss, R.F. (1983). A regional landscape approach to maintain diversity. *BioScience*, **33**, 700–6.

Odum, E.P. (1971). *Fundamentals of Ecology*. Saunders, Philadelphia, Pennsylvania.

Oksanen, L. (1987). Interspecific competition and the structure of bird guilds in boreal Europe: the importance of doing fieldwork in right season. *Trends Ecol. Evol.*, **2**, 376–9.

Oksanen, L., Fretwell S.D., Arruda, J. & Niemelä, P. (1981). Exploitation of ecosystems in gradients of primary productivity. *Amer. Nat.*, **118**, 240–61.

O'Neill, R.V., DeAngelis, D.L., Waide, J.B. & Allen, T.F.H. (1986). *A Hierarchical Concept of Ecosystems*. Princeton University Press, Princeton, New Jersey.

Pacala, S. & Roughgarden, J. (1984). Conrol of arthropod abundance by *Anolis* lizards on St Eustatius (Neth. Antilles). *Oecologia (Berl.)*, **64**, 160–2.

Paine, R.T. (1966). Food web complexity and species diversity. *Amer. Nat.*, **100**, 65–75.

Paine, R.T., Castillo, J.C. & Cancino, J. (1985). Perturbation and recovery patterns of starfish-dominated intertidal assemblages in Chile, New Zealand, and Washington State. *Amer. Nat.*, **125**, 679–91.

Pastor, J., Naiman, R.J., Dewey, B. & McInnes, P. (1988). Moose, microbes and the boreal forest. *BioScience*, **38**, 770–7.

Pelton, M. (1986). Habitat needs of black bears in East. In *Wilderness and Natural Areas in Eastern United States: A Management Challenge*, ed. D. Kulhavy & R. Conner. Center for Applied Studies, School of Forestry, Stephen F. Austin State University, Nacogdoches, Texas.

Peterken, G. (1981). *Woodland Conservation and Management*. Chapman & Hall, London and New York.

Pickett, S.T.A. & White, P.S. (1985). *The Ecology of Natural Disturbance and Patch Dynamics*. Academic Press, New York.

Ranney, J.W. (1977). Forest island edges—their structure, development, and importance to regional forest ecosystem dynamics. EDFB/IBP-77/1. Oak Ridge National Laboratory, Oak Ridge, Tennessee.

Ranney, J.W, Bruner, M.C. & Levinson J.B. (1981). The importance of edge in the structure and dynamics of forest islands. In *Forest Island Dynamics in Man-dominated Landscapes*, ed. R.L. Burgess & D.M. Sharpe. Springer-Verlag, New York, pp. 67–95.

Rasmussen, D.I. (1941). Biotic communities of Kaibab Plateau, Arizona. *Ecol. Monogr.*, **3**, 229–75.

Reese, K.P. & Ratti, J.T. (1988). Edge effect: a concept under scrutiny. *Trans. N. Amer. Wildl. Res. Conf.*, **53**.

Richards, J.F. & Tucker, R.P. (1988). *World Deforestation in the Twentieth Century*. Duke University Press, Durham and London.

Risser, P.G., Karr, J.R. & Forman, R.T.T. (1984). Landscape ecology. Directions and approaches. *III. Nat. Hist. Surv. Spec. Publs*, No. 2.

Robbins, C.S. (1980). Effect of forest fragmentation on breeding bird populations in the Piedmont of the Mid-Atlantic region. *Atlantic Nat.*, **33**, 31–6.

Rodriguez Luengo, J.L. & Rodriguez Pinero, J.C. (1990). Introduced big game: a threat to Canary endemic flora. *19th Trans. Int. Un. Game Biol. Congr.*, pp. 530–5. Norwegian institute for Nature Research, Trondheim, Norway.

Roelofs, J.G.M., Boxman, A.W. & Van Dijk, H.F.G. (1989). Effects on airborne ammonium on natural vegetation and forests. In *Eutrophierung—das gravierendste Problem im Naturschutz*, ed. H. Ellenberg, A. Rüger & G. Vauk. Nordd. Naturschutzakad., *Ber.* **2**, pp. 38–41.

Rolstad, J. (1989). Effects of logging on capercaillie *Tetrao urogallus* leks: cutting experiments in Southcentral Norway. *Scand. J. For. Res.*, **4**, 99–109.

Rolstad, J. & Wegge, P. (1987). Distribution and size of capercaillie leks in relation to old forest fragmentation. *Oecologia* (*Berl.*), **72**, 389–94.

Rolstad, J. & Wegge, P. (1989). Capercaillie, *Tetrao urogallus*, populations and modern forestry—a case for landscape ecological studies. *Finn. Game Res.*, **46**, 43–52.

Roth, R.R. (1976). Spatial heterogeneity and bird species diversity. *Ecology*, **57**, 773–82.

Ruwet, J.C. (1982). Statut et evolution, dans le contexte européen, des populations de tétras lyre *Tetrao tetrix* dans les Ardennes Belges. *Cah. d'Ethol. Appliq.*, **2**, 81–104.

Saastamoinen, O., Hultman, S.-G., Koch, N., Elers, G.M. & Mattson, L. (eds) (1984). Multiple-use forestry in the Scandinavian countries. *Commun. Inst. Forest. Fennici*, **120**, 1–142.

Saunders, D.A. (1989). Changes in the avifauna of a region distinct and remnant as a result of fragmentation of native vegetation; the wheatbelt of western Australia. A case study. *Biol. Conserv.*, **50**, 99–135.

Saunders, D.A. & Hobbs, R.J. (1991). *Nature Conservation: The Role of Corridors*. Surrey Beatty & Sons. Chipping Norton, NSW.

Saunders, D.A., Arnold, G.W., Burbidge, A.A. & Hopkins, A.J.M. (eds) (1987). *Nature Conservation: the Role of Remnants of Native Vegetation*. Surrey Beatty & Sons. Chipping Norton, NSW.

Schoener, T.W. & Spiller, D.A. (1987). Effect of lizards on spider populations: manipulative reconstruction of a natural experiment. *Science, N.Y.*, **236**, 949–52.

Shelford, V.E. (1913). *Animal Communities in Temperate America*. University of Chicago Press, Chicago, Illinois.

Sievert, P.R. & Keith, L.B. (1985). Survival of snowshoe hares at a geographical range boundary. *J. Wildl. Manage.*, **49**, 854–66.

Simberloff, D.S. (1974). Equilibrium theory of island biogeography and ecology. *Ann. Rev. Ecol. Syst.*, **5**, 161–82.

Simberloff, D.S. (1988). The contribution of population and community biology to conservation science. *Ann. Rev. Ecol. Syst.*, **19**, 473–511.

Simberloff, D.S. & Abele, L.G. (1982). Refuge design and island biogeographic theory: effects of fragmentation. *Amer. Nat.*, **120**, 41–50.

Simenstad, C.A., Estes, J.A. & Kenyon, K.W. (1978). Aleuts, sea otters and alternate stable-state communities. *Science. N.Y.*, **200**, 403–11.

Sjörs, H. (1965). Forest regions. *Acta Phytogeogr. Suec.*, **50**, 48–63.

Soulé, M.E. (ed.) (1986). *Conservation Biology: The Science of Scarcity and Diversity*. Sinauer Associates, Sunderland, Massachusetts.

Soulé, M.E. (1987). *Viable Populations for Conservation*. Cambridge University Press, New York.

Soulé, M.E., Wilcox, B.A. & Holtby, C. (1979). Benign neglect: a model of faunal collapse in the game reserves of East Africa. *Biol. Conserv.*, **15**, 259–72.

Soulé, M.E., Bolger, D.T., Alberts, A.C., Wright, J., Sorice, M. & Hill, S. (1988). Reconstructed dynamics of rapid extinctions of chaparall-requiring birds in urban habitat islands. *Conserv. Biol.*, **2**, 75–92.

Söderström, V. (1978). *Ekonomisk Skogsproduktion*. LTs förlag, Stockholm.

Southwood, T.R.E. (1977). Habitat, the templet for ecological strategies? *J. Anim. Ecol.*, **46**, 337–65.

Storaas, T. & Wegge, P. (1985). High nest losses in capercaillie and black grouse in Norway. *Proc. Int. Grouse Symp.*, **3**, 481–98.

Strandgaard, S. (1982). Factors affecting the moose population in Sweden during the 20th century with special attention to silviculture. *Swed. Univ. Agr. Sci., Dep. Wildl. Ecol., Rep.*, No. 8.

Sutherland, J.P. (1974). Multiple stable points in natural communities. *Amer. Nat.*, **108**, 859–73.

Sutherland, W.J. (1988). Predation may link the cycles of lemmings and birds. *Trends Ecol. Evol.*, **3**, 29–30.

Svensson, S. (1978). Size and isolation of natural reserves: some applications of ecological theory. *Anser Suppl.*, **3**, 225–34. (in Swedish with English summary).

Szaro, R.C. & Jakle, M.D. (1985). Avian use of a desert riparian island and its adjacent scrub habitat. *Condor*, **87**, 511–9.

Temple, S.A. (1986). Predicting impacts of habitat fragmentation on forest birds: A comparison of two models. In *Wildlife 2000: Modeling Habitat Relationships of Terrestrial Vertebrates*, ed. J. Verner, M.L. Morrison & C.J. Ralph. University of Wisconsin Press, Madison, Wisconsin, pp. 301–4.

Temple, S. & Cary, J.R. (1988). Modelling dynamics of habitat-interior bird population dynamics in fragmented landscapes. *Conserv. Biol.*, **2**, 340–7.

Terborgh, J. (1986). Keystone plant resources in the tropical forest. In *Conservation Biology: The Science of Scarcity and Diversity*, ed. M.E. Soulé. Sinauer Asociates, Sunderland, Massachusetts, pp. 330–44.

Terborgh, J. & Winter, B. (1980). Some causes of extinction. In *Conservation Biology—an Evolutionary–Ecological Perspective*, ed. M.E. Soulé & B.A. Wilcox Sinauer Associates, Sunderland, Massachusetts, pp. 119–33.

Thomas, J.W. (1979). Wildlife habitats in managed forests: the Blue Mountains of Oregon and Washington. *U.S. Dep. Agric., For. Serv., Agric. Handb.*, No. 553.

Thompson, S.D. (1982). Structure and species composition of desert heteromycid rodent species assemblages: effects of a simple habitat manipulation. *Ecology*, **63**, 1313–21.

Tomialoj'c, L. (1990). *Ptaki Polski, rozmieszczenie i liczebno's'c.* Pa'nstwowe Wydawnictwo Naukowe, Warszawa.
Trimble, G.R. & Tryon, E.H. (1966). Crown encroachment into openings cut in Appalachian hardwood stands. *J. For.*, **64**, 104–8.
Turner, M.G. (1987). *Landscape Heterogeneity and Disturbance.* Springer Verlag, Berlin.
Urban, D.L., O'Neill, R.V. & Shugart, H.H. (1987). Landscape ecology: a hierarchical perspective can help scientists understand spatial patterns. *BioScience*, **37**, 119–27.
Väisänen, R.A., Järvinen, O. & Rauhala, P. (1986). How are extensive, human-caused habitat alternations expressed on the scale of local bird populations in boreal forest? *Ornis Scandinavica*, **17**, 282–92.
Van Ballanberghe, V. (1987). Effects of predaton on moose numbers: a review of recent North American studies. *Swed. Wildl. Res., Suppl.*, **1**, 431–60.
Van Horne, B. (1983). Density as a misleading indicator of habitat quality. *J. Wildl. Manage.*, **47**, 893–901.
Van Wagner, C.E. (1978). Age-class distribution and the forest fire cycle. *Can. J. For. Res.*, **8**, 431–60.
Vauk, G., Prüter, J. & Hartwig, E. (1989). Die aktuelle Bestandszunahme der Seevögel—Ausdruck verbesserter Lebensbedingungen in der Deutschen Bucht. In *Eutrophierung—das gravierendste Problem im Naturschutz*, ed. H. Ellenberg, A. Rüger & G. Vauk. Nordd. Naturschutzakad., *Ber.* **2**, pp. 58–62.
Verner, J., Morrison, M.L. & Ralph, C.J. (1986). *Wildlife 2000: Modeling Habitat Relationships of Terrestrial Vertebrates.* University of Wisconsin Press, Madison, Wisconsin.
Virkkala, R. (1990). Effects of forestry on birds in a changing north-boreal coniferous landscape. PhD thesis, University of Helsinki.
Wales, B.A. (1967). Climate, microclimate and vegetation relationships on northern and southern forest boundaries in New Jersey. *William L. Hutcheson Mem. For. Bull.*, **2**, 1–60.
Wales, B.A. (1972). Vegetation analysis of northern and southern edges in a mature oak–hickory forest. *Ecol. Monogr.*, **42**, 451–71.
Watson, A. & Moss, R. (1979). Population cycles in Tetraonidae. *Ornis Fennica*, **56**, 87–109.
Webb, N.R. (1989). Studies on the invertebrate fauna of fragmented heathland in Dorset, UK, and the implications for conservation. *Biol. Conserv.*, **47**, 153–65.
Western, D. & Ssemakula, J. (1981). The future of the savannah ecosystems: ecological islands or faunal enclaves? *Afr. J. Ecol.*, **19**, 7–19.
Whitcomb, R.F., Robbins, C.S., Lynch, J.F., Whitcomb, B.L., Klimkiewicz, M.K. & Bystrak, D. (1981). Effects of forest fragmentation on avifauna of the eastern deciduous forest. In *Forest Island Dynamics in Man-dominated Landscapes*, ed. R.L. Burgess & D.M. Sharpe. Springer-Verlag, New York, pp. 125–205.
Whitney, G.G. & Runkle, J.R. (1981). Edge versus age effects in the development of a beech–maple forest. *Oikos*, **37**, 377–81.
Wiens, J.A. (1976). Population responses to patchy environments. *Ann. Rev. Ecol. Syst.*, **7**, 81–120.

Wiens, J.A. (1989). *The Ecology of Bird Communities*, Vols 1 and 2. Cambridge University Press, Cambridge.

Wilcove, D.S. (1985). Nest predation in forest tracts and the decline of migratory songbirds. *Ecology*, **66**, 1212–14.

Wilcove, D.S. (1989). Protecting biodiversity in multiple-use lands: lessons from the US Forest Service. *Trends Ecol. Evol.*, **4**, 385–8.

Wilcove, D.S. & May, R.M. (1986). National park boundaries and ecological realities. *Nature, Lond.*, **324**, 206–7.

Wilcove, D. & Whitcomb, R.F. (1983). Gone with the trees. *Nat. Hist.*, **41**, 82–91.

Wilcove, D.S., McLellan, C.H. & Dobson, A.P. (1986). Habitat fragmentation in the temperate zone. In *Conservation Biology*, ed. M.E. Soulé. Sinauer Associates, Sunderland, Massaschusetts, pp. 233–56.

Wilcox, B.A. & Murphy, D.D. (1985). Conservation strategy: the effects of fragmentation on extinction. *Amer. Nat.*, **125**, 879–87.

Williams, M. (1989). *Americans and their forests—a Historical Geography*. Cambridge University Press, Cambridge.

Willis, E.O. (1974). Populations and local extinction of birds on Barro colorado Island, Panama. *Ecol. Monogr.*, **44**, 153–69.

Willmanns, O. & Bruun-Hool, J. (1982). Irish mantel and saum vegetation. *J. Life Sci., Roy. Dublin Soc.*, **3**, 165–74.

Willson, M.F. (1974). Avian community organization and habitat structure. *Ecology*, **55**, 1017–29.

Wing, L.W. (1951). *Practice of Wildlife Conservation*. Wiley, New York.

Wolff, J.O. (1980). The role of habitat patchiness in the population dynamics of snowshoe hares. *Ecol. Monogr.*, **50**, 111–30.

Wright, S.J. (1981). Extinction-mediated competition: the *Anolis* lizards and insectivorous birds of the West Indies. *Amer. Nat.*, **117**, 181–92.

Yahner, R.H. (1984). Effects of habitat patchiness created by a ruffed grouse management plan on avian communities. *Amer. Midl. Nat.*, **111**, 409–13.

Yahner, R.H. (1988). Changes in wildlife communities near edges. *Conserv. Biol.*, **2**, 333–9.

Yahner, R.H. & Scott, D.P. (1988). Effects of forest fragmentation on depredation of artificial avian nests. *J. Wildl. Manage.*, **52**, 158–61.

Yahner, R.H. & Wright, A.L. (1985). Depredation on artificial ground nests: effects of edge and plot age. *J. Wildl. Manage.*, **49**, 508–12.

Yapp, W.B. (1962). *Birds and Woods*. Oxford University Press, Oxford.

Yi, B.G. (1976). Croissance du cedre de l'Atlas *Cedrus atlantica* Manetti en relation avec quelques variables du milieu en Languedoc-Roussillon (France). These Doctorat Ingenieur, Université des Sciences et Techniques du Lanquedoc, Montpellier.

Zimmerman, B.L. & Bierregard, R.O. (1986). Relevance of the equilibrium theory of island biogeography and species–area relations to conservation with a case from Amazonia. *J. Biogeogr.*, **13**, 133–43.

Chapter 3

Conservation of Plant and Animal Populations in Theory and Practice

SVEN G. NILSSON

Department of Ecology, University of Lund, S-223 62 Lund, Sweden

& LARS ERICSON

Department of Ecological Botany, University of Umeå, S-901 87 Umeå, Sweden

3.1 INTRODUCTION

The rapid increase of human populations, together with increasing exploitation of nature per capita, has caused the recent species extinction crisis. The rate of species extinction is now orders of magnitude higher than previously and is expected to increase even more in the near future (Myers, 1976, 1986). Therefore, among other things, we need a deeper knowledge of how to minimize extinction risks of vulnerable populations. Recently, much has been written about genetic diversity of populations and extinction risks (e.g. Soulé & Wilcox, 1980; Schonewald-Cox *et al.*, 1983; Soulé, 1986; O'Brien & Evermann, 1988). In the long run, genetic diversity will presumably be important for the survival of populations. However, population dynamics is generally of more immediate importance than population genetics in determining extinction risk (Shaffer, 1987; Lande, 1988*a*).

Much of the recent debate within conservation biology has focused upon the size and shape of reserves (Simberloff & Abele, 1976) and the significance of population size for survival over a specific time (Shaffer,

1981). Both these branches have their roots in the influential work of MacArthur and Wilson (1967). On the other hand, European botanists have for a long time been more concerned with changes following cessation of management (Pettersson, 1958). During the 1950s and 1960s zoologists were mainly concerned about hunting and the effect of pesticides on animals. Botanists (lichenologists) began to realize the negative effects of air pollution (Ferry *et al.*, 1973; Rose, 1988). Today we are aware of increasing problems with various types of pollution. For example, eutrophication and acidification of waters and soils and the increasing supply of airborne substances seem to constitute an overriding threat to the preservation of biological diversity over large areas in northwestern Europe. However, in Sweden and Finland the main threat is the current change in land use both within forestry and agriculture (Table 3.1), while air pollution seems to have been responsible for the decline of about 15 lichens and a few bryophytes in southern Sweden and Finland (Löfgren & Moberg, 1984; Rassi & Väisänen, 1987; Hallingbäck, 1989*a*,*b*).

Nature conservation has mainly focused upon large mammals, birds, vascular plants and to some extent butterflies. Not until recently has an increased awareness emerged that large groups of invertebrates and cryptogams are severely threatened. This situation has been increasingly stressed during recent years both in Sweden (Databanken för hotade arter och Naturvårdsverket, 1991; Ehnström & Waldén, 1986) and in Finland (Rassi & Väisänen, 1987). A recent example from Sweden is that cryptogams have been used to delimit future forest reserves (Lundqvist, 1990). However, as some of these groups are very species-rich, their proportion of vulnerable species is much lower due to a much more conservative allocation to threat classes (Table 3.2). We also have to admit that species are not treated equally as, for instance, various types of disease are disregarded from Red Data lists (cf. Dobson & May, 1986). However, the Finnish Red Data list includes several species of parasitic fungi (Rassi & Väisänen, 1987).

Another aspect of the ongoing debate within conservation biology is the tendency to draw very broad generalizations from work on a specific species or group of organisms (mainly vertebrates). As shown by Table 3.1, various groups differ strikingly even within Finland and Sweden with regard to the most prominent threat. This calls for an increasing awareness that conservation measures appropriate in one situation may not be so in another.

Table 3.1
The proportion of vulnerable species for various groups threatened due to present changes in land use in forestry, agriculture and other causes in Finland (F) and Sweden (S)

	Vertebrates		*Invertebrates*		*Vascular plants*		*Bryophytes*		*Lichens*		*Fungi*	
	F (69)	*S* (140)	*F* (402)	*S* (764)	*F* (182)	*S* (419)	*F* (117)	*S* (216)	*F* (79)	*S* (212)	*F* (188)	*S* (515)
Forestry	13	27	43	48	21	16	35	33	55	88	71	74
Agriculture	10	25	20	35	35	54	1	18	0	2	11	24
Others	77	48	37	17	44	30	64	49	45	10	18	2

The number of species for each group and country are given in parentheses. Compiled from species lists in Rassi and Väisänen (1987) for Finland and in Ahlén and Tjernberg (1988), Andersson *et al.* (1987), and Databanken för hotade arter och Naturvådsverket (1991).

Table 3.2
Number and proportions of species in different categories of species vulnerability in well-studied organism groups in Sweden (S) and Finland (F)

Group	*Country: total species*	*% in category*[a]			
		0	*1–2*	*3–4*	*0–4*
Mammals	S:67	3·0	14·9	17·9	35·8
	F:61	3·3	11·5	13·1	27·9
Birds	S:250	3·6	6·8	23·6	34·0
	F:235	0·9	6·0	9·4	16·2
Amphibians and reptiles	S:19	0	26·3	36·8	63·2
	F:10	0	20·0	0	20·0
Fish	S:150	0·7	4·0	8·0	12·7
	F:66	1·5	10·6	6·1	18·2
Molluscs	S:110[b]	0	11·8	20·9	32·7
	F:150	0	3·3	6·7	10·0
Lepidoptera	S:2640	0·1	1·2	1·3	2·5
	F:2330	0·4	1·2	3·9	5·5
Coleoptera	S:4350	0·4	2·6	1·2	4·2
	F:3560	0·7	1·8	3·1	5·7
Vascular plants	S:2000	1·8	7·3	10·5	19·6
	F:1350	0·5	5·6	7·3	13·5
Bryophytes	S:1000	1·7	5·1	14·8	21·6
	F:810	1·8	5·4	7·8	15·0
Lichens	S:2100	0·8	3·6	5·8	10·2
	F:1500	1·3	3·5	5·7	10·5

[a] 0, extinct; 1, endangered; 2, vulnerable; 3, rare, 4, at risk. Based on the same sources as Table 3.1.
[b] Excluding Bivalvia.

3.2 SPECIES EXTINCTIONS

3.2.1 Definition of scale

The magnitude of extinction depends upon the scale used. As one can discriminate between the regional, local and within-site distribution pattern of a certain species (Forman, 1964), we can also use a corresponding subdivision when discussing extinction. Thus, global extinction is the total disappearance of a species. For northern areas harbouring only a few endemics, this aspect will be of only minor importance. Extinction from a biotic region may affect several countries. Local extinction comprises the disappearance from a defined area on the basis of dispersal ability of the organism concerned. Thus, for insects local populations may occupy a few hectares (Warren, 1987*a*; Harrison, 1989), while for birds the area may be 10 000–100 000 ha (Pettersson, 1985*a*; Aulén, 1988), or more.

Since in nature a majority of species occur in low densities, local extinctions and recolonizations may be more common than we believe. Most long-term population studies are conducted where a species is common, giving a biased view of the frequency of local extinction. There are also great methodological problems in estimating extinction rates (Nilsson & Nilsson, 1983). Nilsson and Nilsson (1985) suggested a local extinction rate of about 1% among vascular plants on islands in Lake Möckeln, Sweden, but it is not known whether the value is similar on the mainland. Menges (1990) found an extinction rate of 2% for a herb, but many more detailed studies are needed. Naturally high local extinction rates seem to prevail among species confined to naturally fragmented, small biotope patches such as frogs in ponds (Sjögren, 1991) and *Daphnia* in rockpools (Bengtsson, 1989).

We know even less about how common it is for species to become extinct regionally. Presumably, such events were formerly relatively rare but due to the activities of man they are now occurring with increasing frequency (Myers, 1986). When the local extinction rate of a species is higher than the colonization rate this causes concern among conservationists. However, by that time it may be too late to prevent regional extinction due to the inherent non-linearity of population trends when habitat destruction occurs.

3.2.2 Body size and extinction

Large species seem to be especially vulnerable to extinction, mainly due to man (Martin, 1984; Järvinen & Miettinen, 1988). When man is hunting he is presumably foraging optimally (Pyke, 1984), which means concentrating on those species giving the most food per effort, i.e. large species. This is certainly part of the explanation but not all. Large species usually have comparatively low natural mortality rates and low potential rates of increase. Species with such demographic properties are very sensitive to increased adult mortality which can easily cause long-term decreases and extinction (Mertz, 1971; Järvinen & Miettinen, 1988; Lande, 1988*b*). Furthermore, large species usually occur in lower densities than smaller ones (Peters, 1983). Populations with few individuals are usually considered vulnerable, but the intercorrelations between body mass, mortality rate and population density make simple predictions of extinction risks from population numbers unreliable. For example, in North American reserves much smaller populations of large, compared to small, carnivores have survived during the last 75 years (Newmark, 1986 in Soulé, 1987). This is presumably due to recent low mortality rates following legal protection of large carnivores.

However, data from tropical land-bridge islands (Gotteli & Graves, 1990) show that body size may not always be of crucial importance. These authors found that land-bird species did not differ in body mass between the islands and the nearby mainland. Instead, species absent from the islands were those confined to late-successional forest stands on the mainland. This we interpret as species adapted to small-scale disturbances evolving low dispersal abilities and becoming sensitive to fragmentation. Thus the disturbance regime to which a species has adapted is more important than body mass *per se*. A similar case from the plant kingdom is provided by *Plantago cordata* (Meagher *et al.*, 1978), a large—compared to other species of the genus—herb with low reproductive output, confined to shady streams in the eastern United States. Evidently, it is adapted to small-scale disturbances along woodland streams and its decline has followed the recent fragmentation of the woodlands.

3.3 THE IMPORTANCE OF POPULATION SIZE

3.3.1 Minimum viable population sizes

When the range of a species has decreased to a single area or when a population has become isolated with effectively no dispersal from neighbouring local populations, we may ask if this isolated population is viable in the long term. Thus we need to estimate the extinction risk over a specified time period with a given probability of population survival, for example, 99% over 1000 years (Shaffer, 1981). This is called minimum viable population (MVP) analysis. The genetic aspects of such analyses are outside the scope of this chapter (see chapter 4 for references).

Mammal populations isolated on habitat islands in North American deserts for about 8000 years have provided empirical data enabling a comparison with theoretical MVPs. Agreement is fairly good (Belovski, 1987). For large mammalian herbivores 95% population persistence for 100 years requires around 1000 individuals, while for small herbivores numbers are ten times higher. The important point is that these numbers, on the basis of demographic considerations alone, are considerably higher than those based on genetic data (Shaffer, 1987; Lande, 1988*a*).

The emphasis on population size in the prediction of extinction (Terborgh & Winter, 1980; Diamond, 1984; Soulé, 1987) may imply risks for successful species conservation. First, these demographic models have been developed for vertebrates and may not be generally applicable for other groups of organisms. For example, among invertebrates, the high specialization often found may result in even large populations being highly extinction-prone. Thomas (1980, 1983) found that small changes—a few centimetres—in leaf canopy height in calcareous grasslands resulted in large population changes for a butterfly species. This shows that population size may be a misleading indication for species vulnerability of those adapted to successional communities.

The other situation concerns species which demand highly specific substrates for their development. Insect species confined to dead wood may show very high population sizes in a single large, dead tree (Palm, 1959). However, the crucial point is that the suitable substrate must be present in sufficient amounts on a temporal scale. Another case is provided by a forest-gap species. A butterfly, *Mellicta athalia*, only persists in small forest gaps for a few years (Warren, 1987*b*). Thus, the unpredictability of gaps in space and time (Falinski, 1986) implies large fluctuations in population size. Consequently, in the above-mentioned cases it may be better to focus on preserving the appropriate forest stand dynamics within an area large enough to avoid extinction than on counting individuals.

Thus, for habitat specialists the carrying capacity often fluctuates widely. The few long-term studies available for non-pest insect populations show fluctuations in a magnitude of two orders (e.g. Ehrlich & Murphy, 1987; Danell & Ericson, 1990).

Most insect studies have concerned common, usually pest, species. However, fluctuation patterns may differ between common and rare species (Taylor, 1984). A recent examination of population fluctuations in British moths and aphids shows large variability both within and between these two groups (McArdle *et al.*, 1990). The lesson for conservation biology is that we cannot draw any general conclusions based on a few individual species, not to mention applying vertebrate models to other groups of organisms.

Among plants there are several examples which show that even very small populations and minute reserves may be sufficient to prevent local extinctions (Fisher, 1965; Frost, 1981; Elven, 1989). However, establishing reserves while ignoring appropriate disturbance regimes may cause extinctions (Nordhagen, 1963; Ratcliffe, 1973; Høiland, 1981).

3.3.2 Stochastic population variability

In theoretical population models 'species' go extinct much more quickly at a given population size if stochastic environmental variability is introduced or increased in magnitude (Leigh, 1981; Wright & Hubbel, 1983; Goodman, 1987). However, the empirical support for a higher extinction risk of variable populations (Diamond, 1984) is less convincing. For example, Karr (1982*a*) argued that bird species extinctions on Barro Colorado Island in man-made Lake Gatun, Panama could not be predicted from body mass, but from variability in mist-net catches on the nearby mainland. However, catch variability may indicate local movement between dry and wet forest and not population fluctuations (Karr, 1982*b*). Thus, lack of some critical habitat (wet forest) on Barro Colorado rather than population variability *per se* may explain the extinctions. Bird species that vary greatly in density on islands around Great Britain are also those that are most likely to go extinct after taking population size into account (Pimm *et al.*, 1988). However, the result is uncertain since population variability and extinction rate were calculated from the same island census data.

Indirect evidence that variable populations run a higher risk of extinction is given by the observation that among insects scarce, geographically restricted species have populations that fluctuate less than widespread species (Glazier, 1986; Gaston & Lawton, 1988; Spitzer & Lepš, 1988; see, however, McArdle *et al.*, 1990). Among plants, short-lived species are known to show considerable population fluctuations both in space and time (Symonides, 1988). However, due to the seed bank they may rapidly recover provided that suitable conditions for germination prevail. For example, Frost (1981) reported that the annual *Ranunculus ophioglossifolius* was not observed during five of 29 years in a minute nature reserve. However, drastic population fluctuations among annuals may result in increased risks for local extinctions among specialized herbivores (Ehrlich *et al.*, 1980).

More direct tests of the proposed connection between population variability and extinction risk are needed.

3.4 THE IMPORTANCE OF HABITAT AREA

3.4.1 Species with fragmented habitats

Not only true island species but also many continental species have a patchy distribution. In such cases it is *not* sufficient to protect and manage one large viable population and allow the species to go extinct in other patches.

An important concern is whether the habitat is naturally fragmented or whether fragmentation is caused by man. First, some habitats have always been patchy. Vernal pools and rockpools are by definition 'fragmented'. Rich fens provide another example of naturally fragmented habitats and this effect has also been increased by man (Wheeler, 1988). Species of such habitats have evolved dispersal mechanisms to cope with fragmentation (in plants wind-dispersal for small-seeded species such as orchids and plumed species among graminoids, dispersal by birds for species with a hard seed-coat). An interesting example is the annual *Howellia aquatilis*, a species of temporary ponds which did not show any genetic diversity (Lesica *et al.*, 1987). The authors interpret this as a case of long-term isolation. However, this could also be interpreted as due to high dispersal ability.

Secondly, some habitats, such as temperate and boreal forests, have naturally covered large tracts but display considerable differences in patch dynamics. Temperate forests, and some boreal ones, have been characterized by a small-scale gap dynamics with a mean gap size of about 0·01 ha (Runkle, 1985). Many boreal forests have shown a large-scale patch dynamics, up to 10 000 ha (Pickett & Thompson, 1978) due to fire.

Species adapted to habitats which naturally covered large tracts and with small-scale disturbances shoud be highly sensitive to fragmentation. For example, beetles dependent on large, old hollow deciduous trees generally have low dispersal abilities (e.g. Martin, 1989 for click beetles). Such trees are now rare and a high proportion of species confined to old trees are considered as threatened (Ehnström & Waldén, 1986; cf. Table 3.7). Also the vascular flora of temperate deciduous woodlands shows a high proportion of species with low dispersal ability (Peterken & Game, 1984).

Even boreal areas have forest stands characterized by gap dynamics, and which harbour species which today show highly fragmented occurrences due to forestry, for example, the lichen *Usnea longissima* and the beetle *Pytho kolwensis* (see further chapters 7 and 8, respectively).

Forests characterized by gap dynamics also show a large number of species adapted to small, open gaps. The butterfly *Mellicta athalia*, confined to gaps of short duration, is highly sedentary, and less than 2% of the individuals have been shown to disperse between gaps at a maximum distance of 1 km (Warren, 1987*a*). Thus, when long-continuity forest habitats are fragmented we cannot expect their constituent species to possess efficient dispersal mechanisms to cope with an evolutionary novel situation.

Species adapted to fire disturbances in boreal forests should in general be characterized by either good dispersal abilities or the ability to survive in a dormant stage between the disturbances. Consequently, such species do not seem to be particularly sensitive to fragmentation. Instead, it is important that fires occur at regular intervals (see further chapter 7).

The conclusion is that the consequences of habitat fragmentation for species survival will differ, dependent upon the original predictability of the habitat in space and time. Thus, detailed analyses of metapopulation dynamics are most urgent for species with recently fragmented distributions.

3.4.2 Fragmentation models

The most important conclusion from mathematical models by Levins (1970), Carter and Prince (1981) and Hanski (1985) is that species are expected to become extinct even in the presence of a fair amount of suitable, but fragmented, habitat. Furthermore, a proportion of the suitable habitat patches are expected to be uninhabited due to recurring local extinctions. Which patches are empty will also vary over time. These models are conceptually important but are difficult to apply in nature conservation as predictions are qualitative. Recently, Lande (1987) developed these models with a version that makes quantitative predictions for territorial animals with fragmented distributions. The model has only three interrelated parameters:

h = the proportion of suitable habitat in a region;
p = the proportion of suitable habitat in a region currently occupied by females;
k = the proportion of territories that would be occupied by females in a completely suitable region ('the demographic potential of the population').

For an avian life history the equilibrium occupancy of territories by females is:

$$1-(1-k)/h \quad \text{for} \quad h>1-k$$
$$0 \quad \text{for} \quad h\leqslant 1-k$$

Thus, when the proportion of suitable habitat in a region declines the same holds for the proportion that is occupied. Therefore, population

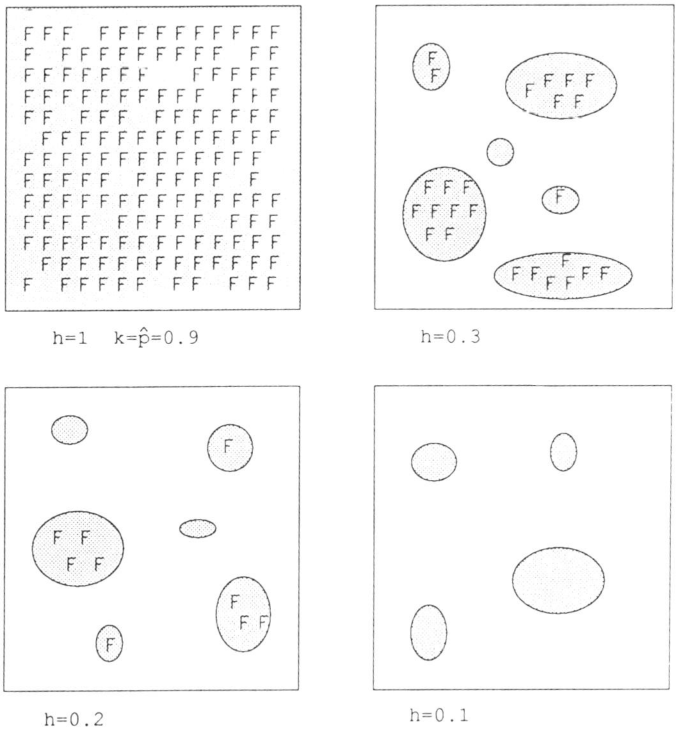

Fig. 3.1. The predictable consequences for a territorial vertebrate (in this case with $k=0{\cdot}9$) following three cases of habitat fragmentation according to Lande's (1987) model. Shaded areas denote suitable habitat. F denotes territorial females. For explanation of parameters used see text.

decrease is not expected to follow habitat decline linearly but to accelerate with diminishing habitat area (Fig. 3.1). Hence, predictions of population decline are likely to be underestimated by projection of current population trends. This model has been applied to the northern spotted owl *Strix occidentalis caurina* (Lande, 1988*b*), a species inhabiting large tracts of very old (>200 years) coniferous forests in western North America. For this species $h=0{\cdot}38$ and $p=0{\cdot}44$, which gives $k=0{\cdot}79$. It appears that the northern spotted owl cannot persist in large regions where less than 21% of the area is composed of suitable habitat. Under the present conservation plans between 7 and 16% of the region is planned to consist of old coniferous forest, and the species is expected to become extinct even if the present, most ambitious conservation plan is implemented. We urgently need empirical tests of Lande's model to examine its validity.

Examples of population decreases stronger than the reduction of habitat area are found for several bird species, dependent upon late successional boreal forests in Finland (Järvinen & Väisinen, 1978). One explanation for this non-linearity is that the species avoid habitat edges near clear-cut forest (Helle & Järvinen, 1986). Another possibility is that the proportion of still suitable habitat occupied has decreased, i.e. according to Lande's model.

Two small (15–25 pairs), isolated populations of middle spotted woodpeckers *Dendrocopos medius* in Denmark and Sweden went extinct despite very small reductions of suitable habitat (Pettersson, 1985*a*). It has been suggested that the final cause of the extinctions was inbreeding depression (Pettersson, 1985*a*). However, it may have been due to the small final reduction in habitat area in combination with unfavourable years. The occupation rate of suitable habitat patches decreased with isolation in the Swedish population (Pettersson, 1985*b*), also according to Lande's model.

The proportion of suitable areas occupied by a species (p) at a given time varies with species, and, because the suitability of a patch also varies over time, the values of p given below are probably minima. For merlins *Falco columbarius* in Wales, about 50% of sites used in any year in 1970–84 were occupied annually (Bibby, 1986). For a nuthatch *Sitta europaea* population studied over a similar time period the corresponding figure was 53% (Fig. 6 in Nilsson, 1987). For local populations of butterflies in England a very high proportion of suitable areas seems to be occupied (see Thomas, 1983; Thomas *et al.*, 1986). The same applies to fungi in infertile grasslands in Sweden (Nitare, 1988). However, these could be transient phenomena caused by recent strong declines of these species. The problem of identifying the amount of suitable habitats is clearly exemplified in the case of five European *Maculinea* butterflies. It has recently been revealed that each of these species is dependent upon a single ant (*Myrmica*) species (Thomas *et al.*, 1989).

Thus, for analyses of minimum viable populations we need very detailed information on the natural history of the actual species in order to estimate the proportion of suitable habitats occupied.

3.4.3 Single large or several small reserves

Obviously, a large nature reserve will support more species than a small one. However, it is not obvious that a single large reserve is better than several small ones with the same combined area—the SLOSS question

(Simberloff & Abele, 1976). In our view, the conservation value (ability to support vulnerable species) of a nature reserve is not linearly related to its area. For very small sites the conservation value will increase rapidly with area as more and more of the minimum area requirements of vulnerable species are satisfied (Nilsson, 1986). Further increases in reserve area will allow a larger number of stenotopic species to reach their minimum viable population sizes (Soulé, 1987). The answer to the SLOSS question may depend on the sizes of the reserves (Nilsson, 1982). Relatively, when dealing with small reserves it is preferable to protect as large an area as possible which will satisfy the requirements of one reproductive group of a local population of the most vulnerable species. However, the size will vary between biotopes. For those with a small area, e.g. rich fens and ravines, a few hectares may be appropriate. For forest biotopes which earlier covered large tracts, areas of 100 ha or more should be considered. However, if one takes into account the risk of catastrophes (disease, fire, etc.), many relatively small reserves may be preferable when dealing with larger preserves. This requires some dispersal (natural or man-made) of vulnerable species between reserves (Wright & Hubbell, 1983). The dispersion and density of habitat patches which meet the area requirements of a species is crucial for the ability of that species to persist in a region (Lande, 1987). Thus, a patch satisfying a species' area requirements will only be occupied if there are other patches within realistic dispersal distances of the species (Pettersson, 1985*b*). Therefore, groups of reserves seem preferable to evenly distributed reserves. However, we need practical recommendations of maximum allowable distances between adjacent reserves. The limited data available indicate that the answer is less than a few kilometres when dealing with vulnerable species (e.g. Müller, 1982; Thomas *et al.*, 1986; Warren, 1987*a*; Harrison, 1989). However, the distance may be lower or higher depending on species and biotope history.

For relatively small biotope areas the proportion of scarce bird species increases with habitat area for several naturally fragmented south Swedish biotopes (Nilsson, 1986). From this it was concluded that both small and large biotope patches must be conserved since they harbour different species assemblages. However, the larger reserves are essential since some species are restricted to large patches of a biotope. On the other hand, there were no bird species occurring on small biotope patches which did not occur on the larger ones. The same pattern was also found for beetles and spiders in fragmented heathlands in England (Hopkins & Webb, 1984). This may be a general pattern with few exceptions (Diamond, 1974).

Two further aspects make the SLOSS issue of less importance in conservation biology at least in northern areas. First, habitat quality and the presence of vulnerable species often seem to override area *per se* when the conservation value of an area is considered (e.g. Boström & Nilsson, 1983; Lahti & Ranta, 1985). Second, many vertebrate species with large home ranges in northern Europe are habitat generalists (for exceptions see Table 3.3). To our knowledge very few big mammals in northern Europe are stenotopic: they are not restricted to reserves but also forage in managed forests and even on agricultural land. Thus, the pessimistic predictions of Belovsky (1987) regarding the extinction of large mammals due to forest fragmentation may not be relevant in northern areas.

Table 3.3
Examples of stenotopic bird species with large breeding area requirements

Species	*Habitat*	*Area*[a] (*ha*)	*References*
Gavia arctica	Oligotrophic lakes	50	Nilsson (1986)
*Tetrao urogallus**	Old coniferous forests	300	Wegge & Rolstad (1986)
Pluvialis apricaria	Bogs and heaths	20	Boström & Nilsson (1983)
Lymnocryptes minimus	Wet sedge-mires	20	Nilsson & Nilsson (1978)
Dendrocopos minor	Old deciduous and mixed forests	20	Wesołowski & Tomialojč (1986)
Dendrocopos leucotos	Old deciduous and mixed forests	100	Wesołowski & Tomialojč (1986)

[a] The size of home-range of one pair or, in a polygamous species (*), the area used by one reproductive group.

3.4.4 Area requirements

The SLOSS question assumes species to be equal while nature conservation focuses on vulnerable species. In some cases species richness can indicate the presence of vulnerable species (Järvinen, 1982; Wheeler, 1988) but this is probably not a general rule. When the goal is to conserve species of special concern, for example large, spectacular or keystone species, the SLOSS question is not the most relevant (Soulé & Simberloff,

1986). Instead we must know the area requirements of species, which obviously differ between taxa (Nilsson, 1978; Soulé & Simberloff, 1986). Area requirements can be determined at different scales dependent on the organism. For birds, a habitat patch may be occupied if it is larger than the home range (Ahlén & Nilsson, 1982) but for insects only when a small population can establish. For example, in unmanaged deciduous forests in Sweden vulnerable species of molluscs (J. Bengtsson & S.G. Nilsson, in prep.) and beetles (R. Baranowski & S.G. Nilsson, unpublished data; cf. Chapter 10) can survive in patches of about 1 ha, while for forest bird species the lower threshold is about 10 ha even in pristine patches (Ahlén & Nilsson, 1982). For the more stenotopic European bird species area requirements range from 20 to 300 ha (Nilsson, 1986; Table 3.3), but this value obviously varies with habitat quality. These differences reflect the higher dispersal distances of most vertebrates, and the large area covered by their local populations, compared to invertebrates. For example, a forest reserve in northern Europe may harbour several local populations of invertebrates but only one or a few individuals of the local population of a large vertebrate. Thus, for the latter many other suitable habitat patches must occur for the local population to persist (see previous section).

Until now forest management has been compatible with the demands of most bird species with higher area requirements, for example, the Ural owl *Strix uralensis* (Ahlén, 1977). However, the situation may change following more intensive forest management and increasing destruction of wetlands.

Area requirements of plants have seldom been studied. Vascular plants in general seem to tolerate a higher degree of habitat fragmentation than birds (Nilsson, 1978; cf. Järvinen, 1982). Nonetheless, in self-incompatible species, pollination success increases rapidly with the number of individuals in a group (e.g. Nilsson & Wästljung, 1987; Widén, 1987; Smith *et al.*, 1988). We need to determine the necessary group size for successful reproduction in such species. Further, species with highly specific demands on microclimate are sensitive to edge-effects. For example, felling of the surrounding forest resulted in extinction of the lichen *Erioderma pedicellatum* in Sweden, in spite of both the species and its richest locality being protected by law (Ingelög *et al.*, 1987; cf. also the example of *Evernia divaricata* in Chapter 8). Another aspect is that we need to preserve very large areas in order to maintain natural species interactions. To our knowledge, no studies in northern areas have documented the importance of this approach for species conservation *per se* (however, cf.

Newmark, 1987). Nonetheless, such areas are essential for our understanding of natural interactions and processes.

3.4.5 Habitat diversity and reserve delimitation

Many species utilize more than one biotope during their life-cycle, for example, many birds feeding in wetlands but breeding in forests. Seasonal migration can also occur within a major biotope between sites with different microclimate or wetness (e.g. Karr, 1982*b*). If a reserve lacks one critical habitat element for a species, this species will disappear if the surroundings are destroyed by exploitation.

Interactions between plants and pollinators involve examples of species shifting between shaded and sunlit sites. For example, several longhorn beetles develop in the wood of old shaded forest trees, while the adults feed on pollen of forest glade herbs (Palm, 1959; Gutowski, 1990). Other pollinating beetles show a shift from early flowering forest edge species to later flowering species in the shade (Pellmyr, 1984).

Edges between habitats are important for species dependent on a warm microclimate. Beetles living in the wood of deciduous trees in temperate and hemiboreal Sweden show a striking dissimilarity with regard to exposure, about twice as many being favoured by sun-exposed sites (Table 3.4). The same pattern also seems to be valid for lichens where

Table 3.4
The numbers of wood-living Swedish beetles (from a total of 820 species) showing preference for, or dependence on, deciduous trees in different exposures (based on Palm, 1959)

	Sun-exposed trees	*Shaded trees*
Preferring	89	32
Dependent on	78	36
Total	167	68

species favoured by light have declined following the closure of earlier open deciduous woodland meadows (Ekman, 1990). In natural forests suitable gaps are created due to windfalls. In managed forests edges bordering arable land or wetlands may provide the only possibility for the existence of such species. The sand-lizard *Lacerta agilis*, a vulnerable species in Britain at the northern limit of its geographic range, uses boundaries between open and closed vegetation for foraging and thermoregulation (House & Spellerberg, 1983).

A third aspect is the spatial variation within a biotope. The butterfly *Euphydryas editha*, dependent on serpentine meadows in western North America, performs best on south-facing slopes in years of ample rainfall but in dry years north-facing slopes are the most suitable sites. Areas with high diversity of slope exposure will improve the chances of survival of this butterfly in the face of a variable climate (Ehrlich & Murphy, 1987; Weiss *et al.*, 1988). Variations in microclimate and edaphic conditions may also result in spatial variation in sensitivity to disease, e.g. parasitic fungi (Matson & Warring, 1983; Wennström & Ericson, 1990).

Edges within and between habitats may have opposite effects on a group of organisms. For example, on raised peat-bogs many waders feed at pool edges but breed elsewhere on the bog. Those with a high proportion of pools also have a higher wader density (Boström & Nilsson, 1983). However, waders seem to avoid bog/forest edges, probably to reduce risks from forest predators (Nilsson, 1977; Boström & Nilsson, 1983; cf. also Chapter 2 for increased predation at field/forest edges).

The conclusion for preservation is that species vary with regard to their demands on habitat diversity. Consequently, knowledge of the vulnerable species present in a planned reserve will help to delimit the size and boundaries necessary for successful species preservation.

3.5 THE IMPORTANCE OF SPECIES INTERACTIONS

3.5.1 Trophic level, keystone species, and species vulnerability

Communities are structured by interactions between organisms, both between and within different trophic levels. Our concern here is to identify the implications for conservation biology and whether changes will result in cascading effects (Carpenter *et al.*, 1985) and linked extinctions (Myers, 1986). Thus, one important aspect is the presence of keystone species (Paine, 1966).

Predators

The role of predators has often been disputed by conservationists in the argument for preserving large carnivores (see also Chapter 2). No doubt the most important predator has been, and still is, man. Being omnivorous, man is able to switch between various types of food, and is thus likely to cause extinctions, profoundly changing biota. During the last

40 000 years the extinction rate of large birds, and especially mammals, has been much higher than that of smaller species of these groups. This pattern is most clear for North American mammals and has been attributed to the direct and indirect effects of hunting by man (Martin, 1984; Owen-Smith, 1987, 1989). Although this explanation is still controversial it is strengthened by the recent extinction of large vertebrates when man arrived on Hawaii, New Zealand and Madagascar (Olson & James, 1982; Järvinen & Miettinen, 1988). In Fennoscandia the four largest carnivores—wolf, brown bear, lynx and wolverine—are on the Red Data lists due to persecution (Anon., 1978).

The long period during which man has altered natural communities by hunting large carnivores and herbivores implies that, even if we acknowledge the significance of predation, our understanding of the consequences of the extinction of many predators is too fragmentary. Nonetheless, there are several examples which indicate that predators often act as keystone species.

The classic example of the effect of the elimination of a single predator is the near extinction of the sea otter *Enhydra lutis* on the North American west coast. As they forage in large kelp forests, on abalones, sea urchins and other herbivores that feed on various algae, their local disappearance resulted both in drastic changes in the grazer and algal communities (Lowry & Pearse, 1973; Dayton, 1975; Estes *et al.*, 1978). However, the present-day situation, with the population having recovered, most probably does not reflect the original situation since one important kelp grazer, the Steller's sea cow *Hydroclamatis gigas*, became extinct in the middle of the 18th century (Dayton, 1975). Other examples of keystone predators are, for example, provided by wolves preying on caribou/reindeer and moose, where elimination of wolves permitted the build-up of excessive herbivore populations (Messier & Crête, 1985; Bergerud, 1988).

Herbivores

Herbivores may have profound effects upon vegetation not only due to grazing or browsing but also other activities such as trampling, digging and burrowing (Crawley, 1983). The consequences of herbivory can be considered by means of the theory of exploitation systems (Oksanen *et al.*, 1981), which states that an ecosystem has *n*-link dynamic structure if the *n*th trophic level is the highest at which the dominant populations strongly influence their resources. Thus in three-link systems, where carnivores constitute the highest trophic level competing for prey (herbivores), the herbivores are not able to build up high population densities. Consequently, their influence on the plants is of minor importance, and plant communities are structured by competition. This implies that excessive hunting of carnivores, e.g. for preserving game species, will result in overgrazing. One example is the overabundant game populations present during 1892–1915 in the Bialowieza National Park in eastern Poland, which has had a long-lasting effect on forest composition and succession (Falinski, 1986). Another example is the present overabundant moose *Alces alces* population in Sweden (Cederlund & Markgren, 1987), restricting the recruitment of several species of deciduous trees (cf. chapters 2 and 10). Another situation is when densities become too high in small game reserves or national parks, as in the case of elephants (Laws *et al.*, 1975).

In two-link systems, which are less productive, the herbivores are close to resource limitation and unable to build up stable population densities. In such systems, plants are adapted to disturbance by grazing. Absence of herbivores (e.g. due to hunting or exclusion) will result in an advantage for more competitive plants. One possible example is provided by mountain and tundra areas with a natural occurrence of reindeer/caribou. Their grazing will result in heaths with a depleted lichen cover (Kelsall, 1968; Helle & Aspi, 1983), strikingly different from the thick greyish *Cladina* carpets in the reindeer-free mountains of southern Scandinavia (Nordhagen, 1943; Dahl, 1956). Circumstantial evidence for the significance of disturbance in these ecosystems is that most plant species are dependent upon seedling establishment (Söyrinki, 1939). The extinction of the only Nordic occurrence of the tundra species *Crepis multicaulis* following fencing to exclude grazers provides another example (Nordhagen, 1963; Høiland, 1981).

For most ecosystems we still do not have the detailed knowledge to understand the importance of herbivory and the consequences of its removal. One possible example of complex interactions is the butterfly *Euphydryas editha*, living in grasslands where digging by gophers *Thomomys bottae* will result in reduced densities of the host plant *Plantago erecta* during winter. However, remaining plants grow larger due to reduced competition on the tilled soil. This results in longer plant survival which allows the larvae to complete their development before plant senescence (Singer & Ehrlich, 1979).

The increased airborne nitrogen deposition over northwestern Europe has during recent years contributed to a change in plant–animal interactions. Outbreaks by the chrysomelid beetle *Lochmaea suturalis* have converted *Calluna* heaths to grasslands (Berdowski & Zeilinga, 1987; Bobbink & Willems, 1987; Berdowski & Siepel, 1988). Earlier this century this beetle experienced regular outbreaks correlated with variations in climate (Overgaard Nielsen, 1986). The present outbreaks seem to be due to the presence of a more nutritious host plant. Another case is provided by the decline of *Fucus vesiculosus* in the Baltic during the 1970s due to overgrazing by *Idothea* species. This has been attributed to increased eutrophication (Haahtela, 1981; Kangas *et al.*, 1982). However, as *Fucus* has not declined in the archipelagoes (Rönnberg, 1981), it is likely that over-fishing also played some role.

Various types of intricate interactions have evolved between herbivores and plants, and the species involved have in several cases been shown to be highly vulnerable. Janzen and Martin (1982) suggested a large number of plant species to be dependent on extinct gomphotheres for their fruit dispersal and subsequent establishment. To what extent this is the case is still open to debate (Hunter, 1989). Temple (1977) suggested that two *Sideroxylon* trees on Mauritius did not become established by seed due to the extinction of dodo but Iverson (1987) believes it was due to the activity of giant tortoises. Cox (1983) reported that the reproduction of a Hawaiian endemic plant declined following the extinction of its endemic bird pollinator. He and co-workers (Cox *et al.*, 1991), working on the rapid extinction of flying foxes throughout the Pacific, suggested that pollinators and seed dispersers may structure the ecosystems on oceanic islands in the same way as predators often do in continental and intertidal communities.

Plants

Plants constitute food for a large number of insects and frugivores and are hosts for a large number of cryptogams. The significance of fruit-producing plants acting as keystone species has been stressed by Gilbert (1980) and Terborgh (1986) for the tropics.

Linked extinctions have been described from the Hawaiian islands where introduced grazers have resulted in the virtual extinction of nectar-producing plants (e.g. the genus *Hibiscadelphus*, comprising six species with a total of 14 individuals present in the wild (Gentry, 1986)), which in turn seems to be a plausible explanation for extinction and rarity of nectarivorous birds (Pimm & Pimm, 1982).

For monophagous species, each plant acts as a keystone species and the extinction of one milkweed, *Asclepias currasavica*, on Barbados has resulted in the extinction of two milkweed bugs unable to feed on other plants (Blakley & Dingle, 1978). The work by Ward and Lakhani (1977) on the insect fauna of juniper *Juniperis communis* is especially instructive for conservation biology. These workers were unable to recommend a suitable population size for juniper in order to preserve its fauna, as a site with 3000 bushes only had a 0·5 probability of including all the 15 more common species (from a total of 38). They also found that species feeding on berries are more likely to go extinct locally (as berries constitute a more limited resource than needles).

Another case is provided by the monophagous geometrids (Lepidoptera) of the forest herb *Actaea spicata* in Sweden. One species, *Baptria tibiale*, is confined to sun-exposed *Actaea* stands. Another, *Acasis appensata*, is confined to stands in late successional shady forests (Ehnström & Waldén, 1986; Hydén, 1990). This illustrates that these two species—both on the Swedish Red Data list—are adapted to feed on one host plant under strikingly different environmental conditions. Another rare geometrid, *Eupithecia groenblomi*, only occurs on its host plant *Solidago virgaurea* in late successional spruce stands in boreal Fennoscandia (Peltonen *et al.*, 1970). This is of specific interest, since *Solidago* is an abundant successional plant, adapted to establish early in the post-fire stage. However, it is able to persist in late successional stages, even if it does not establish itself by seed. This and the above-mentioned cases show that conservation measures for invertebrates must be based on an understanding of the environmental conditions in which the host plant grows.

3.5.2 Species introductions

The outcome of changed interactions both between and within different trophic levels is further illustrated by the consequences of both intentional and unintentional species introductions as well as from biological control (Harper, 1977; Beddington *et al.*, 1978; Ebenhard, 1988).

Animals

Ebenhard (1988) reviewed 1559 animal introductions, including 330 species of mammals and birds. He found clear differences when comparing the effect of introductions on oceanic islands with those on continents and continental-shelf islands. Strict herbivores affected plants and habitat in 50% of the cases on oceanic islands—twice as much as for the other types of introductions (Table 3.5). Consequently isolated oceanic islands have a species-poor fauna, often lacking predators and herbivores. This has frequently resulted in high extinction rates among endemic plants

Table 3.5
The proportion of introduced predators causing shifts in abundance among native species through predation and the proportion of introduced Herbivores which modify plants and habitats on continents, continental shelf islands and oceanic islands (data from Ebenhard, 1988)

Category	*Continents*		*Continental shelf islands*		*Oceanic islands*	
	%	*n*	%	*n*	%	*n*
Strict predators	36	11	25	24	61	83
Omnivorous predators	2	47	5	58	35	202
Strict herbivores	26	86	28	89	50	188
Omnivorous herbivores	11	57	5	58	6	202

following the introduction of generalist vertebrate herbivores (Melville, 1979), such as goats on St Helena (Lucas & Synge, 1978), rabbits on the Kerquelen islands (Werth, 1928) and feral ungulates on Hawaii (e.g. Pimm & Pimm, 1982). Phytophagous invertebrates may also cause drastic declines, as in the case of the Bermuda cedar *Juniperus bermudiana*—earlier the dominant tree species of Bermuda—due to introduced scale insects (Challinor & Wingate, 1971). Bond and Slingsby (1984) report a striking example where an introduced ant displacing a native species impeded the dispersal of a native shrub species.

Ebenhard (1988) also compared the proportion of introductions that resulted in potential competition for resources between species within the same trophic level. He found no major differences between oceanic islands, oceanic-shelf islands and continents (Table 3.6). However, further analysis of how many of these potential competitive situations resulted in effects on other species revealed that they were more likely on oceanic islands (Table 3.6). This, in turn, is analogous with faunas in species-poor island biota being less likely to cope with competitive situations.

However, species introductions may not necessarily cause only negative effects. The spread of the North American muskrat *Ondatra zibethica* in northern Sweden has resulted in the creation of areas with open water in the reeds, which are colonized by a large number of rare aquatic plants (Danell, 1977). Such openings in the reeds are also considered as positive

Table 3.6
Proportion of introductions of mammals and birds competing for resources, and the proportion which may cause changes among native species for different types of introductions (data from Ebenhard, 1988)

Category	*Type of competition*	*Continents*		*Continental shelf islands*		*Oceanic islands*	
		%	*n*	%	*n*	%	*n*
Mammals	Potential	29	84	5	114	5	306
	real	38	24	33	6	80	15
Birds	Potential	28	40	2	41	20	110
	real	46	11	100	1	82	22

for waterfowl (Danell, 1979). On the other hand, coypu *Myocastor coypus* in England created open water and destroyed some rare plants (E. Duffey, pers. comm.).

Plants

The successful dispersal of alien plants over the world has for long attracted botanists. This has resulted in work focused on how to eliminate successful aliens in man-disturbed habitats. However, detailed studies on the consequences for the original flora are comparatively few.

A corresponding survey to that of Ebenhard (1988) for animal introductions has still to be done for plants. That drastic changes may occur is clearly evidenced by numerous examples, for instance from the area of biological control concerning successful aliens in environments disturbed by man (Harper, 1977), the number of native species threatened by aliens (Hall, 1987) and shifts in abundance of native plant communities (Wilson & Belcher, 1989). Even though Sukopp (1962) stated that aliens have been able to establish in most plant communities on the European continent (except for a few more extreme habitats), our understanding of the effects on species and species composition is fragmentary (cf. Mack, 1985). However, there are a few examples illustrating drastic changes, including local extinctions, in plant communities that are only little affected by man (Cwiklinski, 1978; Trepl, 1984; Falinski, 1986).

In Fennoscandia, it is frequently stated that successful introduced aquatic plants (e.g. *Elodea canadensis* and *Glyceria maxima*) have resulted in local extinctions of several rare aquatic plant species (Almquist, 1965; Uotila, 1971; Rørslett, 1981; Martinsson, 1988). However, the relative importance of aliens compared with the eutrophication of water-courses (cf. Wallsten, 1981) often remains unclear. The large-scale plantations in Sweden of the North American *Pinus contorta* have resulted in drastic changes in the fungal flora compared to the situation in stands of the native *P. sylvestris*. One species, the mycorrhiza-forming bolete *Suillus flavidis*, which earlier only occurred in very wet *P. sylvestris* stands, now occurs in abundance in dry *P. contorta* stands (Kardell *et al.*, 1987).

Another consequence of species invasions is that closely related species which earlier were spatially separated may meet. This has resulted in introgressions and the subsequent risk of local extinctions by hybridization. Examples are the endemic *Atriplex tularensis* in California (Freas & Murphy, 1988), *Dryopteris cristata* and *Erica ciliaris* in the United Kingdom (Ratcliffe, 1973), *Polemonium hyperboreum* in its only Nordic locality in Norway (Nilsson, 1986) and the loss of seasonal dimorphism among, for example, some *Euphrasia* species (Karlsson, 1976).

Disease

The significance of disease has until recently largely been neglected in conservation contexts (Dobson & Hudson, 1986; Dobson & May, 1986). The best-known example in the animal kingdom is perhaps the introduction of avian malaria to the Hawaii islands, which has resulted in the remaining endemic land bird fauna becoming restricted to higher elevations where introduced birds, carriers of the pathogen, only rarely occur (Warner, 1968; van Riper *et al.*, 1986). Another case is the extinction of crayfish from most Swedish water-courses due to crayfish disease. There are also examples that disease may result in unexpected extinctions. The large blue butterfly *Maculina arion* declined following the myxomatosis epidemic, which resulted in reduced grazing by rabbits and the subsequent decline of the ant species that house the *Maculina* larvae (Thomas, 1976; Ratcliffe, 1979). In the plant kingdom several examples exist where introduced parasitic fungi have resulted in drastic shifts in abundance of dominant forest trees. This is the case for host-specific pathogens such as *Endothia parasitica* on *Castanea dentata* in North America (Woods & Shanks, 1959; Day & Monk, 1974; Stephenson, 1986) and *Ceratocystis ulmi* on elms in both North America and Europe (Agrios, 1978). The same applies to host-alternating pathogens such as the rust fungus *Cronatium ribicola*, which has eradicated *Pinus strobus* as an economically important forest tree in North America (Agrios, 1978). On the other hand, the generalist pathogen *Phytophthora cinnamomi* has caused both drastic vegetation changes and local extinctions among plant species in Australia

(Weste, 1981). No doubt, such events are likely to remain unexplored in cases other than for dominant species or those of great economic importance.

3.6 DISTURBANCE REGIMES AND HABITAT QUALITY

Fires, windstorms, ice, floods, herbivores, burrowing animals and other agents cause natural disturbances. As a result, a natural landscape is a mosaic of successional patches of various sizes. Disturbed patches vary enormously in size and frequency in different biotopes and regions. Over long periods of time species have become adapted to the characteristic disturbance regime in their range. When man changes the frequency of disturbance and/or the size of disturbance patches he also changes the relative abundance of species.

In fact, most species considered threatened in Sweden and Finland have declined due to an altered disturbance regime (Tables 3.7 and 3.8). However, the causes for decline vary between groups.

3.6.1 Pristine disturbances

Our basic view is that vulnerable species can be identified if we compare present land-use practices with the former disturbance regime of a biotope. This accords with the view of Denslow (1980) that the frequency, extent and nature of disturbance are likely to constitute important selective forces on life-history strategies. For every biotope we need to know the pristine disturbance regime, in order to compare it with the present-day situation and evaluate the response of the constituent species.

Gap dynamics

Temperate, deciduous, and some boreal, forests (in Fennoscandia some spruce forests) show gap dynamics following the death or storm felling of individual old trees. The adaptation of organisms to this type of small-scale disturbance should result in species being present which are characterized by low dispersal abilities. Consequently, such forests are probably more sensitive to habitat fragmentation than those affected by large-scale perturbations such as fire. The high proportion of vulnerable cryptogams and beetles associated with these stands clearly reflects this factor (cf. Tables 3.7 and 3.9).

Many species of temperate deciduous forests seem to demand a long stand continuity for their existence. Peterken and Game (1984) compared

Table 3.7

Vulnerable forest species among vascular plants, bryophytes and lichens in Sweden (based on species lists in Databanken för hotade arter och Naturvådsverket (1991), Floravårdskommittén för Lavar, 1987 and Floravårdskommittén för Mossor, 1988), their swedish distribution and habitat characteristics and main causes of decline

	Disappearance of gap dynamics			*Disappearance of successional stages*		*Small popu-lations*	*Others*	*Total*
	Living trees	*Dead and fallen trees*	*On the ground or on rocks*	*Living[a] trees*	*On the ground*			
Vascular plants								
Temperate-hemiboreal	0	0	26[b]	0	7	3	14	50
Boreal	0	0	13	0	5	0	0	18
Total	0	0	39	0	12	3	14	68
Bryophytes								
Temperate	10	5	20	3	1	8	2	49
Hemiboreal	0	10	7	2	0	0	0	19
Boreal	0	0	3	0	0	0	0	3
Total	10	15	30	5	1	8	2	71
Lichens								
Temperate	83	6	6	9	0	0	1	105
Hemiboreal	10	9	9	13	0	0	0	41
Boreal	25	4	6	5	0	0	0	40
Total	118	19	21	27	0	0	1	186

[a] Deciduous successional trees.
[b] Of which 13 are due to conversion of deciduous to coniferous forests.

Table 3.8
Vulnerable vascular plants, bryophytes and lichens (excluding forest species) in Sweden (for sources see Table 3.7), their habitat characteristics and main causes of decline

	Changed management	*Crop weeds*	*Village weeds*	*Urbanization*	*Small populations*	*Others*	*Eutrophication*	*Changed hydrology*	*Total*
Agricultural areas									
Vascular plants	113	41	25	16	3	33*	0	0	231
Bryophytes	29	0	0	6	4	0	0	0	39
Lichens	0	0	0	0	2	3	0	0	5
Water-courses, shores									
Vascular plants	20	0	0	0	7	11	20	12	70
Bryophytes	21	0	0	0	6	1	0	1	29
Lichens	0	0	0	0	8	0	0	0	8
Rocky ground (outside the mountain range)									
Vascular plants	0	0	0	0	0	14	0	0	14
Bryophytes	12	0	0	0	10	1	0	0	23
Lichens	43	0	0	0	0	0	0	0	43
Mires									
Vascular plants	3	0	0	0	0	0	0	9	12
Bryophytes	0	0	0	0	4	3	0	3	10
Mountain areas									
Vascular plants	0	0	0	0	23	0	0	0	23
Bryophytes	0	0	0	0	42	0	0	0	42
Lichens	0	0	0	0	1	0	0	0	1

[a] Mainly alien species.

Table 3.9
The numbers of vulnerable lichens and beetles and their host-tree species in Sweden (based on data from Floravårdskommittén för Lavar (1987) for lichens and Ehnström and Waldén (1986) for beetles. Lichens category 0–4, beetles category 0–2)

Tree species	*Lichens*	*Beetles*		
		Living trees	*Dead trees*	*Total*
Southern deciduous trees				
Acer	2	1	0	1
Alnus	9	0	3	3
Carpinus	0	1	0	1
Corylus	2	1	0	1
Fagus	15	11	19	30
Fraxinus	9	2	1	3
Quercus	24	33	12	45
Tilia	1	5	6	11
Ulmus	7	5	3	8
Species not host-specific	16	4	1	5
Abundant deciduous trees				
Betula	3	1	12	13
Salix	1	3	0	3
Sorbus	3	1	0	1
Populus	13	0	6	6
Species not host-specific	14	1	3	4
Conifers				
Picea	18	3	7	10
Pinus	4	6	7	13
Species not host-specific	0	0	2	2

89 ancient (existing prior to the year 1600) and 273 recent (established after 1600) temperate, deciduous woodlands in England. They found that a large number of plants were found exclusively in ancient woodlands (often those species which were myrmecochorous, with large seeds and favouring wet places—for the last-mentioned group see also Falinski *et*

al. (1988)). Peterken and Game (1984) also subdivided the recent stands into two groups established prior to and after 1867, and were unable to find any difference in species number. It is remarkable that a species such as *Anemone nemorosa* (myrmecochorous) was confined to ancient stands (Peterken, 1981). This seems strange compared to its occurrence in Fennoscandia, but certainly reflects the much more drastic habitat destruction in England during a long period. Rose (1976), in a corresponding study of the lichen flora, found about 150 taxa/km^2 to be present in stands unaffected during the last 200 years, while there were less than 40 taxa/km^2 in stands clear-felled or replaced during the last 200 years. In order to record complete assemblages of Atlantic bryophyte species Edwards (1986) suggested a woodland continuity of at least 1200 years, although this calculation was based on a rather limited data base.

The conclusion is that the preservation of temperate deciduous forest stands with a long continuity constitutes a prime goal for nature conservation.

Fire disturbances

Fire suppression has caused profound effects in many natural biotopes affecting species adapted to fire. In grasslands, fire and soil disturbance by animals is crucial for the maintenance of many species (Collins & Barber, 1985; McNaughton, 1985; Collins, 1987). The elimination of fires will alter forest stand dynamics. For example, the mixed coniferous forests in the Southern Rocky Mountains were earlier characterized by a patchy disturbance regime including fire and insect outbreaks. Fire elimination, at the beginning of this century, has resulted in more homogeneous, contiguous, closed stands, implying risks for large outbreaks of the western spruce budworm *Choristoneura occidentalis* (Fritts & Swetnam, 1989). The long-term outcome of such changes in dynamics and patchiness for other forest species remains unclear. The active fire suppression in Fennoscandian forests since the end of the 19th century (chapter 7) has resulted in regional extinctions of many insects (Ehnström & Waldén, 1986). The data in Table 3.7 show that a limited number of vascular plants are declining due to lack of fire. However, a considerable number of cryptogams and beetles depend on the deciduous post-fire successional stage (Tables 3.7 and 3.9).

Further, as most late successional stands in boreal Fennoscandia became established after fire during the 19th century, the consequences of recent fire elimination are still unclear. This is highlighted in the case of *Botrychium virginianum*, which was considered as threatened and confined to late successional mire forests. However, it was recently found at several localities in an area hit by extensive storm fellings in 1954. At that time the clear-cuts were burned before planting of pine (Ståhl, 1990).

Animal disturbances

By hunting megaherbivores (body mass more than 1000 kg) to extinction, man has reduced the extent of natural early-successional habitats, apparently causing extinction of additional species (Owen-Smith, 1987). No doubt this has also influenced the temperate deciduous forests. Here, species such as European bison (Borowski & Kossak, 1972), deer (Borowski & Kossak, 1975), and probably the ancestors of domesticated cattle and horses, affected younger deciduous trees by browsing and barking, thus keeping forest gaps open for a longer time. A large proportion of the flora nowadays confined to sun-exposed sites in the agricultural landscape were originally adapted to such habitats. Circumstantial evidence is provided by the high proportion of this flora that is adapted to epizoochorous dispersal (cf. Falinski, 1972). Some domestic animals, for example cattle and horses, may have had a similar effect as extinct herbivores, but firm knowledge is lacking.

Among species causing significant disturbance, beavers *Castor* spp. must be mentioned. They create an abundance of productive marshes, influence succession and increase the amount of dead logs (Naiman *et al.*, 1988). The extinction of beavers over large regions of Europe and America due to hunting is probably one reason for the decline and regional extinction of some species of wood-dwelling beetles and woodpeckers, but detailed studies are lacking.

3.6.2 Man-made disturbances

Some disturbance caused by man may simulate the effects of recently extinct large herbivores. Traditional hay-meadows in Europe are very rich in species of vascular plants and insects (e.g. Willems, 1983; Erhardt, 1985). The reason may be that the practice of opening up forest glades and maintaining a low sward by mowing is sufficiently similar to the effects of large herbivores. In autumn, and sometimes also in early spring, these ancient hay-meadows were grazed by domestic animals. Their deposition of dung and trampling could also be important for the continued existence of some species.

If the rate of succession from small to high plants following disturbance increases, this can also cause local extinction of species. In grasslands, fertilization by nitrogen severely reduces species richness (e.g. Silvertown, 1980; During & Willems, 1986; Tilman, 1987).

It has been argued that few species are adapted to a very high nutrient supply. However, it is more likely that changed disturbance regimes constitute the main reason for the decline of species in nutrient-rich environments. Studies on this problem have mainly been conducted in northwestern Europe and in parts of North America where the original herbivore fauna has been drastically altered by man. The rapid disappearance of poor competitors on Baltic sea-shores have, for example, been the result of the cessation of management and grazing (Luther & Munsterhjelm, 1983).

It is obvious that man has profoundly changed the disturbance regime in most major biotopes, especially very recently. Depending on the amount of change, various proportions of the indiginous species are threatened by regional and perhaps global extinction. We need to restore or simulate the effects of pristine disturbance regimes.

In agricultural landscapes, most species are adapted to various types of disturbance which delay succession. Thus the maintenance of mowing and grazing will enable them to persist, locally, even if they are characterized by low dispersal abilities. Further, a large number of species, e.g. weeds of arable crops, can be maintained by sowing (cf. Svensson & Wigren, 1986*a*,*b*). In forest landscapes we face a more serious problem, namely that the appropriate disturbance regime must be maintained, which implies that large forest stands have to be preserved in order to enable stenotopic species to prevail.

3.7 CONCLUSIONS

Recently, emphasis has been on population size for the prediction of extinction risks (Soulé, 1987). This may be appropriate for species in immediate danger of extinction. However, there are many more species at risk but which still have fairly large populations. For these, other considerations may be more important. We think that restoration of pristine disturbance regimes in nature reserves is in the long run of paramount importance. In addition, we need to develop methods of economic exploitation simulating those disturbance regimes. Even slight but biologically crucial modifications of management methods will considerably reduce the proportion of species restricted to nature reserves.

Finally, we recall the work by Snyder and Snyder (1989) on the

Californian condor. They emphasize that in order to avoid extinction, studies on declining species must be commenced at a much earlier stage than is usual. Too often conservation decisions are made on pure guesses and initiated too late.

REFERENCES

Agrios, G.N. (1978). *Plant Pathology*. Academic Press, New York.

Ahlén, I. (1977). *Faunavård*. Liber, Stockholm.

Ahlén, I. & Nilsson, S.G. (1982). Species richness and area requirements of forest bird species on islands with natural forests in Lake Mälaren and Hjälmaren. *Vår Fågelvärld*, **41**, 161–84 (in Swedish with English summary).

Ahlén, I. & Tjernberg, M. (1988). Hotade och sällsynta ryggradsdjur i Sverige. *Sver. Nat.*, **2**, 32–42.

Almquist, E. (1965). *Flora Upsaliensis. Uppsalatraktens Växter*. Almqvist & Wiksell, Stockholm.

Andersson, H., Coulianos, C.C., Ehnström, B., Hammarstedt, O., Imby, L., Janzon, L.-Å., Lindelöw, Å. & Waldén, H.W. (1987). Hotade evertebrater i Sverige. *Entomol. Tidskr.*, **108**, 65–75.

Anon. (1978). Hotade djur och växter i Norden. Nordiska ministerrådet, NU A 1978 (9), 194 pp. (in Swedish with English summary).

Aulén, G. (1988). Ecology and distribution history of the white-backed woodpecker *Dendrocopos leucotus* in Sweden. *Rep. Depart. Wildl. Ecol., Swed. Univ. Agric. Sci.*, Uppsala, No. 14.

Beddington, J.R., Free, C.H. & Lawton, J.H. (1978). Characteristics of successful natural enemies in models of biological control of insect pests. *Nature, Lond.*, **273**, 513–9.

Belovsky, G.E. (1987). Extinction models and mammalian persistence. In *Viable Populations for Conservation*, ed. M.E. Soulé. Cambridge University Press, Cambridge, pp. 35–57.

Bengtsson, J. (1989). Interspecific competition increases local extinction rate in a metapopulation system. *Nature, Lond.*, **340**, 713–5.

Berdowski, J.J.M. & Siepel, H. (1988). Vegetative regeneration of *Calluna vulgaris* at different ages and fertilizer levels. *Biol. Conserv.*, **46**, 85–93.

Berdowski, J.J.M. & Zeilinga, R. (1987). Transition from heathland to grassland: damaging effects of the heather beetle. *J. Ecol.*, **75**, 159–75.

Bergerud, A.T. (1988). Caribou, wolves and man. *Trends Ecol. Evol.*, **3**, 68–72.

Bibby, C. (1986). Merlins in Wales: site occupancy and breeding in relation to vegetation. *J. appl. Ecol.*, **23**, 1–12.

Blakley, N.R. & Dingle, H. (1978). Competition: butterflies eliminate milkweed bugs from a Caribbean island. *Oecologia (Berl.)*, **37**, 133–6.

Bobbink, R. & Willems, J.H. (1987). Increasing dominance of *Brachypodium pinnatum* (L.) Beauv. in chalk grasslands: a threat to a species-rich ecosystem. *Biol. Conserv.*, **40**, 301–14.

Bond, W. & Slingsby, P. (1984). Collapse of an ant–plant mutalism: the Argentine ant *Iridomyrmex humilis* and myrmecochorous Proteaceae. *Ecology*, **65**, 1031–7.

Borowski, S. & Kossak, S. (1972). The natural food preferences of the European bison in seasons free of snow cover. *Acta theriol.*, **17**, 151–69.

Borowski, S. & Kossak, S. (1975). The food habits of deers in the Bialowieza Primeval Forest. *Acta theriol.*, **20**, 463–506.

Boström, U. & Nilsson, S.G. (1983). Latitudinal gradients and local variations in species richness and structure of bird communities on raised peat-bogs in Sweden. *Ornis Scand.*, **14**, 213–26.

Carpenter, S.R., Kitchell, J.F. & Hogson, J.R. (1985). Cascading trophic interactions and lake productivity. *BioScience*, **35**, 634–9.

Carter, R.N. & Prince, S.D. (1981). Epidemic models used to explain biogeographical distribution limits. *Nature, Lond.*, **293**, 644–6.

Cederlund, G. & Markgren, G. (1987). The development of the Swedish moose population, 1970–1983. *Swed. Wildl. Res. Suppl.*, **1**, 55–62.

Challinor, D. & Wingate, D.B. (1971). The struggle for survival of the Bermuda cedar. *Biol. Conserv.*, **3**, 220–2.

Collins, S.L. (1987). Interaction of disturbance in tallgrass prairie: a field experiment. *Ecology*, **68**, 1243–50.

Collins, S.L. & Barber, S.C. (1985). Effects of disturbance on diversity in mixed-grass prairie. *Vegetatio*, **64**, 87–94.

Cox, P.A. (1983). Extinction of the Hawaiian avifauna resulted in a change of pollinators for the ieie, *Freycinetia arborea*. *Oikos*, **41**, 195–9.

Cox, P.A., Elmqvist, T., Pierson, E.D. & Rainey, W. (1991). Flying foxes as strong indicators in South Pacific island ecosystems: A conservation hypothesis. *Conserv. Biol.*, p. 5.

Crawley, M.J. (1983). *Herbivory. The Dynamics of Animal–Plant Interactions. Studies in Ecology*, **10**. Blackwell, Oxford.

Cwiklinski, E. (1978). Die Einwanderung der synanthropen Art *Impatiens parviflora* DC. in die natürlichen Pflanzengesellschaften. *Acta bot. slov. Acad. Sci. Slov. A*, **3**, 17–30.

Dahl, E. (1956). Rondane: mountain vegetation in south Norway and its relation to the environment. *Nor. Vidensk. Akad. Oslo Skr. I. Mat.-Naturvidensk. Kl.*, **1956**(3), 1–374.

Danell, K. (1977). Short-term plant successions following the colonization of a northern Swedish lake by the muskrat, *Ondatra zibethica*. *J. appl. Ecol.*, **14**, 933–47.

Danell, K. (1979). Reduction of aquatic vegetation following the colonization of a northern Swedish lake by the muskrat, *Ondatra zibethica*. *Oecologia (Berl.)*, **38**, 101–6.

Danell, K. & Ericson, L. (1990). Dynamic relations between the antler moth and meadow vegetation in northern Sweden. *Ecology*, **71**, 1068–77.

Databanken för hotade arter och Naturvårdsverket (1991). *Sweden's Red Lists on Vascular Plants, Bryophytes, Lichens and Macrofungi.*

Day, F.P. & Monk, C.D. (1974). Vegetation patterns on a southern Apalachian watershed. *Ecology*, **55**, 1064–74.

Dayton, P.K. (1975). Experimental studies of algal canopy interactions in a sea otter-dominated kelp community at Amchitka Island, Alaska. *Fish. Bull.*, **73**, 23–237.

Denslow, J.S. (1980). Patterns of plant species diversity during succession under different disturbance regimes. *Oecologia (Berl.)*, **46**, 18–21.

Diamond, J.M. (1974). Colonization of exploded volcanic islands by birds: the supertramp strategy. *Science, N.Y.*, **184**, 803–6.

Diamond, J.M. (1984). 'Normal' extinctions of isolated populations. In *Extinctions*, ed. M.H. Nitecki. University of Chicago Press, Chicago, Illinois, pp. 191–245.

Dobson, A.P. & Hudson, P.J. (1986). Parasites, disease and the structure of ecological communities. *Trends Ecol. Evolut.*, **1**, 11–15.

Dobson, A.P. & May, R.M. (1986). Disease and conservation. In *Conservation Biology. The Science of Scarcity and Diversity*, ed. M.E. Soulé. Sinauer Associates, Sunderland, Massachusetts, pp. 345–65.

During, H.J. & Willems, J.H. (1986). The impoverishment of the bryophyte and lichen flora of the chalk grasslands in the thirty years 1953–1983. *Biol. Conserv.*, **36**, 143–58.

Ebenhard, T. (1988). Introduced birds and mammals and their ecological effects. *Swed. Wildl. Res.*, **13**(4).

Edwards, M.E. (1986). Disturbance histories of four Snowdonian woodlands and their relation to Atlantic bryophyte distribution. *Biol. Conserv.*, **37**, 301–20.

Ehnström, B. & Waldén, H.W. (1986). *Faunavård i Skogsbruket, Del 2. Den Lägre Faunan.* Skogsstyrelsen, Jönköping.

Ehrlich, P.R. & Murphy, D.D. (1987). Conservation lessons from long-term studies of checkerspot butterflies. *Conserv. Biol.*, **1**, 122–31.

Ehrlich, P.R., Murphy, D.D., Singer, M.C., Sherwood, C.B., White, R.R. & Brown, I.L. (1980). Extinction, reduction, stability and increase: the responses of checkerspot butterfly (*Euphydras*) populations to the Californian drought. *Oecologia* (*Berl.*), **46**, 101–5.

Ekman, S. (1990). The lichen flora of Dalby Söderskog National Park. *Svensk Bot. Tidskr.*, **84**, 191–8 (in Swedish with English summary).

Elven, R. (1989). *Aster sibiricus* in Norway. *Blyttia*, **47**, 51–3 (in Norwegian with English summary).

Erhardt, A. (1985). Diurnal Lepidoptera: sensitive indicators of cultivated and abandoned grassland. *J. appl. Ecol.*, **22**, 849–61.

Estes, J.A., Smith, N.S. & Palmisano, J.F. (1978). Sea otter predation and community organization in the western Aleutian Island. *Ecology*, **59**, 822–33.

Falinski, J.B. (1972). Anthropochory in xerothermic grasslands in the light of experimental data. *Acta Soc. Bot. Pol.*, **41**, 357–68.

Falinski, J.B. (1986). Vegetation dynamics in temperature lowland primeval forests. Ecological studies in Bialowieza forest. *Geobotany*, **8**, 1–537.

Falinski, J.B., Canullo, R. & Bialy, K. (1988). Changes in herb layer, litter fall and soil properties under primary and secondary tree stands in a deciduous forest ecosystem. *Phytocoenosis*, **1**, 1–49.

Ferry, B.W., Baddeley, M.S. & Hawksworth, D.L. (eds) (1973). *Air Pollution and Lichens.* Athlone Press, London.

Fisher, F.J.F. (1965). The alpine *Ranunculi* of New Zealand. *Bull. N.Z. Dep. Scient. Ind. Res.*, **165**, 1–192.

Floravårdskommittén för Lavar (1987). Preliminär lista över hotade arter i Sverige. *Svensk Bot. Tidskr.*, **81**, 237–56.

Floravårdskommittén för Mossor (1988). Preliminär lista över hotade mossor i Sverige. *Svensk Bot. Tidskr.*, **82**, 423–45.

Forman, R.T.T. (1964). Growth under controlled conditions to explain the hierarchical distribution of a moss *Tetraphis pellucida*. *Ecol. Monogr.*, **34**, 1–25.

Freas, K.E. & Murphy, D.D. (1988). Taxonomy and the conservation of the critically endangered Bakersfield saltbush, *Atriplex tularensis*. *Biol. Conserv.*, **46**, 317–24.

Fritts, H.C. & Swetnam, T.W. (1989). Dendroecology: a tool for evaluating variations in past and present forest environments. *Adv. Ecol. Res.*, **19**, 111–88.

Frost, L.C. (1981). The study of *Ranunculus ophioglossifolius* and its successful conservation at the Badgeworth Nature Reserve, Gloucestershire. In *The Biological Aspects of Rare Plant Conservation*, ed. H. Synge. Wiley, Chichester, pp. 481–9.

Gaston, K.J. & Lawton, J.H. (1988). Patterns in the distribution and abundance of insect populations. *Nature, Lond.*, **331**, 709–12.

Gentry, A.H. (1986). Endemism in tropical versus temperate plant communities. In Conservation Biology, *The Science of Scarcity and Diversity*, ed. M.E. Soulé. Sinauer Associates, Sunderland, Massachusetts, pp. 153–81.

Gilbert, L.E. (1980). Food web organization and conservation of neotropical diversity. In *Conservation Biology. An Evolutionary Perspective*, ed. M.E. Soulé & B.A. Wilcox. Sinauer Associates, Sunderland, Massachusetts, pp. 11–33.

Glazier, D.S. (1986). Temporal variability of abundance and the distribution of species. *Oikos*, **47**, 309–14.

Goodman, D. (1987). The demography of chance extinction. In *Viable Populations for Conservation*, ed. M.E. Soulé. Cambridge University Press, Cambridge, pp. 11–34.

Gotteli, N.J. & Graves, G.R. (1990). Body size and the occurrence of avian species on land-bridge islands. *J. Biogeogr.*, **17**, 315–25.

Gutowski, J.M. (1990). Pollination of the orchid *Dactylorhiza fuchsii* by longhorn beetles in primeval forests of northeastern Poland. *Biol. Conserv.*, **51**, 287–97.

Haahtela, I. (1981). Probable reasons for the decline of the bladder wrack *Fucus vesiculosus* L. in SW Finland. *Rep. Dep. Biol. Univ. Turku*, **2**, 18–21.

Hall, A.V. (1987). Threatened plants in the fynbos and karoo biomes, South Africa. *Biol. Conserv.*, **40**, 29–52.

Hallingbäck, T. (1989*a*). Occurrence and ecology of the lichen *Lobaria scrobicularia* in southern Sweden. *Lichenologist*, **21**, 331–41.

Hallingbäck, T. (1989*b*). *Neckera pumila*, a moss threatened by acidification in south Sweden. *Svensk Bot. Tidskr.*, **83**, 161–73 (in Swedish with English summary).

Hanski, I. (1985). Single-species spatial dynamics may contribute to long-term rarity and commonness. *Ecology*, **66**, 335–43.

Harper, J.L. (1977). *Population Biology of Plants*. Academic Press, London.

Harrison, S. (1989). Long-distance dispersal and colonization in the bay checkerspot butterfly, *Euphydras editha bayensis*. *Ecology*, **70**, 1236–43.

Helle, T. & Aspi, J. (1983). Effects of winter grazing by reindeer on vegetation. *Oikos*, **40**, 337–43.

Helle, P. & Järvinen, O. (1986). Population trends of North Finnish land birds in relation to their habitat selection and changes in forest structure. *Oikos*, **46**, 107–15.

Høiland, K. (1981). Utryddet i Norge, 1. Altaihavkeskjegg. *Norsk Nat.*, **17**, 4–5.

Hopkins, P.J. & Webb, N.R. (1984). The composition of the beetle and spider faunas on fragmented heathlands. *J. appl. Ecol.*, **21**, 935–46.

House, S.M. & Spellerberg, I.F. (1983). Ecology and conservation of the sand lizard (*Lacerta agilis* L.) habitat in southern England. *J. appl. Ecol.*, **20**, 417–37.

Hunter, J.R. (1989). Seed dispersal and germination of *Enterolobium cyclocarpum* (Jacq.) Griseb. (Legumionosae: Mimosoideae): are megafauna necessary? *J. Biogeogr.*, **16**, 369–78.

Hydén, N. (1990). *Baptria tibiale* och *Eupithecia actaeta* (Lept., Geometridae) i nordvästeuropa—utbredning, biologi, äggläggning och habitatval. *Entomol. Tidskr.*, **111**, 1–15.

Ingelög, T., Thor, G. & Gustafsson, L. (eds) (1987). *Floravård i Skogsbruket, 2nd edn.* Skogsstyrelsen, Jönköping.

Iverson, J.B. (1987). Tortoises, not dodos, and the Tamalaeoque tree. *J. Herpetol.*, **21**, 229–30.

Janzen, D.H. & Martin, P. (1982). Neotropical anachronisms: the fruits the gomphotheres ate. *Science, N.Y.*, **215**, 228–353.

Järvinen, O. (1982). Conservation of endangered plant populations: single large or several small reserves? *Oikos*, **38**, 301–7.

Järvinen, O. & Miettinen, K. (1988). *Sista paret ut? Om naturvårdens biologi.* Painokaari Oy, Helsingfors.

Järvinen, O. & Väisänen, R. (1978). Recent changes in forest bird populations in northern Finland. *Ann. Zool. Fennici*, **15**, 279–89.

Kangas, P., Autio, H., Hällfors, G., Luther, H., Niemi, Å. & Salemaa, H. (1982). A general model of the decline of *Fucus vesiculosus* at Tvärminne, south coast of Finland in 1977–81. *Acta Bot. Fennica*, **118**, 1–27.

Kardell, L., Blomgren, M. & Nitare, J. (1987). Mushroom production and species composition in stands of *Pinus contorta* and *P. sylvestris*. *Svensk Bot. Tidskr.*, **81**, 133–42 (in Swedish with English summary).

Karlsson, T. (1976). *Euphrasia* in Sweden: hybridization, parallelism, and species concept. *Bot. Not.*, **129**, 49–60.

Karr, J.R. (1982*a*). Population variability and extinction in the avifauna of a tropical land bridge island. *Ecology*, **63**, 1975–8.

Karr, J.R. (1982*b*). Avian extinctions on Barro Colorado Island, Panama: a reassessment. *Amer. Nat.*, **119**, 220–39.

Kelsall, J.P. (1968). The migratory barren-ground caribou of Canada. *Can. Wildl. Serv. Monogr. Ser.*, **3**, Queen's Printer, Ottawa.

Lahti, T. & Ranta, E. (1985). The SLOSS principle and conservation practice: an example. *Oikos*, **44**, 369–70.

Lande, R. (1987). Extinction thresholds in demographic models of territorial populations. *Amer. Nat.*, **130**, 624–35.

Lande, R. (1988*a*). Genetics and demography in biological conservation. *Science, N.Y.*, **241**, 1455–60.

Lande, R. (1988*b*). Demographic models of the northern spotted owl (*Strix occidentalis caurina*). *Oecologia (Berl.)*, **75**, 601–7.

Laws, R.M., Parker, I.S.C. & Johnstone, R.C.B. (1975). *Elephants and their Habitat*. Clarendon Press, Oxford.

Leigh, E.G., Jr (1981). The average lifetime of a population in a varying environment. *J. Theoret. Biol.*, **90**, 31–239.

Lesica, P., Leary, R.F., Allendorf, F.W. & Bilderbacke, D.E. (1987). Lack of genetic diversity within and among populations of an endangered plant, *Howellia aquatilis*. *Conserv. Biol.*, **2**, 275–82.

Levins, R. (1970). Extinction. In *Lectures on Mathematics in the Life Sciences*, Vol. 2, ed. M. Gerstenhaber. American Mathematical Society, Providence, RI, pp. 77–107.

Lindström, H. (1980). Hackslått—en försvinnande biotop i mellersta Norrland. *Svensk Bot. Tidskr.*, **74**, 281–306.

Löfgren, O. & Moberg, R. (1984). Oceaniska lavar i Sverige och deras tillbakagång. *Swedish Environment Proctection Board*, SNV PM 1819 (in Swedish with English summary).

Lowry, L.F. & Pearse, J.S. (1973). Abalones and sea urchins in an area inhabited by sea otters. *Mar. Biol.*, **23**, 213–9.

Lucas, G. & Synge, H. (1978). *The IUCN Plant Red Data Book*. IUCN, Morges.

Lundqvist, R. (1990). Transtrandsfjällens urskogar. En naturvårdsinventering av vårt sydligaste fjällområde. *Länsstyrelsen i Kopparbergs län, Miljövårdsenheten*, 1990, **1**. Falun.

Luther, H. & Munsterhjelm, R. (1983). Inverkan av strandbetets upphörande på hydrolittoralens flora i Pojoviken. *Mem. Soc. Fauna Flora Fennica*, **59**, 9–19.

MacArthur, R.H. & Wilson, E.O. (1967). *The Theory of Island Biogeography*. Princeton University Press, Princeton, New Jersey.

Mack, R.N. (1985). Invading plants: their potential contribution to population biology. In *Studies on Plant Demography. A Festschrift for John L. Harper*, ed. J. White. Academic Press, London, pp. 127–42.

Martin, O. (1989). Click beetles (Coleoptera, Elateridae) from old deciduous forests in Denmark. *Entomol. Medd.*, **57** (1–2), 1–107 (in Danish with English summary).

Martin, P.S. (1984). Prehistoric overkill: the global model. In *Quaternary Extinctions*, ed. P.S. Martin, & R.G. Klein. University of Arizona Press, Tucson, Arizona, pp. 354–403.

Martinsson, K. (1988). The decline of *Callitriche hermaphroditica* in freshwater habitats in South and Central Sweden. *Svensk Bot. Tidskr.*, **83**, 243–64 (in Swedish with English summary).

Matson, P.A. & Warring, R.H. (1983). Effects of nutrient and light limitation on mountain hemlock: susceptibility to laminated root rot. *Ecology*, **65**, 1517–24.

McArdle, B.H., Gaston, K.J. & Lawton, J.H. (1990). Variation in the size of animal populations: patterns, problems and artefacts. *J. Anim. Ecol.*, **59**, 439–54.

McNaughton, S.J. (1985). Ecology of a grazing system: the Serengeti. *Ecol. Monogr.*, **55**, 259–94.

Meagher, T.R., Antonovics, J. & Primack, R. (1978). Experimental ecological genetics in *Plantago*, III. Genetic variation and demography in relation to survival of *Plantago cordata*, a rare species. *Biol. Conserv.*, **14**, 243–57.

Melville, R. (1979). Endangered islands floras. In *Plants and Islands*, ed. D. Bramwell. Academic Press, London, pp. 361–77.

Menges, E.S. (1990). Population viability analysis for an endangered plant. *Conserv. Biol.*, **4**, 52–62.

Mertz, D.B. (1971). The mathematical demography of the California condor population. *Amer. Nat.*, **105**, 437–53.

Messier, F. & Crête, M. (1985). Moose–wolf dynamics and the natural regulation of moose populations. *Oecologia (Berl.)*, **65**, 503–12.

Müller, W. (1982). Die Besiedlung der Eichenwälder im Kanton Zürich durch den Mittelspecht, *Dendrocopos medius*. *Orn. Beob.*, **79**, 105–19.

Myers, N. (1976). An expanded approach to the problem of disappearing species. *Science, N.Y.*, **193**, 198–202.

Myers, N. (1986). Tropical deforestation and a mega-extinction spasm. In *Conservation Biology. The Science of Scarcity and Diversity*, ed. M.E. Soulé. Sinauer Associates, Sunderland, Massachusetts, pp. 394–409.

Naiman, R.J., Johnston, C.A. & Kelley, J.C. (1988). Alteration of North American streams by beaver. *BioScience*, **38**, 753–62.

Newmark, W.D. (1986). Mammalian richness, colonization, and extinction in Western North American National Parks. PhD dissertation, University of Michigan, Ann Arbor, Michigan (cited after Soulé, 1987).

Newmark, W.D. (1987). A land-bridge island perspective on mammalian extinctions in western North American parks. *Nature, Lond.*, **325**, 430–2.

Nilsson, S.G. (1977). Composition and density of the bird community on raised peat-bogs in southwestern Sweden. *Fauna och Flora*, **72**, 227–33 (in Swedish with English summary).

Nilsson, S.G. (1978). Fragmented habitats, species richness and conservation practice. *Ambio*, **7**, 26–7.

Nilsson, S.G. (1982) General rules for the design of nature reserves? In *Current Problems in Fauna Conservation*, ed. B. Pettersson. *Rep. Depart. Wildl. Ecol., Swed. Univ. Agric. Sci., Uppsala*, **9**, 25–9 (in Swedish with English summary).

Nilsson, S.G. (1986). Are bird communities in small biotope patches random samples from communities in large patches? *Biol. Conserv.*, **38**, 179–204.

Nilsson, S.G. (1987). Limitation and regulation of population density in the nuthatch *Sitta europaea* (Aves) breeding in natural cavities. *J. Anim. Ecol.*, **56**, 921–37.

Nilsson, S.G. & Nilsson, I.N. (1978). Population, habitat, and display activity of the jack snipe *Lymnocryptes minimus* in southern Sweden. *Vår Fågelvärld*, **37**, 1–8 (in Swedish with English summary).

Nilsson, S.G. & Nilsson, I.N. (1983). Are estimated species turnover rates on islands largely sampling errors? *Amer. Nat.*, **121**, 595–7.

Nilsson, I.N. & Nilsson, S.G. (1985). Experimental estimates of census efficiency and pseudoturnover on island: error trend and between-observer variation when recording vascular plants. *J. Ecol.*, **73**, 65–70.

Nilsson, S.G. & Wästljung, U. (1987). Seed predation and cross-pollination in mast-seeding beech (*Fagus sylvatica*) patches. *Ecology*, **68**, 260–5.

Nilsson, Ö. (1986). *Nordisk Fjällflora*. Bonniers, Stockholm.

Nitare, J. (1988). Earthtongues, a declining group of macrofungi in seminatural grasslands. *Svensk Bot. Tidskr.*, **82**, 341–68 (in Swedish with English summary).

Nordhagen, R. (1943). Sikilsdalen og Norges fjellbeiter. *Bergens Mus. Skr.*, **22**, 1–607.

Nordhagen, R. (1963). Om *Crepis multicaulis* (Led.) og dens utbredelse i Norge, arktisk Russland og Asia. *Blyttia*, **21**, 1–42.

O'Brien, S.J. & Evermann, J.F. (1988). Interactive influence of infectious disease and genetic diversity in natural populations. *Trends Ecol. Evol.*, **3**, 254–59.

Oksanen, L., Fretwell, S.D., Arruda, J. & Niemelä, P. (1981). Exploitation ecosystems in gradients of primary productivity. *Amer. Nat.*, **118**, 240–61.

Olson, S.L. & Janes, H.F. (1982). Fossil birds from the Hawaiian Islands: evidence for wholesale extinction by man before western contact. *Science, N.Y.*, **217**, 633–5.

Overgaard Nielsen, B. (1986). Outbreaks of the heather beetle (*Lochmaea suturalis* Thoms.) in Danish heathland 1900–1984 (Coleoptera: Chrysomelidae). *Entomol. Medd.*, **53**, 99–109 (in Danish with English summary).

Owen-Smith, N. (1987). Pleistocene extinctions: the pivotal role of megaherbivores. *Paleobiology*, **13**, 351–62.

Owen-Smith, N. (1989). Megafaunal extinctions: The conservation message from 11,000 years BP *Conserv. Biol.*, **3**, 405–12.

Paine, R.T. (1966). Food web complexity and species diversity. *Amer. Nat.*, **100**, 65–75.

Palm, T. (1959). Die Holz- und Rinden-Käfer der süd- und mittelschwedischen Laubbäume. *Opusc. Entomol., Suppl.*, **16**.

Pellmyr, O. (1984). Thc pollination ecology of *Actaea spicata* (Ranunculaceae). *Nord. J. Bot.*, **4**, 443–56.

Peltonen, E., Kangas, J. & Kaisila, J. (1970). Zur Bionomie von *Eupithecia groenblomi* Urb. (Lep., Geometridae). *Ann. Entomol. Fenn.*, **36**, 70–5.

Peterken, G.F. (1981). Wood anemone in central Lincolnshire: an ancient woodland indicator? *Trans. Lincs. Nat. Un.*, **20**, 78–82.

Peterken, G.F. & Game, M. (1984). Historical factors affecting the number and distribution of vascular plant species in the woodlands of central Lincolnshire. *J. Ecol.*, **72**, 155–82.

Peters, R.H. (1983). *The Ecological Implications of Body Size.* Cambridge University Press, Cambridge.

Pettersson, B. (1958). Dynamics and constancy in the flora and vegetation of Gotland, Sweden. *Acta phytogeogr. suec.*, **40** (in Swedish with German summary).

Pettersson, B. (1985*a*). Extinction of an isolated population of the middle spotted woodpecker *Dendrocopos medius* (L.) in Sweden and its relation to general theories on extinction. *Biol. Conserv.*, **32**, 335–53.

Pettersson, B. (1985*b*). Relative importance of habitat area, isolation and quality for the occurrence of middle spotted woodpecker *Dendrocopos medius* (L.) in Sweden. *Holarct. Ecol.*, **8**, 53–8.

Pickett, S.T.A. & Thompson, J.N. (1978). Patch dynamics and the design of nature reserves. *Biol. Conserv.*, **13**, 27–37.

Pimm, S.L. & Pimm, J.W. (1982). Resource use, competition and resource availability in Hawaiian honeycreepers. *Ecology*, **63**, 1468–80.

Pimm, S.L., Jones, H.L. & Diamond, J. (1988). On the risk of extinction. *Amer. Nat.*, **132**, 757–85.

Pyke, G.H. (1984). Optimal foraging theory: a critical review. *Ann. Rev. Ecol. Syst.*, **15**, 523–75.

Rassi, P. & Väisänen, R. (eds) (1987). *Threatened Animals and Plants in Finland.* Miljöministeriet, Helsinki.

Ratcliffe, D.A. (1973). Safeguarding wild plants. In *Plants: Wild and Cultivated*, ed. P.S. Green. Botanical Society of the British Isles, London, pp. 18–24.

Ratcliffe, D. (1979). The end of the large blue butterfly. *New Scient.*, **8**, 457–8.

Rönnberg, O. (1981). Traffic effects on rocky-shore algae in the Archipelago Sea, SW Finland. *Acta Acad. Aboensis (B)*, **41**(3), 1–87.

Rørslett, B. (1981). Mykt havfrugras, *Najas flexilis*, i Norge. *Blyttia*, **39**, 1–6.

Rose, F. (1976). Lichenological indicators of age and environmental continuity in woodlands. In *Lichenology: Progress and Problems*, ed. D.H. Brown, D.L. Hawksworth & R.M. Bailey. *Syst. Ass. Spec.* Vol., **8**. Academic Press, London, pp. 279–307.

Rose, F. (1988). Phytogeographical and ecological aspects of Lobarion communities in Europe. *Bot. J. Linn. Soc.*, **96**, 69–79.

Runkle, J.R. (1985). Disturbance regimes in temperate forests. In *The Ecology of Natural Disturbance and Patch Dynamics*, ed. S.T.A. Pickett & P.S. White. Academic Press, New York, pp. 17–33.

Schonewald-Cox, C., Chambers, S.M., McBryde, B. & Thomas, W.L. (eds) (1983). *Genetics and Conservation: A Reference for Managing Wild Animal and Plant Populations.* Benjamin Cummings, Menlo Park, California.

Shaffer, M.L. (1981). Minimum population sizes for species conservation. *Bio-Science*, **31**, 131–4.

Shaffer, M. (1987). Minimum viable populations: coping with uncertainty. In *Viable Populations for Conservation*, ed. M.E. Soulé. Cambridge University Press, Cambridge, pp. 69–86.

Silvertown, J. (1980). The dynamics of a grassland ecosystem: botanical equilibrium in the Park Grass Experiment. *J. appl. Ecol.*, **17**, 491–504.

Simberloff, D.S. & Abele, L.G. (1976). Island biogeography theory and conservation practice. *Science, N.Y.*, **191**, 285–6.

Simenstad, C.A., Estes, J.A. & Kenyon, K.W. (1978). Aleuts, sea otters, and alternate stable state communities. *Science, N.Y.*, **200**, 403–11.

Singer, M.C. & Ehrlich, P.R. (1979). Population dynamics of the checkerspot butterfly *Euphydryas editha. Fortschr. Zool.*, **25**, 53–60.

Sjögren, P. (1991). Extinction and isolation gradients in metapopulations: the case of the pool frog *Rana lessonae. Biol. J. Linn. Soc.*, **42**, 135–47.

Smith, C.C., Hamrick, J.L. & Kramer, C.L. (1988). The effects of stand density on frequency of filled seeds and fecundity in lodgepole pine (*Pinus contorta* Dougl.). *Can. J. For. Res.*, **18**, 453–60.

Snyder, N.F.R. & Snyder, H.A. (1989). Biology and conservation of the Californian condor. In *Current Ornithology*, Vol. 6, ed. D.M. Power. Plenum Press, New York, pp. 175–267.

Soulé, M.E. (ed.) (1986). *Conservation Biology. The Science of Scarcity and Diversity.* Sinauer Associates, Sunderland, Massachusetts.

Soulé, M.E. (ed.) (1987). *Viable Populations for Conservation.* Cambridge University Press, Cambridge.

Soulé, M.E. & Simberloff, D. (1986). What do genetics and ecology tell us about the design of nature reserves? *Biol. Conserv.*, **35**, 19–40.

Soulé, M.E. & Wilcox, B.A. (eds) (1980). *Conservation Biology. An Evolutionary–Ecological Perspective.* Sinauer Associates, Sunderland, Massachusetts.

Söyrinki, N. (1939). Vehrmehrung der Samenpflanzen in der alpinen Vegetation Petsamo Lapplands, II. Spezieller Teil. *Ann. Bot. Soc. 'Vanamo'*, **14**.

Spitzer, K. & Lepš, J. (1988). Determinants of temporal variation in moth abundance. *Oikos*, **53**, 31–6.

Ståhl, P. (1990). *Botrychium virginianum*—a forest species preferring young succession stages. *Svensk Bot. Tidskr.*, **84**, 23–36 (in Swedish with English summary).

Stephenson, S.L. (1986). Changes in a former chestnut-dominated forest after a half century of a succession. *Amer. Midl. Nat.*, **116**, 173–9.

Sukopp, H. (1962). Neophyten in natürlichen Pflanzengesellschaften Mitteleuropas. *Ber. dt. Bot. Ges.*, **75**, 193–205.

Svensson, R. & Wigren, M. (1986*a*). A survey of the history, biology and preservation of some retreating synanthropic plants. *Symb. Bot. Ups.*, **25**(4), 1–74.

Svensson, R. & Wigren, M. (1986*b*). A changing flora—a matter of human concern. *Symb. Bot. Ups.*, **27**(2), 241–51.

Symonides, E. (1988). Population dynamics of annual plants. In *Plant Population Ecology*, ed, A.J. Davy, M.J. Hutchings & A.R. Watkinson. Blackwell, Oxford, pp. 221–48.

Taylor, L.R. (1984). Assessing and interpreting the spatial distribution of insect populations. *Ann. Rev. Entomol.*, **29**, 321–57.

Temple, S.A. (1977). Plant–animal mutualism: coevolution with dodo leads to near extinction of plant. *Science, N.Y.*, **197**, 885–6.

Terborgh, J. (1986). Keystone plant resources in the tropical forest. In *Conservation Biology. The Science of Scarcity and Diversity*, ed. M.E. Soulé. Sinauer Associates, Sunderland, Massachusetts, pp. 330–44.

Terborgh, J. & Winter, B. (1980). Some causes of extinction. In *Conservation Biology*, ed. M.E. Soulé & B.A. Wilcox. Sinauer Associates, Sunderland, Massachusetts, pp. 119–33.

Thomas, J.A. (1976). The ecology of the large blue butterfly. *Ann. Rep. Inst. Terrestr. Ecol.*, pp. 25–7.

Thomas, J.A. (1980). Why did the large blue become extinct? *Oryx*, **15**, 243–7.

Thomas, J.A. (1983). The ecology and conservation of *Lysandra bellargus* (Lepidoptera: Lycaenidae) in Britain. *J. appl. Ecol.*, **20**, 59–83.

Thomas, J.A., Thomas, C.D., Simcox, D.J. & Clarke, R.T. (1986). Ecology and declining status of the silver-spotted skipper butterfly (*Hesperia comma*) in Britain. *J. appl. Ecol.*, **23**, 365–80.

Thomas, J.A., Elmes, G.W., Wardlaw, J.C. & Woyciechowski, M. (1989). Host specificity among *Maculinea* butterflies in *Myrmica* ant nests. *Oecologia (Berl.)*, **79**, 452–7.

Tilman, D. (1987). Secondary succession and the pattern of plant dominance along experimental nitrogen gradients. *Ecol. Monogr.*, **57**, 189–214.

Trepl, L. (1984). Über *Impatiens parviflora* DC als Agriophyt in Mitteleuropa. *Diss. bot.*, **73**, Cramer, Vaduz.

Uotila, P. (1971). Distribution and ecological features of hydrophytes in the polluted lake Vanajavesi, S. Finland. *Ann. Bot. Fennici*, **8**, 257–95.

van Riper, C., III, van Riper, S.G., Goff, M.L. & Laird, M. (1986). The epizootiology and ecological significance of malaria in Hawaiian land birds. *Ecol. Monogr.*, **56**, 327–44.

Wallsten, M. (1981). Changes of lakes in Uppland, central Sweden, during 40 years. *Symb. Bot. Ups.*, **23**(3).

Ward, L.K. & Lakhani, K.H. (1977). The conservation of juniper: The fauna of foodplant island sites in southern England. *J. appl. Ecol.*, **14**, 121–35.

Warner, R.E. (1968). The role of introduced diseases in the extinction of the endemic Hawaiian avifauna. *Condor*, **70**, 101–20.

Warren, M.S. (1987*a*). The ecology and conservation of the heath fritillary butterfly, *Mellicta athalia*, II. Adult population structure and mobility. *J. appl. Ecol.*, **24**, 483–98.

Warren, M.S (1987*b*). The ecology and conservation of the heath fritillary butterfly, *Mellicta athalia*, III. Population dynamics and the effect of habitat management. *J. appl. Ecol.*, **24**, 499–513.

Wegge, P. & Rolstad, J. (1986). The spacing of capercaillie leks in relation to habitat and social organization. *Behav. Ecol. Sociobiol.*, **19**, 401–8.

Weiss, S.B., Murphy, D.D. & White, R.R. (1988). Sun, slope and butterflies: topographic determinants of habitat quality for *Euphydryas editha*. *Ecology*, **69**, 1486–96.

Wennström, A. & Ericson, L. (1990). The interaction between the clonal herb *Trientalis europaea* and the host-specific smut fungus *Urocystis trientalis*. *Oecologia* (*Berl.*), **85**, pp. 283–40.

Werth, E. (1928). Überblick über die Vegetationsgliederung von Kerguelen sowie von Possession-Eiland (Crozet-Gruppe) und Heard-Eiland. In *Deutsche Südpolar-Expedition 1901–1903*. Part 8, ed. E. von Drygalski. De Gruyter, Berlin, pp. 127–76.

Wesołowski, T. & Tomialojč, L. (1986). The breeding ecology of woodpeckers in a temperate primeval forest—preliminary data. *Acta Orn.*, **22**, 1–21.

Weste, G. (1981). Changes in the vegetation of sclerophyll shrubby woodland associated with invasion by *Phytophthora cinnamomi*. *Austr. J. Bot.*, **29**, 261–76.

Wheeler, B.D. (1988). Species richness, species rarity and conservation evaluation of rich-fen vegetation in lowland England and Wales. *J. appl. Ecol.*, **25**, 331–53.

Widén, B. (1987). Population biology of *Senecio integrifolius* (Compositae), a rare plant in Sweden. *Nord. J. Bot.*, **7**, 687–704.

Willems, J.H. (1983). Species composition and above ground phytomass in chalk grassland with different management. *Vegetatio*, **52**, 171–80.

Wilson, S.D. & Belcher, J.W. (1989). Plant and bird communities of native prairie and introduced Eurasian vegetation in Manitoba, Canada. *Conserv. Biol.*, **3**, 39–44.

Woods, F.W. & Shanks, R.E. (1959). Natural replacement of chestnut by other species in the Great Smoky Mountains National Park. *Ecology*, **40**, 349–61.

Wright, S.J. & Hubbell, S.P. (1983). Stochastic extinction and reserve size: a focal species approach. *Oikos*, **41**, 466–76.

Chapter 4

Conservation of Genetic Variation in Plants— The Importance of Population Size and Gene Flow

Björn Widén & Linus Svensson

Department of Systematic Botany, University of Lund, Ö. Vallgatan 18–20, S-223 61 Lund, Sweden

4.1 INTRODUCTION

Recently, many authors have focused on the role of population genetics in the management and conservation of threatened species (see Soulé & Wilcox, 1980; Frankel & Soulé, 1981; Schonewald-Cox *et al.*, 1983; Soulé, 1985, 1987 for reviews). There is one concept in the current theoretical framework of conservation biology that includes genetical aspects—Minimum Viable Population size (hereafter MVP). Conservation biologists have for a long time known that the smaller the population, the more susceptible it is to extinction by various causes. The MVP concept constitutes one of the major foci of current conservation biology (Simberloff, 1988) because the increasing fragmentation of the landscape caused by man will in the future reduce population size. Many of the factors that affect population dynamics, and thus, potentially, the likelihood of extinction, contain elements of uncertainty. Shaffer (1981) summarized the sources of uncertainty into four classes: demographic and environmental stochasticity, natural catastrophes and genetic stochasticity. Different sources of uncertainty interact in extremely complex ways (Shaffer, 1987). We will in this chapter review the current knowledge of the conservation biology of plants with the focus on genetic stochasticity, i.e. genetics of small populations, and gene flow among populations.

Most models of genetic effects connected to MVP have been developed for vertebrates. Although basic genetic processes in plant populations are similar to those in vertebrates, there exist important biological differences between plants and vertebrates that influence genetic diversity and gene flow among populations. Specifically, plants are sedentary and can have significant genetic differentiation on a scale of a few metres; they have a much greater variation in life histories; and they have a much greater diversity in breeding system than vertebrates. As a consequence, the variety and complexity of the genetic structures of plant populations are far greater than those found among vertebrates. Here we adopt a plant-biased approach (especially conifers and flowering plants), addressing the following issues:

(1) What do we know about the distribution of genetic variation within and among plant populations?
(2) What role(s) does gene flow among plant populations play in creating and maintaining genetic diversity?
(3) What are the ecological and evolutionary effects of loss of genetic variation?
(4) What are the implications for plant conservation biology?

4.2 SOME BASIC POPULATION GENETICS

Before we can discuss these issues, and before we can evaluate the importance of different aspects of genetic diversity for the conservation of plant species, we have to consider some basic theoretical models and concepts in population genetics. A number of factors influence the distribution of genetic variation, and the gene frequencies in a population are brought about by a number of evolutionary forces: mutation, gene flow, recombination, natural selection and genetic drift. Here we can only briefly discuss some of the more important aspects of this complex subject.

4.2.1 Population size and genetic variation

Fisher's (1930) fundamental theorem of natural selection implies that the rate of evolution at a single locus by selection is limited by the amount of genetic variation available. Theoretical population genetics predicts that in finite populations (i.e. without immigration) the presence of only a small number of individuals, maintained for several generations, will lead to the depletion and rearrangement of genetic variation by inbreeding and genetic drift (Wright, 1931). Thus, the number of breeding individuals is a crucial measure in determining the amount of genetic variation that is accessible in a population. The population size influences the probability of long-term survival of a population because genetic variation is needed as raw material for evolution in a changing environment.

Genetic variation is recovered by mutation which, according to Lande and Barrowclough (1987), requires very large population sizes (10^5–10^6) for neutral single-locus variation to reach equilibrium, but rather small population sizes (≈ 500) for quantitative variation under neutral or stabilizing selection. The recovery time is considerable for single-locus genes (10^5–10^7 generations), but much shorter for quantitative variation (10^2–10^3).

Effective population size, and neighbourhood size

Considering only the number of individuals in the population is not appropriate for population genetic models because factors such as unequal sex ratio, overlapping generations, non-random distribution of offspring and non-random mating make the 'effective' population size, N_e, less than the censused population size. N_e, represents the size of an idealized population in which individuals contribute equally to the gamete pool and have the same variation in allele frequencies and level of inbreeding as the observed population. If factors such as mating system, sex ratio, and variation in fertility are known, effective population size can be estimated from the actual number of breeding individuals. However, very few estimates exist in plants.

Wright (1946) introduced another useful parameter, the 'neighbourhood', which he defined as a group equivalent to a panmictic unit within a continuous array of individuals. Because of the sedentary habit of plants and the low mobility of many animal species, mating does not usually occur over large distances. Therefore, a distance parameter must be incorporated in the estimation of the neighbourhood. The neighbourhood area, N_a, is an area from which the parents of central individuals may be treated as if drawn at random (Wright, 1943*a*, 1946). The effective size of a neighbourhood, N_s, is then equivalent to the number of reproducing individuals in N_a. A circle of radius 2σ, twice the standard deviation of the gene flow distance, would include 86·5% of the parents of an individual at the centre. The neighbourhood size is described by the formula $N_s = 4\pi d\sigma^2$, where d = density.

Because gene flow in plants occurs via pollen and seeds, the standard deviation of pollen flow (σ_p) and of seed flow (σ_s) must be calculated separately: $N_s = 4\pi d(\sigma_s^2 + \frac{1}{2}t\sigma_p^2)$ where t is the outcrossing rate (cf. Crawford, 1984 for details of the model). The neighbourhood size is not exactly the same as the effective population size but may be considered as an upper limit. Depending on deviations from idealized conditions, N_e will be less than N_s (Crawford, 1984; cf. Govindaraju, 1988*b*).

4.2.2 Genetic drift—the random variation in allele frequencies

Genetic drift, the random fluctuation in allele frequencies between generations, tends to diminish genetic variation and may result in the fixation of one allele (homozygosity) at each locus. The prerequisite in the model for genetic drift to occur is a finite, diploid outbreeding population where no mutation, natural selection or gene flow occurs. An idealized population, with N diploid, hermaphroditic individuals reproducing at random (panmixis), will produce an infinite gamete pool to which each individual contributes equally. For an allele with initial frequency p in the population, the variance in allelic frequency among repeated samples of N offspring is $p(1-p)/2N$. It can be shown that the expected heterozygosity, H_e, declines by a factor of $1–1/2N$ per generation (Wright, 1931).

The effective size of a population, N_e (see above), can then be defined as the size of an idealized population which shows the same amount of distribution of allele frequencies under random genetic drift or the same amount of inbreeding (see below) as in the actual population. When sizes fluctuate between generations, N_e is limited by the harmonic mean of the population numbers. Thus, small population sizes, or bottlenecks, reduce genetic diversity but the final reduction in heterozygosity depends on how rapidly the original population size is restored (Nei *et al.*, 1975).

4.2.3 Inbreeding

Two alleles may be *identical in state* simply if they are functionally equivalent, or they may be *identical by descent* if they are copies of a single ancestral gene. If inbreeding occurs, the latter increases in frequency, and the inbreeding coefficient F is then the probability that the two alleles at any locus are identical by descent. If we assign F to an individual, then it expresses the degree of relationship between the individual's parents. The degree of relationship expressed in F is essentially a comparison between the population under study and an original theoretical population a number of generations earlier, which, by definition, has an inbreeding coefficient of zero. The level of inbreeding increases with time at a rate dependent on the effective population size N_e: $\Delta F = \frac{1}{2}N_e$, per generation (Falconer, 1981).

In general, the inbreeding coefficient of individuals in generation t is

$$F_t = \Delta F + (1 - \Delta F)F_{t-1}$$

The inbreeding coefficient can also be expressed by the formula:

$$F = 1 - H_o/H_e$$

where H_o is the observed and H_e is the expected frequency of heterozygotes.

A convenient property of the F-statistics is that it can be partitioned among subgroups in a substructured population (Wright, 1951, 1965). Three parameters are used to describe the correlation between uniting gametes within groups, between groups and the overall correlation:

F_{IS} is the correlation between two gametes within a subgroup;
F_{ST} is the correlation between uniting gametes taken at random from two different subpopulations;
F_{IT} is the correlation between uniting gametes, relative to the whole population.

The relationship among the parameters is, according to Wright (1951): $(1 - F_{IT}) = (1 - F_{IS})(1 - F_{ST})$.

4.2.4 Gene flow

Many factors, in addition to population size, influence the amount of genetic variation maintained in natural populations: the level of gene flow among populations, the breeding system, the sex-ratio, overlapping generations and generation time. Organisms may differ in many of these respects. A very central factor often considered in conservation biology is gene flow, i.e. movement of genes within or among populations or subpopulations. Gene flow counteracts depletion of genetic variation in small populations, and has long been considered an important evolutionary force. Mayr (1963) coined the concept of a 'biological species', a group of actually or potentially interbreeding organisms. He considered the potential for interbreeding to be often realized and gene flow to be important in maintaining the genetic and phenotypic homogeneity of a species. This widely accepted view was questioned by Ehrlich and Raven

(1969) and others, mainly because observations of dispersal showed that, for most species, the actual movement of gametes and individuals between populations was too rare and restricted (cf. Levin & Kerster, 1974) to tie up the genetic variation within a species. Slatkin (1985*a*) has recently reviewed the subject and concluded that the level of gene flow differs greatly among taxa and that the documented gene flow cannot generally account for morphological homogeneity within species. Based on gene flow measurements by indirect methods (see below), he also claimed that 'gene flow in some species appears to be sporadic, with dispersal over short distances occurring most of the time but with dispersal over much longer distances occurring frequently enough to produce widespread genetic homogeneity'. Recent studies of long-distance gene flow (cf. below) have, however, further increased the difficulties in interpreting the relationship between gene flow and the homogeneity of species. Theoretical estimates (Slatkin, 1985*a*, 1987) have revealed that one individual exchanged every other generation between local populations, without respect to their sizes, is sufficient to prevent genetic divergence through the effect of genetic drift alone. The extinction and recolonization of local populations is probably an important form of gene flow, but few reliable studies under natural conditions exist (e.g. Ehrlich, 1983; McCauley, 1987). This excludes the possibility to draw general conclusions about this mechanism of gene flow (Slatkin, 1985*a*). According to MacArthur and Wilson (1967), extinction and recolonization are important mechanisms responsible for patterns of species diversity in islands and island-like habitats, but it is unclear if colonization is a gene flow mechanism distinct from migration (Wade & McCauley, 1988).

How to measure gene flow

Gene flow can be estimated by direct and indirect methods. The former are based on measurements of potential and actual gene flow while the latter use spatial distribution of alleles, chromosomal segments, or phenotypic traits to calculate the level of gene flow in a population.

Direct methods. Studies of dispersal of individuals at particular life-history stages can be assessed by direct observations and can be a measure of potential gene flow. The actual gene flow will, however, also be dependent on the subsequent establishment of individuals and their reproductive success. This information is not easy to obtain. Actual gene flow can be measured directly by studying the spread of marker alleles in a population. Direct methods cannot detect occasional changes in the pattern of gene flow due to the limited time scale of observation and experimentation, and are thus assumed to underestimate long-distance gene flow. The main components of gene flow in plants are pollen and seed dispersal. Their movements in natural populations are hard to trace because of the difficulties in determining the source of different pollen grains or seeds. Dispersal can be estimated by placing traps for seeds and pollen around isolated plants, but experimental manipulations are usually needed. Many techniques have been developed in order to study actual pollen flow or to simulate pollen flow (Table 4.1). The study of seed dispersal is less well-developed.

Table 4.1

Summary of different methods to study gene flow in plants. Direct methods of seed dispersal are not considered. Indirect methods often estimate *Nm* (except spatial autocorrelation)

Simulation methods	
Artificial 'pollen grains' applied to stamens are allowed to be transferred to recipient flowers	Waser & Price (1982) Svensson (1985)
Chemical labelling of pollen grains	Handel (1976) Reinke & Bloom (1979)
Pollinator movement recordings	Schmitt (1980) Levin (1981)
Direct methods	
Recording grains on the stigma	Thomson & Plowright (1980) Waser & Price (1982)
Genetic markers	Schaal (1980) Handel (1985) Bos *et al.* (1986)
Paternity analysis	Ellstrand & Marshall (1985)
Indirect methods	
Recessive lethal alleles	Dobzhansky & Wright (1941)
Inbreeding coefficient F_{ST} $F_{ST}=1/(4Nm+1)$	Wright (1969)
Rare or 'private' alleles	Slatkin (1980, 1981, 1985*b*)
Genetic distance $D^a = -\ln(J_{XY}/(J_X J_Y)^{1/2})$	Nei (1972)
Spatial autocorrelation	Sokal & Wartenberg (1983)

[a] D is $J_X=\Sigma p_i^2$, $J_Y=\Sigma q_j^2$ and $J_{XY}=\Sigma p_i q_j$, and p_i and q_j are the frequencies of alleles at a locus in population X and Y.

Most information about gene flow by direct methods in plants comes from estimates of pollen flow, especially studies of pollinator movements. Pollinators tend to move between near neighbours, but occasionally longer flights occur. There are serious problems with extrapolating gene flow from pollinator movements. One of these is pollen carryover, in which only a fraction of the pollen present on a pollinator is deposited on the stigma at each visit (Schaal, 1980; Svensson, 1985). By using genetic markers Schaal (1980) and Levin (1981) found pollen carryover to increase estimates of gene flow by a factor 1·5–1·9 compared with estimates from pollinator movements.

Paternity analysis is a new and rewarding method to measure gene flow in plants (Ellstrand & Marshall, 1985). If multi-locus genotypes can be assayed in seeds and maternal parents, then the potential paternal parents in the local population can be estimated. If the paternal gene combination does not fit any of the possible multi-locus gametes generated by the local population, then that gene combination represents gene flow between populations.

Indirect methods. Most indirect methods estimate Nm, where N is the effective population size and m is the fraction of genes exchanged between populations each generation. Assumptions are that a certain probability of identity between two alleles (they are identical by descent) randomly chosen from the population, and that the population is in genetic equilibrium. The most frequent methods used are listed in Table 4.1 (for more details see Slatkin, 1985*a*).

4.2.5 Selection

In simple genetical models of small populations it is often assumed that the alleles are selectively neutral. However, the natural selection, and type of selection, is important for the discussion of the maintenance of genetic variation. The genetic variation within a population is influenced by a balance between gene flow and selection. If an allele is favoured by selection s, them immigration of other alleles at a rate m would not prevent a high frequency of the favoured allele if, roughly, $m<s$ (Haldane, 1930). Pronounced spatial differences in selection pressure have been shown to counteract strong gene flow by pollen (McNeilly, 1968). Very strong selection pressures have actually been documented in natural populations; $s>0{\cdot}5$ in *Agrostis tenuis*, *Anthoxanthum odoratum*, and *Dryas octopetala* (Levin, 1988 and references therein; Endler, 1986).

4.2.6 Local differentiation

Sedentary organisms, like plants growing in a fine-scale environment, have two general methods of adapting to environmental heterogeneity: *phenotypic plasticity* and *genetic micro-differentiation*. Phenotypic plasticity refers to the degree to which the phenotypic expression of a genotype varies under different environmental conditions, its 'norm of reaction' (Johannsen, 1911). Much of the temporal and spatial variation in the plant's environment is too fine-scaled and too unpredictable to be tracked genetically. The expression of plasticity is under genetic control and plasticity is a widespread and important phenomenon in plants. Many studies have demonstrated differences in plasticity among populations within a species, and some have provided evidence for variation among genotypes within populations. The most important aspect of individual plasticity, in an evolutionary context, is its buffering effect against selection, which may prevent selective elimination of individuals and their genes. This buffering can passively maintain genetic diversity in a population rather than doing so on the basis of ongoing selection (Bradshaw, 1965; Sultan, 1987). Local genetic differentiation in morphological and phenological traits over short distances is well-documented in plants (see references in Levin, 1988). For example, Linhart (1974) has described a case of genetic differentiation between central and peripheral stands of a self-pollinating annual (*Veronica peregrina*) in vernal pools. The patches were only 2–5 m apart, but were subject to quite diverse selection pressures due to different stand densities, the availability of water and the presence or absence of tall competitors.

This micro-differentiation may not be the result of selection acting at the loci examined. Hamrick and Allard (1972) found different multilocus allozyme genotypes of the annual *Avena barbata* in xeric and mesic sites on a hillside, and Nevo *et al.* (1986) found significant allozyme frequency–microniche association for six microniches in *Hordeum spontaneum*. The genotype–environment covariance in both species could be the result of diversifying selection, but Hedrick and Holden (1979) have shown that strong genotype–environment association could be a transient pattern generated by genetic hitch-hiking of selectively neutral allozymes with alleles at loci other than those on which selection acts.

4.2.7 Measures of genetic diversity

The genetic variation within a species can be described at two fundamental levels: genetic variation among individuals within a local population, and genetic variation between different local populations. The best method to assess the genetic variation is by allozymes because measurements can be standardized and comparison between species can easily be made. For this purpose, several parameters of genetic diversity are used (Table 4.2).

Table 4.2
Different parameters for measuring genetic diversity

A	=	mean number of alleles per locus
A_P	=	mean number of alleles per polymorphic locus
P	=	the proportion of loci which are polymorphic
P_A	=	the proportion of the total number of alleles found within each population
H_o	=	observed heterozygosity
H_e	=	expected heterozygosity
H_T	=	total allelic diversity
H_S	=	mean allelic diversity within populations
G_{ST}	=	the ratio of the allelic diversity among populations to the total allelic diversity

The most frequent methods used to estimate genetic structures are Wright's (1951) F-statistics (see above), Nei's (1973) gene diversity statistics, and Nei's (1972) genetic distance (Table 4.1).

Nei's (1973) gene diversity statistics can be calculated from allele frequency data. The total genetic diversity for a species is estimated from the formula

$$H_T = 1 - \sum p_i^2$$

where p_i is the mean frequency of the i-th allele at a locus. This statistic can be partitioned into genetic diversity within populations (H_S) and genetic diversity among populations D_{ST}. Of special interest is the proportion of the genetic diversity among populations (G_{ST}): $G_{ST} = D_{ST}/H_T$.

Other important statistics are the number of alleles per polymorphic locus (A_P) and the proportion of alleles found in each population (P_A).

4.3 EFFECTS OF GENETIC SUBSTRUCTURING OF PLANT POPULATIONS

4.3.1 Inbreeding depression

In an ideal population, all matings are supposed to be between randomly chosen parents, i.e. one individual has the same probability of mating with all other compatible individuals. Most plant populations do, however, deviate significantly from that ideal scheme in being spatially structured. The spatial structure is established and maintained by plants being sedentary, restricted gene flow, heterogeneity in the environment and differentiated selection pressure (see above). This spatial structure will increase the probability of mating between close neighbours, which, in turn, have a high probability of common descent (Turner *et al.*, 1982; Campbell & Waser, 1987). Plants also show a wide variety of breeding systems that influence the mating system and population structure. Self-compatible species vary in selfing rate depending on floral biology and local environmental conditions. Some species are predominantly selfing, some have a mixed mating system, while others are highly outcrossing (Schemske & Lande, 1985).

A high level of self-fertilization or crossing with close relatives will reduce the effective population size and hence increase inbreeding. Inbreeding will increase the expression of recessive deleterious alleles, the sublethal ones eventually becoming fixed. The loss of fitness during inbreeding is known as *inbreeding depression*. This can be defined as the relative reduction in fitness of selfed offspring compared to outcrossed offspring.

The phenomenon complementary to inbreeding depression is *heterosis*, which includes the relative increase in fitness when crossing two inbred (i.e. homozygous) lines. Heterosis is regarded as either the heterozygotic offspring having a real fitness advantage over that of the homozygotic offspring ('over-dominance hypothesis') or just countering inbreeding depression ('dominance hypothesis') (Falconer, 1981; Ledig, 1986). Heterozygotes have two partially different genomes, showing a more pronounced developmental homeostasis (e.g. stability and consistency) in several characters. Observations of homeostasis are commonly explained in terms of a higher 'buffering' capacity perhaps at the enzyme level (Powers & Place, 1978). The characteristics of a higher buffering capacity are thought to be adaptive in fluctuating or stressful environments, and especially so for sedentary organisms (Shields, 1982).

Very few studies have concluded that real heterosis over-dominance exists, and only in animals (Watt, 1985). In plants heterosis is usually explained by the dominance hypothesis, although over-dominance has been found to be of some importance. Most of the studies of heterosis have focused on growth rate, usually that of pine trees (see Mitton & Grant, 1984 for a review), but other species and fitness parameters have been used (Schaal & Levin, 1977; Strauss, 1987; Waser *et al.*, 1987;

Svensson, 1988, 1990), yielding different conclusions. Relating allozyme heterozygosity to growth rate has resulted in positive relationship (Knowles & Grant, 1981; Mitton & Grant, 1980), no relationship (Ledig *et al.*, 1983; but see Bush *et al.*, 1987), or negative relationship (Knowles & Mitton, 1980). Recently, Lande and Barrowclough (1987) claimed that little evidence, in fact, has been found for heterozygote advantage.

The extent of inbreeding depression in natural populations can be assessed in experiments that compare the relative fitness of progeny obtained from artificial self- and cross-pollination, or from electrophoretic data (Ritland, 1989). The magnitude of inbreeding depression can be considerable in natural populations of both plants and animals (Charlesworth & Charlesworth, 1987), but the frequency and level of inbreeding in plants is often far greater than in most animal groups. Charlesworth and Charlesworth (1987) and Barrett and Kohn (in press) have recently reviewed the empirical evidence for inbreeding depression in natural plant populations. There has been a widely accepted view that inbreeding depression is, as a rule, low in partially inbreeding populations, while it is high in outbreeders (Wright, 1977). However, the available data (Fig. 4.1) indicate that even inbred species show strong inbreeding depression, and this supports the idea that recessive lethals are not the sole cause of inbreeding depression in plants.

Inbreeding depression can be expressed in different stages of the plant's

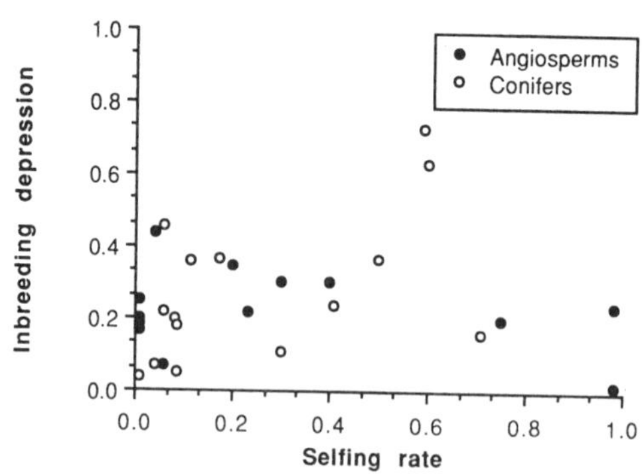

Fig. 4.1. The relationship between inbreeding depression ($1 - w_{selfed}/w_{outcrossed}$) and selfing rate in non-domesticated plants (from Charlesworth and Charlesworth, 1987). Characters used to estimate inbreeding depression were size, fertility or survival of selfed and outcrossed progenies. When their Tables 2 and 3 show range of values for inbreeding depression or selfing rate we use the average.

life-cycle. For example, numerous studies have shown deleterious effects of inbreeding on seed number and size, and embryo development particularly in conifers (Charlesworth & Charlesworth, 1987). However, the few studies that have employed a multiplicative fitness function (seed set, germination, survival, fecundity) indicate that inbreeding effects are most strongly manifest late in the life-cycle (Barrett & Kohn, in press). Most studies have been conducted in gardens or in glasshouses, suggesting that our estimates of inbreeding depression are underestimates, since measures of inbreeding depression in the field tend to be higher than those in cultivation (Charlesworth & Charlesworth, 1987).

The current empirical evidence from plant populations indicates that the relationship between inbreeding depression and rate of selection varies considerably from species to species (Fig. 4.1; Barrett & Kohn, in press). It would be premature to draw far-reaching conclusions of the conservation biology implications from the restricted material available. If both outcrossing and selfing plants show inbreeding depression the relevant comparisons to be made are among populations within species. The few available studies of this type indicate great variation within species (Barrett & Kohn, in press). To our knowledge, however, no study has yet compared inbreeding depression between genetically depleted and non-depleted populations within a species. There is an urgent need for well-designed experiments to test the relevant questions in natural populations.

4.3.2 Outbreeding depression

Mating between two individuals which are separated by large spatial distances may result in a relative decrease in fitness. Observations showing matings over greater spatial distances to have a decreased fitness compared with matings occurring at smaller distances are collectively termed 'outbreeding depression' ('The phrase "outbreeding depression" makes no judgement as to why hybridization leads to a fitness reduction; it merely refers to the fact that it occurs.' Templeton, 1986).

An individual plant separated from other plants by a certain distance will have a coadapted genome that may not be fully equivalent to the other individuals within or outside the population. The variation in the genomes of different individuals is dependent upon their degree of isolation and the rate of change in the genome. Mixing two genomes, by breeding plants with not completely identical evolutionary histories, will decrease the fitness of the progeny compared with inbred progeny displaying a coadapted genome.

The intensity of outbreeding depression would be related to spatial distance. That prediction is based on the hypothesis that genetic distance is correlated with, or at least related to, spatial distance. The relationship between spatial and genetic distances in plants has received increasing interest during recent years, due in large part to the work by Schaal (1974), Schaal and Levin (1977) and by Price and Waser (1979), Waser (1987), Waser & Price (1989) and Waser (in press). The existence of a relationship between spatial and genetic distances has, however, been questioned. Several investigators have not found any significant relationship within populations or between populations (e.g. Bos *et al.*, 1986; Waser, 1987).

Consider the following three assumptions: (i) mating between two individuals which are too similar genetically may give rise to inbreeding depression; (ii) mating between individuals which are genetically too dissimilar may give rise to outbreeding depression; (iii) there is a correlation between geographic and genetic distance. When all the specific assumptions have been fulfilled, the combination could imply an *optimal outcrossing distance*, the intermediate distance at which offspring are produced having a superior fitness compared with the offspring of

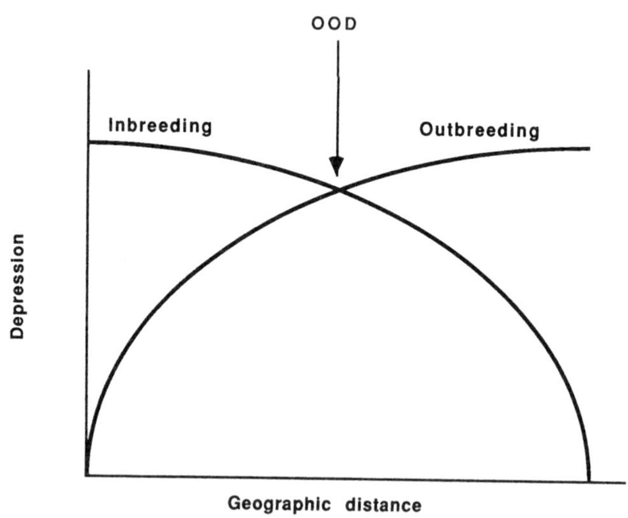

Fig. 4.2. The relation between outbreeding and inbreeding depression and geographic distance. Arrow indicates a hypothetical optimal outcrossing distance (OOD). For details of hypothesis, see text. Redrawn after Shields (1982).

long- and short-distance crosses (Ritland & Ganders, 1987; Waser & Price, 1989) (cf. Fig. 4.2). Several investigators have performed experiments in order to test this hypothesis. Waser and Price (1989) found an optimal outcrossing distance for the plants *Delphinium nelsonii* and *Ipomopsis aggregata* by measuring an increase in seed set in the progeny of individuals at an intermediate crossing distance. Subsequently, this result was confirmed by using other parameters such as pollen tube length and seedling survival under natural conditions (Waser *et al.*, 1987; Waser & Price, 1989). Other investigations have also shown a distance-related response in fitness parameters to crossing (e.g. Banyard & James, 1979; Kephart, 1981; Koptur, 1984; Park *et al.*, 1984; Schemske & Pautler, 1984; Harder *et al.*, 1985; Svensson, 1988, 1990). Additional research has shown a linear relationship between geographic distance between parents and offspring fitness (e.g. Kruckeberg, 1957; Levin, 1983, 1984; Stratton *et al.*, 1985), or no relationship at all (Motten, 1983; Molau *et al.*, 1989; Newport, 1989). These differing conclusions may be due to a variety of reasons—among them the different breeding systems among the taxa studied, different magnitude and scale of local selection pressures, different genetic structures for the populations under study and/or the problem of choosing the relevant character associated with fitness.

4.3.3 Genetic substructuring of plant populations—Conclusions

From what has been discussed above it is clear that plant populations are usually spatially structured. The background for substructuring is diverse, and dependent upon abiotic as well as biotic factors. Local adaptation, restricted gene flow, differentiated selection pressure and genetic drift all act together towards a genetic subdivision of plant populations, thereby increasing the probability of non-random matings. A genetic subdivision of a plant population will, in turn, result in an increased probability for inbreeding depression effects. Enforced inbreeding has, in normally outbreeding organisms (i.e. self-incompatible plant species or species with a mixed mating system), generally been considered to be a drawback.

When turning from theory to the real world, more complicated patterns emerge. Obviously a substantial number of plant species exist in situations which, theoretically, should be less optimal, e.g. extremely small populations, species which are obligate-selfing or surviving under extremely low gene flow.

4.4 'THE REAL WORLD'—MEASURES OF POPULATION STRUCTURE AND GENE FLOW

In the light of the common substructuring of plant populations, one question emerges: does fragmentation of a species' distribution area by human activities result in genetic depletion? No careful experiments under natural conditions exist, so it is not possible to draw general conclusions from direct observations. Most models of genetic diversity assume that the population under study is in genetic equilibrium, which, for many species, is probably not true. For different models, the rate of approach to equilibrium is of the same magnitude as $1/m$ and $1/\alpha$, the inverted value of migration or mutation rates, respectively (Slatkin, 1985*b*). This is on a scale of hundreds of generations, or even more (Lande & Barrowclough, 1987). Therefore, we do not expect experimental studies to provide a realistic estimate of the genetic effects of habitat fragmentation caused by human impact. Instead we have to draw conclusions from the distribution of genetic variation in natural systems with different spatial structures. However, very few studies have addressed questions which are important for the establishment of the MVP for any plant species. For instance, very few studies have yet been carried out to study the correlation between population size and genetic diversity within a plant species (but see Husband & Barret, 1989). In this section we will discuss published estimates of population size, gene flow and partitioning of genetic diversity. However, very few estimates of plant population sizes are published. Even if the population can be defined and the number of individuals counted, it is usually difficult to assess the number of *breeding* individuals in a natural population. Even in strictly annual plants, generations are overlapping if there is a dormant seed bank, another problem in determining population size. In perennial plants individuals may not reproduce each year (Widén, 1987), or may survive some years as subterranean parts. Individuals may differ greatly in progeny number, a very common feature of natural plant populations, and this can cause significant reduction in effective population size (Crawford, 1984; Heywood, 1986). Although the genetic structure within a population has been described in a few cases, the relative influence of selection, genetic drift and migration on the observed pattern is poorly understood. Another problem in evaluating patterns of genetic structure in natural populations is the usual lack of information about the history of the population.

The effective size of a population determines if genetic drift will occur. Wright (1969) suggested that a moderate amount of drift would be expected if $N_e < 200$, and a considerable amount would be expected if $N_e < 20$ (given no selection and gene flow). Based on experiences from captive animal breeding, Franklin (1980) suggested that 50 breeding individuals are needed for short-term survival and 500 for long-term survival of a population. This 50/500 rule has been applied as a guideline in some management plans for animals (Simberloff, 1988). We will first compare these figures with estimates of neighbourhood sizes obtained from natural plant populations, and we will then discuss the role of gene flow in counteracting genetic depletion, referring to published data. Finally, we will discuss the relationship between genetic diversity and geographic distribution in a few conspecific taxa.

4.4.1 Neighbourhood size

The estimates of neighbourhood size are based on both direct and indirect measures of gene flow (Table 4.3). These produce very different estimates depending on the method used, because many factors are known to influence the estimates (Levin & Kerster, 1969; Levin, 1981). For instance, neighbourhood structure varies with the density of plants since pollen dispersal due to pollinator activity is inversely related to plant density (e.g. Pyke, 1978). Temporal patterns in pollinator visitation and flowering phenology have been shown to influence the estimates of N_s in several studies (Schmitt, 1983; Palmer *et al.*, 1988; Molau *et al.*, 1989; Olesen & Warncke, 1989). In age-structured populations it is important to consider demographic factors when estimating neighbourhood size (Govindaraju, 1988*b*).

The effective population sizes of the species in Table 4.3 are probably smaller than the estimates of N_s given, due to the great variation in progeny number often found in natural plant populations. We expect the distribution of progeny numbers to follow a Poisson distribution, with the ratio of the variance and mean progeny number to be 1. However, Heywood (1986) showed that the ratio (measured for seed number = progeny) for 34 annual species exceeded 10 in most cases, and even 100 in some species. According to Kimura and Crow (1963*a*) such pronounced deviation from 1 will considerably decrease the effective population size. On the other hand, Levin (1981, 1988) concluded that actual neighbourhood sizes for plants in general are probably larger than that estimated from pollen and seed dispersion, because (1) pollen flow distances based on pollinator flight distances do not consider pollen carryover and flight directionality; (2) pollen–pistil compatibility may be lower among near-neighbours than among moderately spaced individ-

Table 4.3
Estimates of neighbourhood, N_s, taken from different published sources

Species	N_s	*Reference*	*Species*	N_s	*Reference*
Conifers			Perennial herbs		
Abies concolor	3619	Govindaraju (1988*b*)	*Bartsia alpina*	1·5–185·3	Molau *et al.* (1989)
Cedrus atlantica	208	Wright (1953)	*Borrichia frutescens*	20–30	Antlfinger (1982)
Picea glauca	50118	Govindaraju (1988*b*)	*Delphinium nelsonii*	28	Waser & Price (1983)
P. mariana	7819	Govindaraju (1988*b*)	*Ipomopsis aggregata*	69	Waser & Price (1983)
P. rubens	5897	Govindaraju (1988*b*)	*Liatris cylindracea*	165	Schaal & Levin (1978)
Pinus cembrioides	11	Wright (1953)	*L. aspera*	30–191	Levin & Kerster (1969)
P. elliotti	2023	Govindaraju (1988*b*)	*Phlox pilosa*	75–282	Levin & Kerster (1968)
P. monticola	44662	Govindaraju (1988*b*)	*Plantago lanceolata*	14–20	Bos *et al.* (1986)
P. palustris	1244	Govindaraju (1988*b*)	*Primula vulgaris*	1–38	Cahalan & Gliddon (1985)
P. ponderosa	3587	Govindaraju (1988*b*)	*P. veris*	7.4	Richards & Ibrahim (1978)
P. radiata	1–3200	Bannister (1965)	*Saxifraga hirculus*	970–1532	Olesen & Warncke (1989)
P. strobus	29529	Govindaraju (1988*b*)	*Senecio* spp. (bees)	8–24	Schmitt (1980)
P. taeda	6111	Govindaraju (1988*b*)	*Senecio* spp. (butterflies)	933–6154	Schmitt (1980)
Pseudotsuga taxifolia	26	Wright (1953)	*Thermopsis montana*	132	Zimmerman (1982)
			Thymus vulgaris	7–311	Belhassen *et al.* (1987)
			Viola pedata	432	Beattie & Culver (1979)
			V. rostrata	167	Beattie & Culver (1979)
Deciduous trees			Annual/biennial herbs		
Fraxinus pennsylvanica	16	Wright (1953)	*Avena barbata*	40–400	Rai & Jain (1982)
F. americana	4·4	Wright (1953)	*Linanthus bicolor*	121.6–253·4	Schmitt (1983)
Populus deltoides	230	Wright (1953)	*Linanthus parryae*	14–27	Wright (1943*b*)
Ulmus americana	253	Wright (1953)	*L. parryae*	10–100	Wright (1978)
			Lithospermum caroliniense	2–7	Kerster & Levin (1968)
			Lupinus texensis	42–95	Schaal (1980)
			L. amplus	36	Zimmerman (1982)

uals; (3) seed-abortion may be higher, seed-set lower, and fitness of progeny lower in crosses between neighbours than more spaced plants; (4) seed and seedling mortality may be higher near the seed parent because of competition, disease and predation; and (5) the tails of pollen and seed distributions may be under-represented when direct methods are used.

Even if the values in Table 4.3 are only rough approximations of the neighbourhood sizes, it is instructive to compare them with the critical sizes suggested by Wright (1969) and Franklin (1980). Of the species concerned, 44% ($n = 29$) had at least some of the estimates of $N_s \leqslant 20$, 65% had $N_s \leqslant 50$, 83% had estimates of $N_s \leqslant 200$, and only 3% had estimates of $N_s > 500$.

4.4.2 Estimates of gene flow

Estimates of the potential gene flow by direct methods indicate leptocurtic distributions in plants (Levin & Kerster, 1974). Pollen movement can be substantial in wind-pollinated trees (Silén, 1962), but usually gene flow measured by direct methods is confined. Bos *et al.* (1986) estimated mean gene transport per generation to be 0·2–1·4 m in the wind-pollinated herb *Plantago lanceolata*. Similar results have been reported by Schaal (1980) for an insect-pollinated *Lupinus* species. Widén and Widén (1990) found average pollen dispersal distances to be 5·3–5·9 m in a natural population and 1·1 m in an experimental population of *Glechoma hederacea*. Paternity analysis has revealed a substantial interpopulation gene flow in a few studies, however. Ellstrand and Marshall (1985) found the rate of interpopulation gene flow by pollen to vary from 8·2% to 17·9% per population of wild radish *Raphanus sativus*, isolated by 100 m to 1 km from conspecifics. In the tropical tree *Platypodium elegans* in Panama, gene flow by pollen was as high as 40% between trees separated more than 0·5 km (J. Hamrick, pers. comm.).

Estimates of gene flow by indirect methods (Nm) for 124 plant species are shown in Fig. 4.3(A). Wright (1931) established a rule of thumb when he showed that there can be significant genetic differentiation due to drift when $Nm < 1$. In Fig. 4.3, 51% of the plants show a gene flow level of less than 1. Thus, the low amount of gene flow among populations observed in a considerable number of plant species may induce genetic depletion due to drift. However, there are great differences among life forms and mating systems (Fig. 4.3(B)). Generally speaking, herbs show the least gene flow and trees the most. Specifically, the lowest gene flow level is found in self-pollinated herbs and the highest in wind-pollinated conifers.

The present knowledge of gene flow in plants has been summarized in

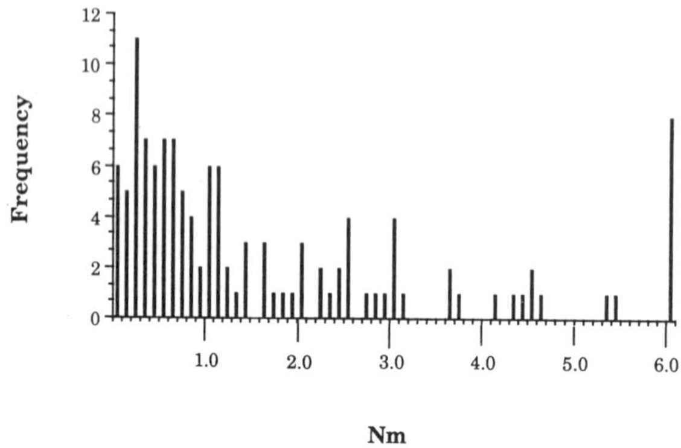

(a)

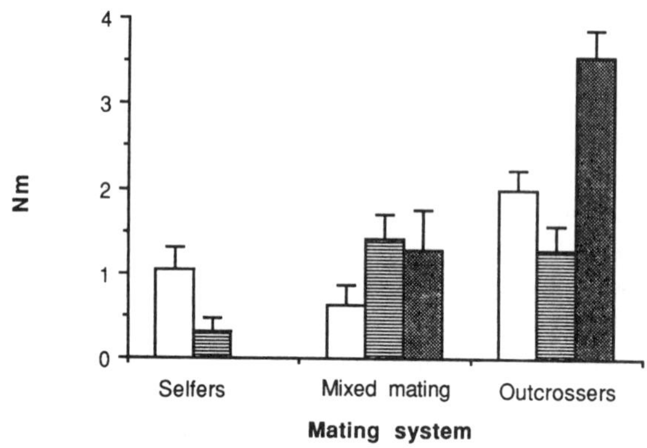

(b)

Fig. 4.3. Estimates of gene flow in plants by indirect methods. (A) *Nm*-estimates from 124 plant species. (B) The distribution of gene flow (*Nm*) among different types of mating systems. □, annual and biannual life-histories; ▤, short-lived perennials; ▦, long-lived perennials. Error bars indicate standard error (SE). The calculation of *Nm* has been performed according to Crow and Aoki (1984) using data from Bayer *et al.* (1987), Bayer (1988), Crawford and Whitkus (1988), Loveless and Hamrick (1988), and from references in Govindaraju (1988*a*).

Fig. 4.4. Gene flow within populations is leptocurtic and confined, and gene flow between populations constitutes a more or less constant background level. Within-population gene flow differs among species, but also among populations, depending on population density, temporal variation

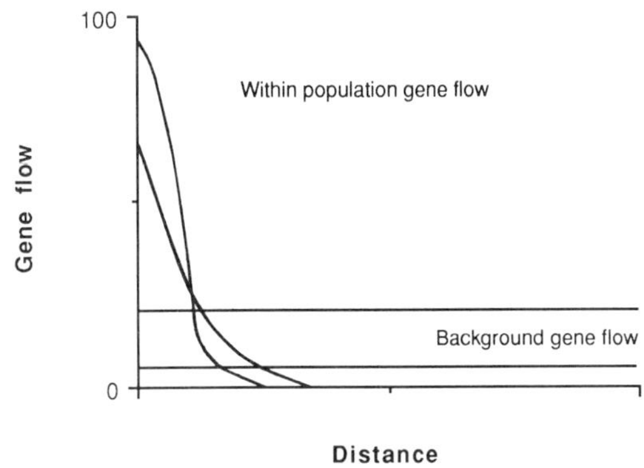

Fig. 4.4. Gene flow within and between plant populations (after an idea of J.L. Hamrick, see Hamrick & Murawski, 1990). See text for further details.

in flowering phenology and variation in breeding system. The level of background gene flow differs considerably among species, being completely absent in some species and very high in others.

4.4.3 Genetic structure

Species distributed over large, continuous areas are expected to exhibit greater genetic resources than species which have patchy distributions. The prediction is that species with limited ranges and few individuals will exhibit low levels of genetic polymorphism compared to widespread species. The factors responsible for reduction of allelic variation could be either genetic drift, strong directional selection toward genetic uniformity in a limited array of environments, or high levels of inbreeding in small populations, or a combination of these factors. The genetic variation among widespread and restricted species can thus yield important information for conservation biologists studying the genetic effects of habitat fragmentation. Several surveys of allozyme variation in plants support the prediction that species with restricted ranges will exhibit low levels of genetic diversity. However, it is important that comparisons are made among closely related species because levels of genetic diversity are known to be correlated with breeding system and life history characters (Hamrick *et al.*, 1979; Hamrick, 1983; Loveless & Hamrick, 1984; Hamrick & Godt, 1990) which could make comparison among unrelated species misleading.

In Table 4.4 we show within-population and within-species estimates

Table 4.4
Genetic variation in widespread (W) and restricted (R) taxa

	Genus	*Species*	*Pop*	*Sample size*	*No. loci*
Ancestor-derivate[a]					
A1	*Capsicum*	*eximum*	·	·	23
A1	*Capsicum*	*cardenasii*	·	·	24
A2	*Cirsium*	*canescens*	16	24	10
A2	*Cirsium*	*pitcheri*	21	24	14
A3	*Clarkia*	*biloba*	3	33–67	8
A3	*Clarkia*	*lingulata*	2	33–67	8
A4	*Erythronium*	*albidum*	7	2–24	37
A4	*Erythronium*	*propullans*	7	4–19	37
A5	*Gaura*	*longiflora*	3	30–40	18
A5	*Gaura*	*demareei*	2	30–40	18
A6	*Lastenia*	*minor*	16	8–32	20
A6	*Lastenia*	*maritima*	18	31251	20
A7	*Layia*	*glandulosa*	2	30	21
A7	*Layia*	*discoidea*	2	30	21
A8	*Stephanomeria*	*exigua* ssp. *coronaria*	1	31	13
A8	*Stephanomeria*	*malheurensis*	1	40	13
Congeners					
B1	*Astragalus*	*pattersoni*	3	25	12
B1	*Astragalus*	*pectinatus*	3	25	12
B1	*Astragalus*	*linifolius*	3	25	12
B1	*Astragalus*	*osterhouti*	3	25	12
B2	*Eucalyptus*	*grandis*	12	·	20
B2	*Eucalyptus*	*saligna*	7	·	20
B2	*Eucalyptus*	*cloeziana*	17	·	20
B2	*Eucalyptus*	*delegatensis*	23	·	20
B2	*Eucalyptus*	*caesia* ssp. caesia	7	31181	18
B2	*Eucalyptus*	*caesia* ssp. magna	6	31181	18

B2	*Eucalyptus*	*lateritica*	2	·	16
B2	*Eucalyptus*	*pendens*	7	·	16
B2	*Eucalpytus*	*johnsoniana*	3	·	16
B2	*Eucalyptus*	*suberea*	3	·	16
B3	*Layia*	*platyglossa*	6	·	17
B3	*Layia*	*jonesii*	3	·	17
B3	*Layia*	*leucopappa*	2	·	17
B3	*Layia*	*munzii*	2	·	17
B4	*Limnanthes*	*douglassi* 'inland'	14	15–40	19
B4	*Limnanthes*	*bakeri*	3	15–40	19
B4	*Limnanthes*	*macounii*	8	15–40	18
B4	*Limnanthes*	*vinculans*	4	15–40	17
B5	*Lisianthius*	*skinneri*	3	25	12
B5	*Lisianthius*	*aurantiacus*	1	25	12
B5	*Lisianthius*	*habuensis*	1	25	12
B5	*Lisianthius*	*jefensis*	1	25	12
B5	*Lisianthius*	*peduncularis*	1	25	12
B6	*Lupinus*	*texensis*	10	14–45	8
B6	*Lupinus*	*subcarnosus*	8	13–36	8
B7	*Oenothera*	*biennis I*	57	·	20
B7	*Oenothera*	*organensis*	10	3–51	15
B8	*Pinus*	*coulteri*	8	·	33
B8	*Pinus*	*sabiniana*	8	·	29
B8	*Pinus*	*contorta*	·	·	25
B8	*Pinus*	*ponderosa*	·	·	20
B8	*Pinus*	*resinosa*	·	·	27
B8	*Pinus*	*rigida*	·	·	21
B8	*Pinus*	*torreyana*	2	51–106	59
B9	*Pinus*	*attenuata*	4	15–53	32
B9	*Pinus*	*radiata*	5	10–36	32
B9	*Pinus*	*muricata*	7	10–36	32

[a]Cases A1–A8 represent ancestor–derivative species-pairs, B1–B10 represent taxa with unknown or unclear taxonomic relationships.

Table 4.4—*contd.*

				Diversity Parameters		
	Average Ho	*P-species*	*AP-species*	*A-species*	*P-pop average*	*P-pop range*
Ancestor-derivate[a]						
A1	0·02	0·44	2·18	1·57	·	·
A1	0·01	0·25	2·17	1·29	·	·
A2	0·17	0·90	3·20	3·00	0·83	·
A2	0·02	0·29	2·50	1·43	0·11	·
A3	0·15	0·63	3·60	2·63	0·58	0·50–0·62
A3	0·08	0·63	3·20	2·37	0·53	0·50–0·62
A4	0·42*	0·38	3·10	1·80	0·34	0·30–0·38
A4	0·18*	0·24	2·30	1·32	0·12	0·08–0·19
A5	0·07	0·33	3·17	1·72	0·25	0·22–0·29
A5	0·05	0·28	3·00	1·56	0·22	0·22–0·22
A6	0·07	0·50	·	·	0·26	0·10–0·40
A6	0·01	0·45	·	·	0·10	0·00–0·35
A7	0·28	0·81	3·53	3·05	0·76	0·71–0·81
A7	0·27	0·91	3·11	2·91	0·86	0·86–0·86
A8	0·16	0·62	2·88	2·15	·	·
A8	0·00	0·15	2·50	1·23	·	·
Congeners						
B1	·	0·25	2·35	1·33	0·20	0·17–0·25
B1	·	0·33	3·00	1·67	0·33	0·33–0·33
B1	·	0·33	2·25	1·42	0·25	0·25–0·25
B1	·	0·17	2·50	1·25	0·11	0·08–0·17
B2	0·17*	0·65	·	·	·	·
B2	0·24*	0·65	·	·	·	·
B2	0·24*	0·30	·	·	·	·
B2	0·24*	0·85	·	·	·	·
B2	0·07*	0·44	2·50	1·83	0·28	0·17–0·44

B2	0·06*	0·39	2·18	1·72	0·30	0·22–0·39
B2	0·28*	·	·	2·56	·	·
B2	0·16*	·	·	2·88	·	·
B2	0·08*	·	·	1·88	·	·
B2	0·17*	·	·	2·44	·	·
B3	0·25	0·82	3·70	3·24	·	·
B3	0·18	0·65	3·00	2·29	·	·
B3	0·17	0·59	3·20	2·29	·	·
B3	0·14	0·53	3·20	2·18	·	·
B4	0·17*	0·95	4·17	4·00	0·65	·
B4	0·07*	0·37	2·29	1·47	0·22	·
B4	0·002*	0·11	2·50	1·17	0·02	·
B4	0·11*	0·41	2·71	1·71	0·33	·
B5	0·00	0·00	–	0·00	·	·
B5	0·08	0·33	2·00	1·33	·	·
B5	0·02	0·08	2·00	1·08	·	·
B5	0·06	0·17	2·00	1·17	·	·
B5	0·08	0·33	2·00	1·33	·	·
B6	0·34	0·88	3·74	3·12	·	·
B6	0·10	0·88	2·08	1·84	·	·
B7	0·09	0·30	2·33	1·40	·	·
B7	0·02	0·07	2·00	1·07	·	·
B8	·	0·49	·	·	·	·
B8	·	0·93	·	·	·	·
B8		0·92	–	3·20	·	·
B8		0·60	–	2·30	·	·
B8		0·11	–	2·00	·	·
B8		0·77	–	3·10	·	·
B8	0·00	0·00	–	1·00	0·00	0–0
B9	0·13*	0·75	2·83	2·38	0·55	0·50–0·59
B9	0·14*	0·97	3·03	2·97	0·58	0·44–0·91
B9	0·12*	0·94	2·83	2·72	0·47	0·28–0·66

Table 4.4—*contd.*

	Diversity Parameters[b]				
	Ap-pop average	*Ap-pop range*	*A-pop average*	*A-pop range*	*BS*[c]
Ancestor-derivate[a]					
A1	·	·	·	·	HSC
A1	·	·	·	·	SI
A2	·	·	·	·	Mixed
A2	·	·	·	·	Mixed
A3	2·77	2·50–3·00	2·04	1·75–2·25	Outcrossed
A3	2·90	2·80–3·00	2·06	2·00–2·12	Outcrossed
A4	2·50	2·17–2·64	1·52	1·39–1·62	Sex and asex
A4	2·21	2·00–2·67	1·14	1·05–1·27	Asex
A5	2·60	2·25–2·80	1·40	1·28–1·53	SI
A5	2·75	2·75–2·80	1·39	1·39–1·39	SI
A6	2·60	2·00–3·40	1·42	1·15–1·60	SI
A6	2·17	2·00–3·00	1·12	1·00–1·53	SC
A7	3·03	3·00–3·06	2·60	2·52–2·67	SI
A7	2·73	2·67–2·78	2·48	2·43–2·52	SI
A8	·	·	·	·	SI
A8	·	·	·	·	SC
Congeners					
B1	2·17	2·00–2·50	1·22	1·17–1·25	·
B1	2·75	2·50–3·00	1·58	1·50–1·67	SI
B1	2·23	2·00–2·35	1·31	1·25–1·34	HSC
B1	2·33	2·00–3·00	1·14	1·08–1·17	SSC
B2	·	·	·	·	·
B2	·	·	·	·	·
B2	·	·	·	·	·
B2	·	·	·	·	·
B2	2·06	2·00–2·29	1·30	1·17–1·50	·
B2	2·07	2·00–2·25	1·31	1·22–1·44	·

B2	·	·	·	·	·
B2	·	·	·	·	·
B2	·	·	·	·	·
B2	·	·	·	·	·
B3	·	·	·	·	SI
B3	·	·	·	·	SI
B3	·	·	·	·	SI
B3	·	·	·	·	SI
B4	·	·	2·20	·	·
B4	·	·	1·20	·	HSC
B4	·	·	1·02	·	·
B4	·	·	1·50	·	·
B5	·	·	·	·	SC
B5	·	·	·	·	SC
B5	·	·	·	·	SC
B5	·	·	·	·	SC
B5	·	·	·	·	SC
B6	·	·	·	·	·
B6	·	·	·	·	·
B7	·	·	·	·	SC
B7	·	·	·	·	SI
B8	·	·	·	·	·
B8	·	·	·	·	·
B8	·	·	·	·	·
B8	·	·	·	·	·
B8	·	·	·	·	·
B8	·	·	·	·	·
B8	·		1·00	1·00–1·00	·
B9	2·35	2·19–2·47	1·70	1·60–1·90	·
B9	2·64	2·53–2·83	1·90	1·70–2·40	·
B9	2·25	2·06–2·47	1·60	1·40–1·90	·

[b]Diversity parameters are defined in Table 4.2, * denotes expected heterozygosity.
[c]BS (breeding system); SI, self-incompatible; SC, self-compatible; Mixed, mixed mating; Sex, sexually reproducing; Asex, asexually reproducing.

Table 4.4—*contd.*

	Distribution	*Polymorphic*	*Life time*[d]	*Source*
Ancestor-derivate[a]				
A1	W	0·01	·	McLeod *et al.* (1983)
A1	R	0·01	·	McLeod *et al.* (1983)
A2	W	0·05	P	Loveless & Hamrick (1988)
A2	R	0·05	P	Loveless & Hamrick (1988)
A3	W	0·01	A	Gottlieb (1974)
A3	R	0·01	A	Gottlieb (1974)
A4	W	0·01	P	Pleasants & Wendel (1989)
A4	R	0·01	P	Pleasants & Wendel (1989)
A5	W	0·01	A	Gottleib & Pilz (1976)
A5	R	0·01	A	Gottleib & Pilz (1976)
A6	W	0·01	A	Crawford *et al.* (1985)
A6	R	0·01	A	Crawford *et al.* (1985)
A7	W	0·01	A	Gottlieb *et al.* (1985)
A7	R	0·01	·A	Gottlieb *et al.* (1985)
A8	W	0·05	A	Gottlieb (1973)
A8	R	0·05	A	Gottlieb (1973)
Congeners				
B1	W	0·05	P	Karron *et al.* (1988)
B1	W	0·05	P	Karron *et al.* (1988)
B1	R	0·05	P	Karron *et al.* (1988)
B1	R	0·05	P	Karron *et al.* (1988)
B2	W	·	P	Moran & Bell (1983); Moran & Hopper (1987)
B2	W	·	P	Moran & Bell (1983); Moran & Hopper (1987)
B2	W	·	P	Moran & Bell (1983); Moran & Hopper (1987)
B2	W	·	P	Moran & Bell (1983); Moran & Hopper (1987)
B2	R	0·01	P	Moran & Hopper (1983)
B2	R	0·01	P	Moran & Hopper (1983)

B2	R	·	P	Moran & Hopper (1987)
B2	R	·	P	Moran & Hopper (1987)
B2	R	·	P	Moran & Hopper (1987)
B2	R	·	P	Moran & Hopper (1987)
B3	W	0·01	A	Warwick & Gottlieb (1985)
B3	R	0·01	A	Warwick & Gottlieb (1985)
B3	R	0·01	A	Warwick & Gottlieb (1985)
B3	R	0·01	A	Warwick & Gottlieb (1985)
B4	W	·	A	Kesseli & Jain (1984)
B4	R	·	A	Kesseli & Jain (1984)
B4	R	·	A	Kesseli & Jain (1984)
B4	R	·	A	Kesseli & Jain (1984)
B5	W	0·02	P	Sytsma & Schaal (1985)
B5	R	0·02	P	Sytsma & Schaal (1985)
B5	R	0·02	P	Sytsma & Schaal (1985)
B5	R	0·02	P	Sytsma & Schaal (1985)
B5	R	0·02	P	Sytsma & Schaal (1985)
B6	W	0·05	A	Babbel & Selander (1974)
B6	R	0·05	A	Babbel & Selander (1974)
B7	W	0·05	·	Levy & Levin (1975); Levy *et al.* (1975)
B7	R	0·05	·	Levin *et al.* (1979)
B8	W	0·01	P	Ledig (1986)
B8	W	0·01	P	Ledig (1986)
B8	W	0·01	P	Yen & Layton (1979)
B8	W	0·01	P	O'Malley *et al.* (1979)
B8	W	0·01	P	Allendorf *et al.* (1982)
B8	W	0·01	P	Guries & Ledig (1982)
B8	R	0·01	P	Ledig & Conckle (1983)
B9	W	0·01	P	Millar *et al.* (1988)
B9	R	0·01	P	Millar *et al.* (1988)
B9	R	0·01	P	Millar *et al.* (1988)

[d]P, perennial; A, annual.

of the proportion of polymorphic loci (P), average number of alleles per locus (A) and average number of alleles per polymorphic locus (A_P) in rare and widespread conspecifics taken from published papers. The within-population P ranges from 0% in *Pinus torreyana* to 86% in *Layia*

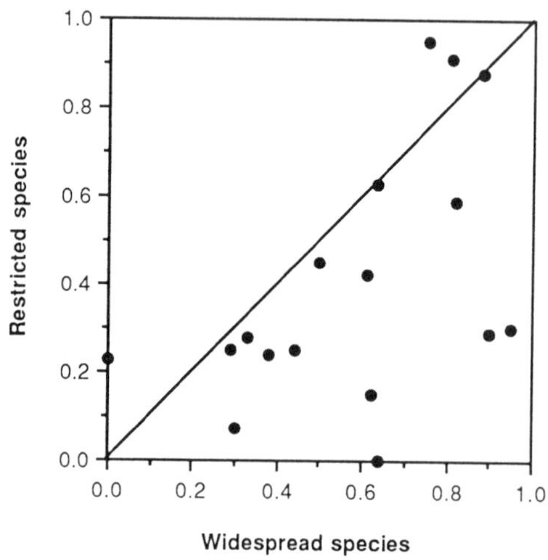

Fig. 4.5A. The percentage of polymorphic loci for species with wide and restricted distributions. Data from Table 4.4.

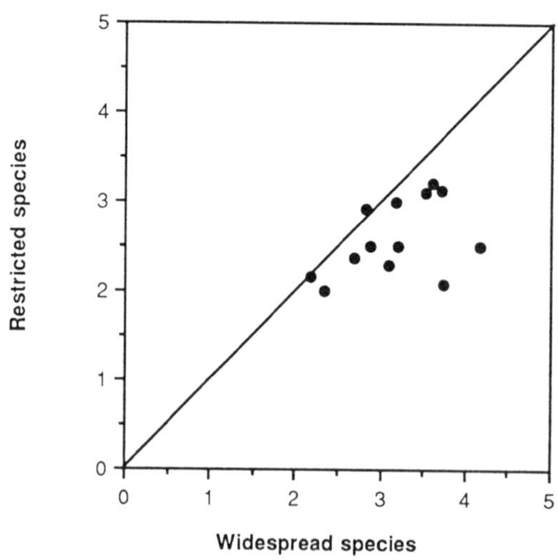

Fig. 4.5B. The mean number of alleles at the polymorphic loci for species with wide and restricted distributions. Data from Table 4.4.

discoidea, and the within-population A_P varies from 2·00 (e.g. *Eucalyptus*) to 3·1 in *Layia*. Populations of widespread taxa are not clearly more polymorphic than populations of restricted taxa, but there seems to be a great variation among populations (Fig. 4.5(B) and 4.5(C)). At the species level widespread taxa are more polymorphic than restricted taxa, both for

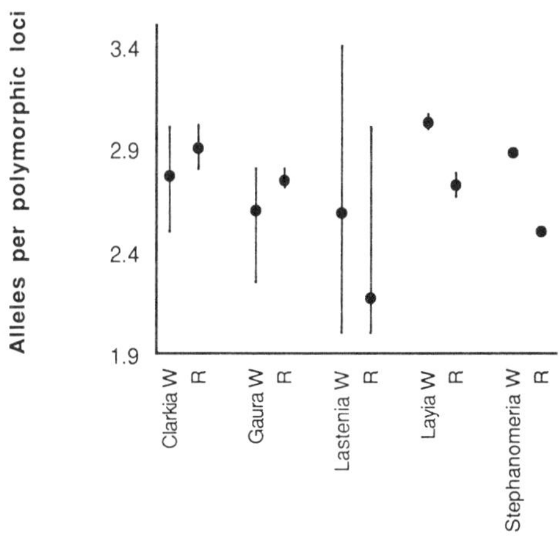

Fig. 4.5C. The proportion of polymorphic loci (mean and range) among populations of ancestor-derivative species-pairs with wide (W) and restricted (R) distributions. Data from Table 4.4.

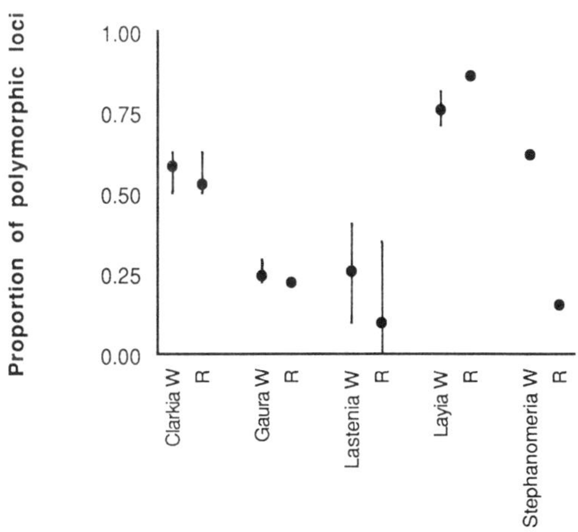

Fig. 4.5D. Average number of alleles per polymorphic loci (mean and range) among populations of ancestor-derivative species-pairs with wide (W) and restricted (R) distributions. Data from Table 4.4.

P and A_P (Fig. 4.5(A)). Unfortunately, the size of the database is quite limited.

The lack of knowledge about the past history of most floras is a serious drawback for the conclusions drawn from comparison between widespread and rare species. We do not know if population sizes have gone through bottlenecks or if the populations have been established from few individuals. In very few cases we have enough well-documented evidence to show that rare species can be either young or ancient. In Table 4.5 we show seven species-pairs where the rare species has been suggested to represent a recent derivative from an ancient, more widespread species.

Table 4.5
Average heterozygosity within widespread (W) and restricted (R) ancestor–derivate species-pairs (cf. Table 4.4)

Species	*W*	*R*
Capsicum	0·021	0·010
Cirsium	0·174	0·018
Clarkia	0·150	0·080
Erythronium	0·420	0·180
Gaura	0·074	0·050
Lasthenia	0·074	0·007
Layia	0·280	0·270
Stephanomeria	0·159	0·001

The rare and widespread species have very high genetic identity in these seven species groups, which is consistent with the hypothesis that the rare species had originated from only a few individuals of the ancestral species. The lower polymorphism in the rare species may, therefore, be an effect of a *founder* event instead of a long-lasting small population size. Significant losses in genetic variation have been recorded after founder events (Schwaegerle & Schaal, 1979; Berlocher, 1984). A naturalized population of *Sarracenia purpurea*, known to be established from only a single founder, had an average heterozygosity of 0·042 compared with a mean of 0·095 for ten other populations (Schwaegerle & Schaal, 1979).

Pinus torreyana is a narrow endemic restricted to two populations along the Californian coast, with estimated population sizes of 7000 and 2000 mature trees, respectively (Ledig & Conkle, 1983). Both populations were monomorphic for 59 loci, a remarkable condition among conifers, which usually have a within-population *P* of 88 to 97% (Ledig & Conkle, 1983). The within-species *P* was only 3·4%, which could be compared with 49% and 93%, respectively, for the two close relatives, *P. coulteri* and *P. sabiniana* (Ledig, 1986). Ledig and Conkle (1983) suggested that genetic drift is the most likely explanation for the lack of variability in *P. torreyana* due to a reduction in population size some 10 000 to 5000 years ago. Despite the lack of genetic differentiation, measured in terms of allozymes, between the two populations they exhibit a great many morphological differences; they have recently been recognized as two subspecies (Haller, 1986).

Several other narrow endemics have also been shown to be genetically depauperate, including *Silene diclinis* (Prentice, 1984), *Pedicularis furbishiae* (Waller *et al.*, 1987) and three species of *Sullivantia* (Soltis, 1982).

The difficulties in interpreting the pattern of genetic diversity can be illustrated with the narrow endemic Californian fan palm *Washingtonia filifera* (McClenaghan & Beauchamp, 1986). Here the average within-population *P* was 9·8% for 16 allozyme loci, and the proportion of polymorphic loci, but not the heterozygosity level, was significantly correlated with population size for 16 populations ranging from one to 82 mature individuals. The authors hypothesized that stochastic processes reduced the level of genetic variability during formation of these populations. The observed levels of low among-population genetic diversity are consistent with a recent colonization or a high level of gene flow among populations ($Nm = 9{\cdot}2$ calculated according to Crow and Aoki, 1984). However, the authors could not discriminate between these and alternative hypotheses regarding uniform selection pressures favouring a narrow array of specialized genotypes in either relict or colonizing populations. Lack of genetic variation has also been reported in widespread species. Mashburn *et al.* (1978) reported no intraspecific polymorphism within populations of *Typha latifolia* and *T. domingensis* surveyed for variation at 10 enzyme loci.

4.5 DISCUSSION

Before discussing genetic diversity in fragmented habitats we should consider some exampls of population size and its relevance to the MVP concept in plants. After that we shall stress some problems in the current use of rare and widespread species as models for MVP in plants. Finally, we shall consider the level of gene flow in plants with different breeding systems and life-histories, and its consequences for the MVP and the metapopulation concepts in plants.

There are few reliable estimates of actual population sizes over long periods of time, but there are some data which indicate the existence of very small sizes over long periods. In the Aegean archipelago (in the Mediterranean sea), 68% of the censused populations of *Erysimum* contained $\leqslant 100$ individuals (Fig. 4.6). In this study, Snogerup (1967) estimated the size of 53% of all known populations ($n = 107$) of eight taxa of the self-compatible but habitually outcrossing, perennial chasmophytes of the genus *Erysimum* sect. *Cheiranthus*. In the more widespread *Brassica* sect. *Brassica*, Snogerup *et al.* (1990) found similar patterns of distribution of population sizes (Fig. 4.6). These species occur on more or less vertical

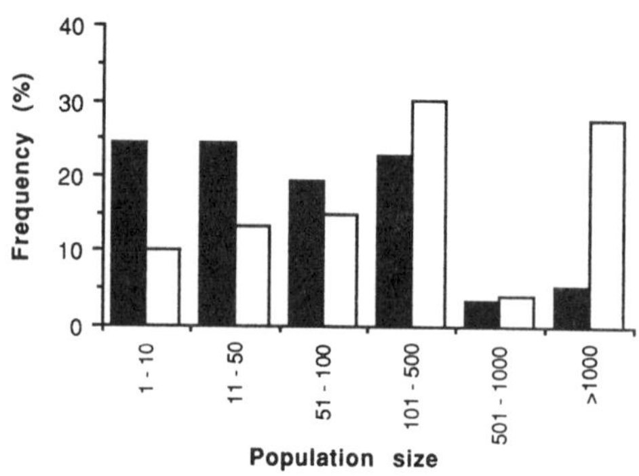

Fig. 4.6. Frequency of different population sizes among species of the plant genus *Erysimum* sect. *Cheiranthus* (■) among the Aegean islands and the more widespread *Brassica* sect. *Brassica* (□). Data from Snogerup (1967) and Snogerup *et al.* (1990). The number of populations are 57 and 149, respectively.

cliffs, often in ravines—a stable environment. Snogerup (1967) assumed that the populations of *Erysimum* have been isolated for thousands of generations, and that it is likely that these chasmophytes have existed in small populations for a considerable length of time. These species exhibited a great deal of morphological variation, even among progeny derived from a single plant, which the author ascribed to high levels of heterozygosity. Unfortunately no allele frequencies are available, so we cannot estimate the levels of heterozygosity and the genetic structures of these species. Snogerup (1967) found greenhouse-grown progeny derived from seeds collected from natural populations to exhibit numerous morphological abnormalities, which may indicate inbreeding depression. There are, however, several explanations for the morphological variations in these species. The populations may have evolved well-developed plastic responses as an alternative to genetic variation. This may allow the populations to persist for long periods in the stabile chasmophytic vegetation despite severe inbreeding. Anyhow, it seems that the population system in the Aegean chasmophytes would provide suitable material for the testing of basic predictions regarding MVP in plants.

The effective population sizes in plants (as indicated by the estimates of neighbourhood sizes in Table 4.3) are usually much smaller than the actual number of mature individuals in local populations. The small N_s suggests that the MVP for plant species might be small. However, the difficulties of relating MVP to neighbourhood size could be shown by the genetic structure of *Liatris cylindracea*, an outbreeding perennial herb in an undisturbed population. Schaal (1975) divided the population into a number of 3 × 3 m plots—approximately the area of a neighbourhood in this species. There was a significant degree of local differentiation among these neighbourhoods as assessed by allozyme frequencies (Schaal, 1975). However, Sokal and Oden (1978) applied a spatial autocorrelation analysis to Schaal's material and found no consistent spatial pattern, indicating that the pattern does not exist on a larger scale. The individuals within each neighbourhood exhibited 93% of the total genetic diversity found in the local population (Schaal, 1976). This is probably the result of a certain level of gene flow among neighbourhoods within the population ($Nm = 3{\cdot}4$, calculated according to Wright, 1969). For a plant species in a fragmented world, however, we expect gene flow among local populations to be much more restricted than gene flow among neighbourhoods in a continuous population. Thus, an MVP must probably be larger than a neighbourhood size.

Despite the very few estimates of N_s in plant populations, the available information has some implications for conservation biology. If the actual number of individuals in local populations of rare or threatened species is smaller than the general level of N_s estimated for plants with comparable life-histories and breeding systems, this could be a warning signal. For instance, more than 50% of the actual populations sizes of the rare *Eucalyptus caesia* (Table 4.4) were composed of fewer than 100 individuals, and no population size of the fan palm *Washingtonia filifera*, discussed above, exceeded 90 individuals.

According to present theoretical models, the small N_s recorded in many plants, and the small actual population sizes found in some cases, will ultimately lead to genetic drift and inbreeding. One consequence is inbreeding depression. In addition, Charlesworth and Charlesworth (1987) have recently suggested that inbreeding depression has generally been underestimated, rather than overestimated, in natural populations. However, many plant species have evolved mechanisms to overcome enforced inbreeding (Jain, 1976; Lloyd, 1979). In many plant taxa predominantly self-fertilizing species exist which are closely related to primarily outcrossing species (Stebbins, 1957). The origin of self-fertilization is polyphyletic and may be a frequent event within a species (Lloyd, 1988; Wyatt, 1988). Lande and Schemske (1985) hypothesized that species with a long history of occasional population bottlenecks are expected to exhibit relatively little inbreeding depression and to be selected for self-fertilization, and species with historically large, outcrossing populations are expected to show much inbreeding depression and to be selected for mechanisms which prevent inbreeding. If this is true, such adaptive mechanisms may suggest different conservation strategies for plants with different breeding systems. However, empirical evidence of differences in inbreeding depression from natural population is still sparse (cf. Fig. 4.1).

Knowledge gained from the comparisons of widespread and rare species-pairs generally cannot be transferred directly to models of the genetic effects of habitat fragmentation. Many studies have been carried out on progenitor-derivative species-pairs on what are probably very recent speciation events. Here we could expect the progenitor to have all the allelic variation present in the derivative, while the derivative should contain only part of this variation and possess few unique alleles. These predictions are confirmed for many of the species-pairs in Table 4.4. Thus, the lower genetic variability in these derivative species is probably an effect of a founder event in the speciation processes and not a result of small population sizes *per se*. In some of the species-pairs (e.g. *Cirsium*), the rare, less variable species is restricted to areas which were covered by ice during the last glacial period, while the more widespread species is found beyond the glacial boundary. Marginal populations of conifers tend to have lower heterozygosity than central populations (Ledig, 1986). Cwrynar and McDonald (1987) compared the time since populations of *Pinus contorta* were founded, based on pollen records, with genetic diversity along a north–south transect in northern Canada. They found a significant reduction in the average number of alleles per population from the centre of the distribution to its northern periphery. Similarly, species of large mammals with northern latitudinal distribution have been shown to have lower heterozygosity than species with a more southern distribution in the Northern Hemisphere (Sage & Wolff, 1986). Serial colonization (*stepping-stone model*) from southern non-glaciated to northern glaciated latitudes, with only a part of the species' gene pool involved in the dispersal or bottlenecks and genetic drift during colonization, is the likely explanation for these patterns.

We do not think that available studies of genetic diversity in rare and widespread species provide unambiguous answers to our questions concerning fragmented habitats. We suggest that further studies of the genetic effects of habitat fragmentation must be designed carefully in order to minimize the uncertainty caused by historical factors (see above).

Several authors have argued that a metapopulation of small subpopulations with occasional gene flow among them will better preserve genetic variation (Kimura & Crow, 1963*b*; Chesser, 1983) and maximize the rate of evolution than would a single large population of equal total size. However, the level of gene flow in natural populations of some species is still unclear; application of indirect estimation methods in both plant and animal studies produces much higher measurements than direct methods (Slatkin, 1985*a*, 1987). This suggests that direct methods underestimate the long-distance dispersal (see above), or perhaps that such species have undergone large-scale demographic changes both relatively frequently and recently. Colonization and migration among existing populations both contribute to gene flow (Slatkin, 1977; Wade & McCauley, 1988), but indirect methods do not clearly identify their respective effects.

Most of the species with $Nm > 1$ in Fig. 4.3 are long-lived, outcrossing perennials. These species may conform to the metapopulation structure with a homogenizing gene flow among existing populations. Conifers from glaciated areas with a short genetic history have an especially high Nm, and an alternative explanation for the observed level of genetic identity among populations could be an effect of recent colonization. A majority of plants exhibit very little gene flow, indicated by $Nm < 1$ (Fig. 4.3). Thus, in these cases we expect genetic drift and inbreeding to be important evolutionary forces. This group of plants consists predominately of self-fertilizing annuals (Fig. 4.3) with low levels of heterozygosity. It is precisely those plants which have evolved mechanisms to handle inbreeding (Lande & Schemske, 1985). We, therefore, do not expect self-fertilizing annuals to be the kind of species which are likely to suffer from inbreeding in a fragmented world. These plants are often colonizers and demographic or environmental stochasticity may be more important in determining the MVP for this group.

4.6 CONCLUSIONS AND FURTHER RESEARCH NEEDS

In this chapter we have shown that, according to theory, small population size will ultimately lead to genetic depletion, and the presently available information on genetic diversity in natural populations seems to confirm this assumption by suggesting an ongoing genetic depletion in small populations. However, we claim that our empirical knowledge about genetic diversity and population size in plants is still insufficient. The interpretation of empirical evidence in the context of fragmentation is not unambiguous for most plant species. There are several reasons for this:

(1) the present knowledge is not a result of studies designed to understand the effect of fragmentation;
(2) the correlation among genetic variation and breeding system and life-history makes comparisons between species difficult to interpret;
(3) the lack of knowledge of the past history of natural populations is a further complication when interpreting patterns of genetic variation;
(4) plants with different breeding systems and life-histories may be vulnerable to fragmentation in different ways.

Self-fertilizing annuals have been selected for inbreeding and may not be expected to suffer genetically from habitat fragmentation. Outbreeding perennials, on the other hand, may suffer from inbreeding, and should, therefore, have priority in the further study of the genetic effects of fragmentation.

The immediate requirements for conservation biology in order to predict the long-term genetic survival of fragmented plant populations are twofold:

(1) The level of genetic depletion in a fragmented landscape must be established.
(2) The ecological effects of inbreeding and loss of genetic diversity must be understood in detail.

An optimal design for increasing our knowledge of genetic depletion and habitat fragmentation would be:

(A) In order to eliminate the founder-effect present in the current comparisons of species-pairs, and other problems arising with inter-species comparison (e.g. variation in breeding systems), an ideal type of study should fulfil one or several of the following criteria:

(1) Limiting the scope of research to a single species, with samples drawn from populations of varying size and not making inter-species comparisons.
(2) Locating a species with a long history and a stable population size.
(3) A study species selected from a glaciated area should be one with a well-documented and previously more extensive distribution (e.g. documented by pollen depositions), and now occurring as a fragmented species. The time elapsed since the onset of fragmentation should be countable in hundreds of generations.
(4) And, finally, it would be ideal to find a study species with a distribution covering both glaciated and non-glaciated areas.

(B) To study the demographic effects of inbreeding in natural populations in carefully designed experiments. Outbreeding perennials should be used and, ideally, populations found to be genetically depleted due to isolation should be compared with non-depleted populations.

ACKNOWLEDGEMENTS

This work was part of a research project financially supported by the National Swedish Environmental Protection Agency. We are grateful for discussions and comments on earlier versions of the manuscript by S. Andersson, A. Carlström, H. Runemark and H. Tegelström. We thank E. Schippert for linguistic advice and G. Andersson and M. Widén for technical assistance.

REFERENCES

Allendorf, F.W., Knudsen, K.L. & Blake, G.M. (1982). Frequencies of null alleles at enzyme loci in natural populations of ponderosa and red pine. *Genetics*, **100**, 497–504.

Antlfinger, A.E. (1982). Genetic neighborhood structure of the salt marsh composite *Borrichia frutescens*. *J. Hered.*, **73**, 128–32.

Babbel, G.R. & Selander, R.K. (1974). Genetic variability in edaphically restricted and widespread plant species. *Evolution*, **28**, 619–30.

Bannister, M.H. (1965). Variation in the breeding system of *Pinus radiata*. In *The Genetics of Colonizing Species*, ed. H.G. Baker & G.L. Stebbins. Academic Press, New York, pp. 353–72.

Banyard, B.J. & James, S.H. (1979). Biosystematic studies in the *Stylidium crassifolium* species complex (Stylidiaceae). *Aust. J. Bot.*, **27**, 27–37.

Barrett, S.C.H. & Kohn, J.R. (in press). Genetic and evolutionary consequences of small population size in plants: implications for conservation. In *The Conservation of Rare Plants: Biology and Genetics*, ed. D.A. Falk & H.E. Holsinger. Oxford University Press, Oxford.

Bayer, R.J. (1988). Patterns of isozyme variation in western North American *Antennaria* (Asteraceae: Inuleae), I. Sexual species of sect. *Dioicae*. *Syst. Bot.*, **13**, 525–37.

Bayer, R.J., La Duke, J.C. & Crawford, D.J. (1987). Isozyme variation in *Trillium nivale* (Liliaceae). *Can. J. Bot.*, **65**, 2250–4.

Beattie, A.J. & Culver, D.C. (1979). Neighbourhood size in *Viola*. *Evolution*, **33**, 1226–9.

Belhassen, E., Dockes, A.C., Gliddon, C. & Gouyon, P.H. (1987). Gene dispersal and neighborhood in a gynodioecious species, the case of *Thymus vulgaris* L. *Genet. Sel. Evol.*, **19**, 307–20.

Berlocher, S.H. (1984). Genetic changes coinciding with the colonization of California by the walnut husk fly *Rhagoletis complete*. *Evolution*, **38**, 906–18.

Bos, M., Harmens, H. & Vrieling, K. (1986). Gene flow in *Plantago*, I. Gene flow and neighbourhood size in *P. lanceolata*. *Heredity*, **56**, 43–54.

Bradshaw, A.D. (1965). Evolutionary significance of phenotypic plasticity in plants. *Adv. Genet.*, **13**, 115–55.

Bush, R.M., Smouse, P.E. & Ledig, F.T. (1987). The fitness consequences of multiple-locus heterozygosity: the relationship between heterozygosity and growth rate in pitch pine (*Pinus rigida* Mill.). *Evolution*, **41**, 787–98.

Cahalan, C. & Gliddon, C. (1985). Genetic neighbourhood size in *Primula vulgaris*. *Heredity*, **54**, 65–70.

Campbell, D.R. & Waser, N.M. (1987). The evolution of plant mating systems: multilocus simulations of pollen dispersal. *Amer. Nat.*, **129**, 593–609.

Charlesworth, D. & Charlesworth, B. (1987). Inbreeding depression and its evolutionary consequences. *Ann. Rev. Ecol. Syst.*, **18**, 237–68.

Chesser, R.K. (1983). Isolation by distance: relationship to the management of genetic resources. In *Genetics and Conservation: A Reference for Managing Wild Animal and Plant Populations*, ed. C.M. Schonewald-Cox, S.M. Chambers, B. MacBryde & L. Thomas. Benjamin Cummings, Menlo Park, California, pp. 66–77.

Crawford, D.J. & Whitkus, R. (1988). Allozyme divergence and the mode of speciation for *Coreopsis gigantea* and *C. maritima* (Compositae). *Syst. Bot.*, **13**, 256–64.

Crawford, D.J., Ornduff, R. & Vasey, M.C. (1985). Allozyme variation within and between *Lasthenia minor* and its derivate species, *L. maritima* (Asteraceae). *Amer. J. Bot.*, **75**, 1177–84.

Crawford, T.J. (1984). What is a population? In *Evolutionary Ecology*, ed. B. Shorrocks. *Symp. Brit. Ecol. Soc.*, *23rd*. Blackwell Scientific Publications, Oxford, pp. 135–73.

Crow, J.F. & Aoki, K. (1984). Group selection for a polygenic behavioral trait: estimating the degree of population subdivision. *Proc. Natl Acad. Sci. USA*, **81**, 6073–77.

Cwynar, L.C. & McDonald, G.M. (1987). Geographical variation of lodgepole pine in relation to population history. *Amer. Nat.*, **123**, 463–9.

Dobzhansky, T. & Wright, S. (1941). Genetics of natural populations, Vol. V. Relations between the mutation rate and the accumulation of lethals in populations of *Drosophila pseudoobscura*. *Genetics*, **26**, 23–51.

Ehrlich, P.R. (1983). Genetics and the extinctions of butterfly populations. In *Genetics and Conservation: A Reference for Managing Wild Animal and Plant Populations*, ed. C.M. Schonewald-Cox, S.M. Chambers, B. MacBryde & L. Thomas. Benjamin Cummings, Menlo Park, California, pp. 152–63.

Ehrlich, P.R. & Raven, P.H. (1969). Differentiation of populations. *Science, N.Y.*, **165**, 1228–32.

Ellstrand, N.C. & Marshall, D.L. (1985). Interpopulation gene flow by pollen in wild radish, *Raphanus sativus*. *Amer. Nat.*, **126**, 606–16.

Endler, J.A. (1986). *Natural Selection in the Wild*. Princeton University Press, Princeton, New Jersey.

Falconer, D.S. (1981). *Introduction to Quantitative Genetics*. Longman, London.

Fisher, R.A. (1930). *The Genetical Theory of Natural Selection*. Clarendon, Oxford.

Frankel, O.H. & Soulé, M.E. (1981). *Conservation and Evolution*. Cambridge University Press, Cambridge.

Franklin, I.R. (1980). Evolutionary changes in small populations. In *Conservation Biology. An Evolutionary – Ecological Perspective*, ed. M.E. Soulé & B.A. Wilcox. Sinauer Associates, Sunderland, Massachusetts, pp. 135–49.

Gottlieb, L.D. (1973). Genetic differentiation, sympatric speciation, and the origin of a diploid species of *Stephanomeria*. *Amer. J. Bot.*, **60**, 545–53.

Gottlieb, L.D. (1974). Genetic configuration of the origin of *Clarkia lingulata*. *Evolution*, **28**, 244–50.

Gottlieb, L.D. & Pilz, G. (1976). Genetic similarity between *Gaura longiflora* and its obligately outcrossing derivative *G. demareei*. *Syst. Bot.*, **1**, 181–7.

Gottlieb, L.D., Warwick, S.I. & Ford, V.S. (1985). Morphological and electrophoretic divergence between *Layia discoidea* and *L. glandulosa*. *Syst. Bot.*, **10**, 484–95.

Govindaraju, D.R. (1988*a*). Relationship between dispersal ability and level of gene flow in plants. *Oikos*, **52**, 31–5.

Govindaraju, D.R. (1988*b*). Life history, neighbourhood sizes, and variance structure in some North American conifers. *Biol. J. Linn. Soc.*, **35**, 69–78.

Guries, R.P. & Ledig, F.T. (1982). Genetic diversity and population structure in pitch pine (*Pinus rigida* Mill.). *Evolution*, **36**, 387–402.

Haldane, J.B.S. (1930). A mathematical theory of natural and artificial selection, Part IV. Isolation. *Proc. Camb. Phil. Soc.*, **26**, 220–30.

Haller, J.R. (1986). Taxonomy and relationships of the mainland and island populations of *Pinus torreyana* (Pinaceae). *Syst. Bot.*, **11**, 39–50.

Hamrick, J.L. (1983). The distribution of genetic variation within and among natural plant populations. In *Genetics and Conservation: A Reference for Managing Wild Animal and Plant Populations*, ed. C.M. Schonewald-Cox,

S.M. Chambers, B. MacBryde & L. Thomas. Benjamin Cummings, Menlo Park, California, pp. 501–8.

Hamrick, J.L. & Allard, R.W. (1972). Microgeographical variation in allozyme frequencies in *Avena barbata*. *Proc. Natl Acad. Sci. USA*, **69**, 2100–4.

Hamrick, J.L. & Godt, M.J. (1990). Allozyme diversity in plant species. In *Plant Population Genetics, Breeding and Genetic Resources*, ed. A.H.D. Brown, M.T. Clegg, A.L. Kahler & B.S. Weir. Sinauer Associates, Sunderland, Massachusetts, pp. 43–63.

Hamrick, J.L. & Murawski, D.A. (1990). The breeding structure of tropical tree populations. *Plant Species Biol.*, **5**, 157–60.

Hamrick, J.L., Linhart, Y.B. & Mitton, J.B. (1979). Relationships between life history characteristics and electrophoretically detectable genetic variation in plants. *Ann. Rev. Ecol. Syst.*, **10**, 173–200.

Handel, S.N. (1976). Restricted pollen flow of two woodland herbs determined by neutron-activation analysis. *Nature, Lond.*, **260**, 422–3.

Handel, S.N. (1985). Pollen flow patterns and the creation of local genotypic variation. In *Structure and Functioning of Plant Populations, 2. Phenotypic and Genotypic Variaton in Plant Populations*, ed. J. Haeck & J.W. Woldendorp. North-Holland Publishing Company, Amsterdam, pp. 251–65.

Harder, L.D., Thomson, J.D., Cruzan, M.B. & Unnash, R.S. (1985). Sexual reproduction and variation in floral morphology in an ephemeral vernal lily, *Erythronium americanum*. *Oecologia* (*Berl.*), **67**, 286–91.

Hedrick, P.W. & Holden, L. (1979). Hitch-hiking: An alternative to coadaptation for barley and slender wild oat examples. *Heredity*, **43**, 79–86.

Heywood, J.S. (1986). The effect of plant size variation on genetic drift in populations of annuals. *Amer. Nat.*, **127**, 851–61.

Husband, B.C. & Barrett, S.C.H. (1989). Population size and genetic diversity in populations of *Eichhornia paniculata*. *Amer. J. Bot.*, **76**, 148.

Jain, S.K. (1976). The evolution of inbreeding in plants. *Ann. Rev. Ecol. Syst.*, **7**, 469–95.

Johannsen, W. (1911). The genotypic concept of heredity. *Amer. Nat.*, **45**, 129–59.

Karron, J.D., Linhart, Y.B., Chaulk, C.A. & Robertson, C.A. (1988). Genetic structure of populations of geographically restricted and widespread species of *Astragalus* (Fabaceae). *Amer, J. Bot.*, **75**, 1114–9.

Kephart, S. (1981). Breeding systems in *Asclepias incarnata* L., *A. syriaca* L. *Amer. J. Bot.*, **68**, 226–32.

Kerster, H.W. & Levin, D.A. (1968). Neighbourhood size in *Lithospermum carolinense*. *Genetics*, **60**, 577–87.

Kesseli, R.V. & Jain, S.K. (1984). New variation and biosystematic patterns detected by allozyme and morphological comparisons in *Limnanthes* sect. *Reflexae* (Limnanthaceae). *Pl. Syst. Evol.*, **147**, 133–65.

Kimura, M. & Crow, F.J. (1963*a*). The measurement of effective population number. *Evolution*, **17**, 279–88.

Kimura, M. & Crow, F.J. (1963*b*). On the maximum avoidance of inbreeding. *Genet. Res.*, **4**, 399–415.

Knowles, P. & Grant, M.C. (1981). Genetic patterns associated with growth variability in ponderosa pine. *Amer. J. Bot.*, **68**, 942–6.

Knowles, P. & Mitton, J.B. (1980). Genetic heterozygosity and radial growth variability in *Pinus contorta*. *Silvae Genet.*, **29**, 114–8.

Koptur, S. (1984). Outcrossing and pollinator limitation of fruit set: breeding systems of neotropical *Inga* trees (Fabaceae: Mimosoideae). *Evolution*, **38**, 1130–43.
Kruckeberg, A.R. (1957). Variation in fertility of hybrids between isolated populations of serpentine species, *Streptanthus glandulosus* Hook. *Evolution*, **11**, 185–211.
Lande, R. & Barrowclough, G.F. (1987). Effective population size, genetic variation, and their use in population management. In *Viable Populations for Conservation*, ed. M.E. Soulé. Cambridge University Press, Cambridge, pp. 87–123.
Lande, R. & Schemske, D.W. (1985). The evolution of self-fertilization and inbreeding depression in plants, I. Genetic models. *Evolution*, **39**, 24–40.
Ledig, F.T. (1986). Heterozygosity, heterosis, and fitness in outbreeding plants. In *Conservation Biology. The Science of Scarcity and Diversity*, ed. M.E. Soulé. Sinauer Associates, Sunderland, Massachusetts, pp. 77–104.
Ledig, F.T. & Conkle, M.T. (1983). Gene diversity and genetic structure in a narrow endemic Torrey pine (*Pinus torreyana* Parry ex. Carr). *Evolution*, **37**, 79–85.
Ledig, F.T., Guries, R.P. & Bonefeld, B.A. (1983). The relation of growth to heterozygosity in pitch pine. *Evolution*, **37**, 1227–38.
Levin, D.A. (1981). Dispersal vs gene flow in plants. *Ann. Missouri Bot. Gard.*, **68**, 233–53.
Levin, D.A. (1983). Plant parentage: an alternative view of the breeding structure of populations. In *Population Biology–Retrospect and Prospect*, ed. C.E. King & P.S. Dawson. Columbia University Press, New York, pp. 171–88.
Levin, D.A. (1984). Inbreeding depression and proximity-dependent crossing success in *Phlox drummondii*. *Evolution*, **38**, 116–27.
Levin, D.A. (1988). Local differentiation and the breeding structure of plant populations. In *Plant Evolutionary Biology*, ed. L.D. Gottlieb & S.K. Jain. Chapman & Hall, London, pp. 305–29.
Levin, D.A. & Kerster, H.W. (1968). Local gene dispersal in *Phlox*. *Evolution*, **22**, 130–9.
Levin, D.A. & Kerster, H.W. (1969). Density-dependent gene dispersal in *Liatris*. *Amer. Nat.*, **103**, 61–74.
Levin, D.A. & Kerster, H.W. (1974). Gene flow in seed plants. *Evol. Biol.*, **7**, 139–220.
Levin, D.A., Ritter, K. & Ellstrand, N.C. (1979). Protein polymorphism in the narrow endemic *Oenothera organensis*. *Evolution*, **33**, 534–42.
Levy, M. & Levin, D.A. (1975). Genetic heterozygozity and variation in permanent translocation heterozygotes of the *Oenothera biennis* complex. *Genetics*, **79**, 493–512.
Levy, M., Steiner, E.E. & Levin, D.A. (1975). Allozyme genetics in permanent translocation heterozygotes in the *Oenothera biennis* complex. *Biochem. Genet.*, **13**, 487–500.
Linhart, Y.B. (1974). Intrapopulation differentiation in annual plants, I. *Veronica peregrina* raised under non-competitive conditions. *Evolution*, **28**, 232–43.
Lloyd, D.G. (1979). Some reproductive factors affecting the selection of self-fertilization in plants. *Amer. Nat.*, **113**, 67–79.
Lloyd, D.G. (1988). Benefits and costs with biparental and uniparental reproduc-

tion in plants. In *The Evolution of Sex. An examination of Current Ideas*. ed. R.E. Michod & B.R. Levin. Sinauer Associates, Sunderland, Massachusetts, pp. 233–52.

Loveless, M.D. & Hamrick, J.L. (1984). Ecological determinants of genetic structure in plant populations. *Ann. Rev. Ecol. Syst.*, **15**, 65–95.

Loveless, M.D. & Hamrick, J.L. (1988). Genetic organization and evolutionary history in two North American species of *Cirsium*. *Evolution*, **42**, 254–65.

MacArthur, R.H. & Wilson, E.O. (1967). *The Theory of Island Biogeography*. Princeton University Press, Princeton, New Jersey.

Mashburn, S.J., Sharitz, R.R. & Smith, M.H. (1978). Genetic variation among *Typha* populations of the southeast United States. *Evolution*, **32**, 681–5.

Mayr, E. (1963). *Animal Speciation and Evolution*. Harvard Unversity Press, Cambridge, Massachusetts.

McCauley, D.E. (1987). Population genetic consequences of local colonization: evidence from the milkweed beetle *Tetraopes tetraopthalmus*. *Flor. Entomol.*, **70**, 21–30.

McClenaghan, L.R. Jr & Beauchamp, A.C. (1986). Low genetic differentiation among isolated populations of the California fan palm (*Washingtonia filifera*). *Evolution*, **40**, 315–22.

McLeod, M.J., Guttmanm S.I., Eshbaugh, W.H. & Rayle, R.E. (1983). An electrophoretic study of evolution in *Capsicum* (Solanaceae). *Evolution*, **37**, 562–74.

McNeilly, T. (1968). Evolution in closely adjacent plant populations, III. *Agrostis tenuis* on a small copper mine. *Heredity*, **23**, 99–118.

Millar, C.I., Strauss, S.H., Conkle, M.T. & Westfall, R.D. (1988). Allozyme differentiation and biosystematics of the Californian closed-cone pines (*Pinus* sect. *Oocarpae*). *Syst. Bot.*, **13**, 351–70.

Mitton, J.B. & Grant, M.C. (1980). Observation on the ecology and evolution of quaking aspen, *Populus tremuloides*, in the Colorado front range. *Amer. J. Bot.*, **67**, 202–9.

Mitton, J.B. & Grant, M.C. (1984). Associations among protein heterozygosity, growth rate and developmental homeostasis. *Ann. Rev. Ecol. Syst.*, **15**, 479–99.

Molau, U., Carlsson, M., Dahlgren, A. & Hill, Ö. (1989). Mating system and pollen-mediated gene flow in *Bartsia alpina*. *Oikos*, **55**, 409–19.

Moran, G.F. & Bell, J.C. (1983). *Eucalyptus*. In *Isozymes in Plant Genetics and Breeding*, ed. S.D. Tanksley & T.J. Orton. Elsevier, Amsterdam, pp. 423–41.

Moran, G.F. & Hopper, S.D. (1983). Genetic diversity and the insular population structure of rare granite rock species, *Eucalyptus caesia* Benth. *Aust. J. Bot.*, **31**, 161–72.

Moran, G.F. & Hopper, S.D. (1987). Conservation of the genetic resources of rare and widespread eucalypts in remnant vegetation. In *Nature Conservation: The Role of Remnants of Native Vegetation*, ed. D.A. Saunders, G.W. Arnold, A.A. Burbidge & J.M. Hopkins. Surrey Beatty and Sons, London, pp. 151–62.

Motten, A.F. (1983). Reproduction of *Erythronium umbilicatum* (Liliaceae): pollination success and pollinator effectiveness. *Oecologia* (*Berl.*), **59**, 351–9.

Nei, M. (1972). Genetic distance between populations. *Amer. Nat.*, **106**, 283–92.

Nei, M. (1973). Analysis of gene diversity in subdivided populations. *Proc. Natl Acad. Sci. USA*, **70**, 3321–3.

Nei, M., Mauyama, T. & Chakraborty, R. (1975). The bottleneck effect and genetic variability in populations. *Evolution*, **29**, 1–10.

Nevo, E., Beiles, A., Kaplan, D., Goldenberg, E.M., Olsvig-Whittaker, L. & Naveh, Z. (1986). Natural selection of allozyme polymorphisms: a microsite test revealing ecological differentiation in wild barley. *Evolution*, **40**, 13–20.

Newport, M.E.A. (1989). A test for proximity-dependent outcrossing in the alpine skypilot, *Polemonium viscosum*. *Evolution*, **43**, 1110–3.

Olesen, J.M. & Warncke, E. (1989). Temporal changes in pollen flow and neighbourhood structure in a population of *Saxifraga hirculus* L. *Oecologia (Berl.)*, **79**, 205–11.

O'Malley, D.M., Allendorf, F.W. & Blake, G.M. (1979). Inheritance of isozyme variation and heterozygosity in *Pinus ponderosa*. *Biochem. Genet.*, **17**, 233–50.

Palmer, M., Travis, J. & Antonovics, J. (1988). Seasonal pollen flow and progeny diversity in *Amianthum muscaetoxicum*: ecological potential for multiple mating in a self-compatible, hermaphroditic perennial. *Oecologia (Berl.)*, **77**, 19–24.

Park, Y.S., Fowler D.P. & Coles, F.J. (1984). Population studies of white spruce, II, Natural inbreeding and relatedness among neighboring trees. *Can J. For. Res.*, **14**, 909–13.

Pleasants, J.M. & Wendel, J.F. (1989). Genetic diversity in a clonal narrow endemic *Erythronium propullans*, and its widespread progenitor, *Erythronium albidum*. *Amer. J. Bot.*, **76**, 1136–51.

Powers, D.A. & Place, A.R. (1978). Biochemical genetics of *Fundulus heteroclitus* (L.), I. Temporal and spatial variation in gene frequencies of LDH-B, MDH-A, GPI-B and PGM-A. *Biochem. Genet*, **16**, 593–607.

Prentice, H.C. (1984). Enzyme polymorphism, morphometric variation and population structure in a restricted endemic, *Silene diclinis* (Caryophyllaceae). *Biol. J. Linn. Soc.*, **22**, 125–44.

Price, M.V. & Waser, N.M. (1979). Pollen dispersal and optimal outcrossing in *Delphinium nelsonii*. *Nature, Lond.*, **277**, 294–6.

Pyke, G.H. (1978). Optimal foraging: movements of bumblebees between inflorescences. *Theoret. Popul. Biol.*, **13**, 72–98.

Rai, K.N. & Jain, S.K. (1982). Population biology of *Avena*, IX. Gene flow and neighbourhood size in relation to micrgeographic variation in *Avena barbata*. *Oecologia (Berl.)*, **53**, 399–405.

Reinke, D.C. & Bloom, W.L. (1979). Pollen dispersal in natural populations: a method for tracking individual pollen grains. *Syst. Bot.*, **4**, 223–9.

Richards, A.J. & Ibrahim, H. (1978). Estimation of neighbourhood size in two populations of *Primula veris*. In *The Pollination of Flowers by Insects*, ed. A.J. Richards. Academic Press, London, pp. 165–74.

Ritland, K. (1989). Gene identity and the genetic demography of plant populations. In *Plant Population Genetics, Breeding, and Genetic Resources*, ed. A.H.D. Brown, M.T. Clegg, A.L. Kahler & B.S. Weir, Sinauer Associates. Sunderland, Massachusetts, pp. 181–99.

Ritland, K. & Ganders, F.R. (1987). Crossability of *Mimulus guttatus* in relation to components of gene fixation. *Evolution*, **41**, 772–86.

Sage, R. & Wolff, J.O. (1986). Pleistocene glaciations, fluctuating ranges, and low genetic variability in a large mammal (*Ovis dalli*). *Evolution*, **40**, 1092–5.
Schaal, B.A. (1974). Isolation by distance in *Liatris cylindracea*. *Nature, Lond.*, **252**, 703.
Schaal, B.A. (1975). Population structure and local differentiation in *Liatris cylindracea*. *Amer. Nat.*, **109**, 511–28.
Schaal, B.A. (1976). Genetic diversity in *Liatris cylindracea*. *Syst. Bot.*, **1**, 163–8.
Schaal, B.A. (1980). Measurement of gene flow in *Lupinus texensis* *Nature, Lond.*, **284**, 450–1.
Schaal, B.A. & Levin, D.A. (1977). The demographic genetics of *Liatris cylindracea* Michx. (Compositae). *Amer. Nat.*, **110**, 191–206.
Schaal, B.A. & Levin, D.A. (1978). Morphological differentiation and neighborhood size in *Liatris cylindracea*. *Amer. J. Bot.*, **65**, 923–8.
Schemske, D.W. & Lande, R. (1985). The evolution of self-fertilization and inbreeding depression, II. Emperical observations. *Evolution*, **39**, 41–52.
Schemske, D.W. & Pautler, L.P. (1984). The effect of pollen composition on fitness components in a neotropical herb. *Oecologia (Berl.)*, **62**, 31–6.
Schmitt, J. (1980). Pollinator foraging behavior and gene dispersal in *Senecio* (Compositae). *Evolution*, **34**, 934–43.
Schmitt, J. (1983). Density-dependent pollinator foraging, flowering phenology, and temporal pollen dispersal patterns in *Linanthus bicolor*. *Evolution*, **37**, 1247–57.
Schonewald-Cox, C.M., Chambers, S.M., MacBryde, B. & Thomas, L. (eds) (1983). *Genetics and Conservation. A reference for Managing Wild Animal and Plant Populations*. Benjamin Cummings, Menlo Park, California.
Schwaegerle, K.E. & Schaal, B.A. (1979). Genetic variability and founder effect in the pitcher plant *Sarracenia purpurea* L. *Evolution*, **33**, 1210–8.
Shaffer, M.L. (1981). Minimum population sizes for species conservation. *BioScience*, **31**, 131–4.
Shaffer, M.L. (1987). Minimum viable population: coping with uncertainty. In *Viable Populations for Conservation*, ed. M.E. Soulé. Cambridge University Press, Cambridge, pp. 69–86.
Shields, W.M. (1982). *Philopatry, Inbreeding, and the Evolution of Sex*. State University of New York Press, Albany, New York.
Silén, R.R. (1962). Pollen dispersal considerations for Douglas-fir. *J. For.*, **60**, 790–5.
Simberloff, D. (1988). The contribution of population and community biology to conservation science. *Ann. Rev. Ecol. Syst.*, **19**, 473–511.
Slatkin, M. (1977). Gene flow and genetic drift in a species subject to frequent local extinctions. *Theoret. Popul. Biol.*, **12**, 253–62.
Slatkin, M. (1980). The distribution of mutant alleles in a subdivided population. *Genetics*, **95**, 503–23.
Slatkin, M. (1981). Estimating levels of gene flow in natural populations. *Genetics*, **99**, 323–35.
Slatkin, M. (1985*a*). Gene flow in natural populations. *Ann. Rev. Ecol. Syst.*, **16**, 393–430.
Slatkin, M. (1985*b*). Rare alleles as indicators of gene flow. *Evolution*, **39**, 53–65.

Slatkin, M. (1987). Gene flow and geographic structure of natural populations. *Science, N.Y.*, **236**, 787–92.
Snogerup, S. (1967). Studies in the Aegean flora, IX. *Erysimum* sect. *Cheiranthus*, B. Variation and evolution in the small-population system. *Opera Bot.*, **14**.
Snogerup, S., Gustafsson, M. & von Bothmer, R. (1990). *Brassica* sect. *Brassica* (Brassicaceae), I. Taxonomy and variation. *Willdenowia*, **19**, 271–365.
Sokal, R.R. & Oden, N.C. (1978). Spatial autocorrelation in biology, 2. Some biological implications and four applications of evolutionary and ecological interest. *Biol. J. Linn. Soc.*, **10**, 229–49.
Sokal, R.R. & Wartenberg, D.E. (1983). A test of spatial autocorrelation analysis using an isolation-by-distance model. *Genetics*, **105**, 219–37.
Soltis, D.E. (1982). Allozyme variability in *Sullivantia* (Saxifragaceae). *Syst. Bot.*, **7**, 26–34.
Soulé, M.E. (1985). What is conservation biology? *BioScience*, **35**, 727–34.
Soulé, M.E. (ed.) (1987). *Viable Populations for Conservation.* Cambridge University Press, Cambridge.
Soulé, M.E. & Wilcox, B.A. (eds) (1980). *Conservation Biology. An Evolutionary–Ecological Perspective.* Sinauer Associates, Sunderland, Massachusetts.
Stebbins, G.L. (1957). Self-fertilization and population variability in the higher plants. *Amer. Nat.*, **91**, 337–54.
Stratton, D.A., Cruzan, M.B. & Thomson, J.D. (1985). Pollen tube growth rate and outcrossing distance in *Erythronium grandiflorum. Amer. J. Bot.*, **72**, 866.
Strauss, S.H. (1987). Heterozygosity and developmental stability under inbreeding and crossbreeding in *Pinus attenuata. Evolution*, **41**, 331–9.
Sultan, S.E. (1987). Evolutionary implications of phenotypic plasticity in plants. *Evol. Biol.*, **21**, 127–78.
Svensson, L. (1985). An estimate of pollen carryover by ants in a natural population of *Scleranthus perennis* L. (Caryophyllaceae). *Oecologia* (*Berl.*), **66**, 373–7.
Svensson, L. (1988). Inbreeding crossing and variation in stamen number in *Scleranthus annuus* (Caryophyllaceae), a selfing annual. *Evolutionary Trends in Plants*, **2**; 31–7.
Svensson, L. (1990). Distance-dependent regulation of stamen number in crosses of *Scleranthus annuus* from a discontinuous population. *Amer. J. Bot.*, **77**, 889–96.
Sytsma, K.J. & Schaal, B.A. (1985). Genetic variation, differentiation, and evolution in a species complex of tropical shrubs based on isozymic data. *Evolution*, **39**, 582–93.
Templeton, A.R. (1986). Coadaptation and outbreeding depression. In *Conservation Biology, The Science of Scarcity and Diversity*, ed. M.E. Soulé. Sinauer Associates, Sunderland, Massachusetts, pp. 105–16.
Thomson, J.D. & Plowright, R.C. (1980). Pollen carryover, nectar rewards and pollinator behavior with special reference to *Diervilla lonicera. Oecologia* (*Berl.*), **46**, 68–74.
Turner, M.E., Stephens, J.C. & Anderson, W.W. (1982). Homozygosity and patch structure in plant populations as a result of nearest-neighbor pollination. *Proc. Natl Acad. Sci. USA*, **79**, 203–7.

Wade, M.J. & McCauley, D.E. (1988). Extinction and recolonization: their effects on the genetic differentiation of local populations. *Evolution*, **42**, 995–1005.

Waller, D.M., O'Malley, D.M. & Gawler, S.C. (1987). Genetic variation in the extreme endemic *Pedicularis furbishiae* (Scrophulariaceae). *Conserv. Biol.*, **1**, 335–40.

Warwick, S.I. & Gottlieb, L.D. (1985). Genetic divergence and geographic speciation in *Layia* (Compositae). *Evolution*, **39**, 1236–41.

Waser, N.M. (1987). Spatial genetic heterogeneity in a population of the montane perennial plant *Delphinium nelsonii*. *Heredity*, **58**, 249–56.

Waser, N.M. (in press). Population structure, optimal outbreeding, and assortative mating in angiosperms. In *The Natural History of Inbreeding and Outbreeding*, ed. N.W. Thornhill. University of Chicago Press, Chicago, Illinois.

Waser, N.M. & Price, M.V. (1982). A comparison of pollen and fluorescent dye carry-over by natural pollinators of *Ipomopsis aggregata* (Polemoniaceae). *Ecology*, **63**, 1168–72.

Waser, N.M. & Price, M.V. (1983). Optimal and actual outcrossing in plants, and the nature of plant pollinator interaction. In *Handbook of Experimental Pollination Biology*, ed. C.E. Jones & R.J. Little. Scientific and Academic Editions, Van Nostrand, New York, pp. 341–59.

Waser, N.M. & Price, M.V. (1989). Optimal outcrossing in *Ipomopsis aggregata*: seed set and offspring fitness. *Evolution*, **43**, 1097–109.

Waser, N.M., Price, M.V., Montalvo, A.M. & Gray, R.N. (1987). Female mate choice in a perennial herbaceous wildflower, *Delphinium nelsonii*. *Evolutionary Trends in Plants*, **1**, 29–33.

Watt, W.B. (1985). Bioenergetics and evolutionary genetics: opportunities for new synthesis. *Amer. Nat.*, **125**, 118–43.

Widén, B. (1987). Population biology of *Senecio integrifolius* (Compositae), a rare plant in Sweden. *Nord. J. Bot.*, **7**, 687–704.

Widen, B. & Widén, M. (1990). Pollen limitation and distance-dependent fecundity in females of the clonal gynodioecious herb *Glechoma hederacea* (Lamiaceae). *Oecologia* (*Berl.*), **83**, 191–6.

Wright, J.W. (1953). Pollen-dispersion studies: some practical applications. *J. For.*, **51**, 114–8.

Wright, S. (1931). Evolution in Mendelian populations. *Genetics*, **16**, 97–159.

Wright, S. (1943*a*). Isolation by distance. *Genetics*, **28**, 114–38.

Wright, S. (1943*b*). An analysis of local variability of flower color in *Linanthus parryae*. *Genetics*, **28**, 139–56.

Wright, S. (1946). Isolation by distance under diverse systems of mating. *Genetics*, **31**, 39–59.

Wright, S. (1951). The genetical structure of populations. *Ann. Eugenics*, **15**, 323–54.

Wright, S. (1965). The interpretation of population structure by *F*-statistics with special regard to systems of mating. *Evolution*, **10**, 395–420.

Wright, S. (1969). *Evolution and the Genetics of Populations, Vol. 2. The Theory of Gene Frequencies*. University of Chicago Press, Chicago, Illinois.

Wright, S. (1977). *Evolution and the Genetics of Natural Populations, Vol. 3.*

Experimental Results and Evolutionary Deductions. University of Chicago Press, Chicago.
Wright, S. (1978). *Evolution and the Genetics of Populations, Vol. 4. Variability Within and Among Natural Populations.* University of Chicago Press, Chicago, Illinois.
Wyatt, R. (1988). Phylogenetic aspects of the evolution of self-pollination. In *Plant Evolutionary Biology*, ed. L.D. Gottlieb & S.K. Jain. Chapman & Hall, London, pp. 109–31.
Yeh, F.C. & Layton, C. (1979). The organisation of genetic variability in central and marginal populations of lodgepole pine *Pinus contorta* ssp. *latifolia. Can. J. Genet. Cytol.*, **21**, 487–503.
Zimmerman, M. (1982). The effect of nectar production on neighborhood size. *Oecologia* (*Berl.*), **52**, 104–8.

Chapter 5

The Ecology of Dispersal in Relation to Conservation

LENNART HANSSON

Department of Wildlife Ecology, Swedish University of Agricultural Sciences, S-750 07 Uppsala, Sweden

LARS SÖDERSTRÖM

Department of Plant Ecology, University of Umeå, S-901 83 Umeå, Sweden

CHRISTER SOLBRECK

Department of Plant and Forest Protection, Swedish University of Agricultural Sciences, S-750 07 Uppsala, Sweden

5.1 INTRODUCTION

Small populations easily become extinct (Goodman, 1987). Small populations may also be affected by genetic drift and inbreeding (Schonewald-Cox *et al.*, 1983), although the importance of these latter effects on population dynamics is less clear (Ehrlich, 1983; Shaffer, 1987; see also chapter 4). The adverse demographic and genetic processes in small populations are both counteracted by immigration. Such immigration should keep the density over a certain threshold level (Brown & Kodric-Brown, 1977) or supply at least one reproducing individual per generation (Schonewald-Cox *et al.*, 1983) to circumvent extinction or loss of genetic variability. However, immigration of foreign competing or predating species may have negative effects on local and circumscribed communities (Helle & Järvinen, 1986).

Most boreal and temperate plant and animal species live in habitats which change over short or moderate time-scales. As old habitats deteriorate relative to the needs of the organism, new ones have to be colonized. In many species, movements and colonization of new habitats have to take place in every generation, and in some cases more than one dispersal period is necessary during the life cycle. This chapter deals with movements which may occur on one or more occasions during the life span and which result in a permanent change of habitat site or home range. Such movements are usually called 'dispersal', while the term 'migration' is used for two-way movements. Some ecologists consider dispersal as one type or effect of migration.

For animals, we are thus concerned with the movements which occur outside the organisms' normal daily excursions to feeding stations and other points of vital needs. For the purposes of definition, the absolute length of the movements or direction or means of transport are thus not of prime interest. In animals the dispersal may be a matter of a few metres' transport in all directions in some species while other species may move intercontinentally.

In plants, these movements may involve anything from diaspores landing a few centimetres from the mother colony to transoceanic dispersal. The scale of dispersal studies varies in the same way. If diaspores are dispersed only short distances, they fall within the limits of the population under study and contribute mainly to population growth and persistence. Distance dispersal should be defined as when diaspores are transported over greater distances than the population under study extends. The scale of the populations (e.g. on a decaying log or within a forest) will then determine the limits between within-population dispersal and distance dispersal. This chapter deals mainly with distance dispersal.

Plant and animal species display many kinds of adaptations that facilitate dispersal and the spatial tracking of favourable habitat patches. Adaptive between-habitat movements have been described under various terms, such as migration or dispersal (Rankin, 1985). Unfortunately, the vocabulary is often confusing, with different terminologies employed for different groups of organisms and emphasis put on different aspects of the process of movement (see, for example, Taylor & Taylor, 1983; Kennedy, 1985; Taylor, 1986). Part of the confusion stems from not distinguishing dispersal or migration as an individual trait (characterized physiologically and behaviourally) from dispersal or migration as a population process (Fig. 5.1). The terms 'migration' (a concept often used by entomologists also for one-way movements) and 'distance dispersal' (for plants) demonstrate the semantic confusion; we will here instead compare

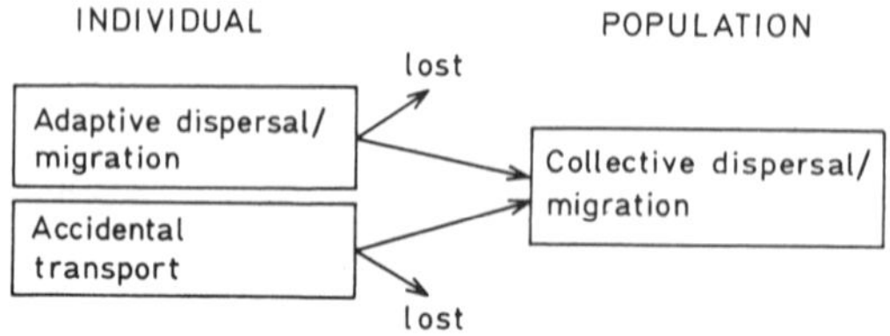

Fig. 5.1. Relationships between dispersal or migration as individual and population concepts.

mechanisms and effects in population dynamics of various groups of organisms in order to approach some kind of general understanding of the importance of dispersal in conservation.

At the individual level, dispersal is a process usually confined to a specific phase in the life cycle of an organism. Thus, most plants disperse in the seed or spore stage, many sessile marine organisms as planktonic larvae, and most insects by flight as adults prior to reproduction. Other groups, such as mammals, may disperse both as young (natal dispersal) and as adults (breeding dispersal). The tendency to migrate or disperse often varies widely between individuals. This variation may be due to genetic factors as well as to responses to environmental signals. The relative roles of genes and environment differ considerably between species, as shown, for example, by numerous studies of insects (see below).

Dispersal as a concept in population biology differs somewhat from that of individual dispersal (Fig. 5.1). Not all presumptive dispersers (e.g. as genetically defined) will be able to move at all. Others that start a journey will fail to colonize suitable habitats. Furthermore, colonization will sometimes occur as a result of accidental transport of organisms. Hence population dispersal is not only performed by individuals according to adaptive behaviour. The immigrants recorded by a population biologist are thus a subset of the presumptive migrants often mixed with accidentally transported 'non-migrants'. In some cases 'non-adaptive' dispersal may be an important (or the only) component of collective dispersal. In most cases, however, it seems as though dispersal in populations is mainly the result of adaptive behaviour (individual dispersal). The individual and population concepts thus often overlap to a large extent.

A further consequence of migration usually being adaptive is that dispersing individuals are often different from non-moving conspecifics, both those left behind and those joined. These differences (whether genotypic or only phenotypic) may pertain, for example, to age, size, reproductive potential or health. Thus, at the collective (population) level, dispersal may not only have immediate effects on numbers in local patches due to departures and arrivals. When migrants are qualitatively different, the future birth and death schedules of local populations will also be affected. Adaptive dispersal behaviour will thus have both immediate and delayed effects on local populations, and it will connect population processes in different habitat patches.

In population studies emphasis has long been placed on examining birth and death processes in local habitat patches. Although the possible importance of interpatch movements has long been realized (e.g. Andrewartha & Birch, 1954), these problems have only recently been the focus of more intense studies. Studies have proceeded at a very different pace in various organism groups and with emphasis on different aspects of dispersal or migration. As it is difficult to present a unified view of dispersal covering all organism groups, we concentrate here on three types, in which dispersal studies have been performed with some success. They are bryophytes and lichens, which disperse passively both as sexually and asexually produced diaspores, insects with usually distinct ontogenetic dispersal phases and small mammals where population dynamics has appeared to play a fundamental role in releasing dispersal and permitting settlement. Seed dispersal in vascular plants has been reviewed by Howe and Smallwood (1982) and dispersal in many other animal groups in Rankin (1985).

Due to the heterogeneity of approaches we will make only very broad comparisons. The topics covered are primarily:

(1) Morphological, physiological, behavioural and life history adaptations to dispersal.
(2) Variability in dispersal within and between species, and ultimate and proximate factors determining this variability.
(3) The role of dispersal in population processes.
(4) The importance of considering dispersal for conservation purposes.

We provide specific examples on the importance of dispersal for nature conservation in the three organism groups reviewed. Finally we discuss some common dispersal features to be pursued in conservation management and future research.

5.2 BRYOPHYTES AND LICHENS

5.2.1 General background

Plants are sessile organisms and dispersal occurs usually in connection with reproduction (asexual or sexual). Exceptions are, for example, some species floating on the water surface (such as *Lemna* spp., *Ricciocarpos natans* and some *Riccia* spp.) which may move along river systems. Dispersal is mostly passive. However, many plants have developed special tissue by which they disperse better with wind (e.g. wings on maple seeds, pappus on many Asteraceae, etc.), attach better to animals (e.g. hooked spines in *Galium aparine*, sticky substances in *Plantago major*) or may be internally transported in animals (many berries). In some plants, mechanisms of more or less active ejection occur, e.g. the sudden rupture of capsules in *Sphagnum* species. Dispersal devices (diaspores) may be produced either sexually or asexually.

The distance dispersal ability in plants depends on (1) the number of diaspores produced; (2) how efficiently diaspores are transported to new localities; and (3) how easily diaspores will establish in new localities.

5.2.2 Diaspore production

Bryophytes and lichens produce both spores and several types of asexual propagules. In bryophytes the number of spores per capsule generally varies between 20 000 and 1 600 000 but values up to 50–80 millions spores/capsule and down to 16 spores/capsule have been recorded. Within species, there are substantial variations in spore numbers, e.g. between localities, between years (Longton & Miles, 1982), between colonies (Jonsson & Söderström, 1988) and in populations in relation to capsule size (Kreulen, 1972).

A vast majority of all hepatics and a large proportion of mosses have asexually produced diaspores, e.g. gemmae, bulbils and tubers. Many bryophytes, especially mosses, also have the possibility to regrow from fragments (Longton & Schuster, 1983). The importance of asexually produced diaspores is considerable. First, populations unable to produce gametangia may, theoretically, maintain clonal populations indefinitely. Eventually, favourable conditions for sexual reproduction may arise. Secondly, the production of asexual diaspores may maintain unisexual taxa indefinitely. Unisexual colonies may disperse by gemmae and eventually come in contact with a clone of the other sex and be able to form sporophytes. A large number of unisexual taxa rely principally or entirely on asexual reproductive devices for dispersal.

Lichens consist of two different organisms living in symbiosis, a fungus and an alga. The sexually produced diaspores are fungal spores. To be able to form a new lichen thallus, these spores have to establish contact with an appropriate alga. Lichens also produce several types of asexual diaspores, which consist both of the fungus and the alga so that the entire lichen symbiosis is dispersed. The most common asexual propagules are soredia and isidia. Thallus fragments (differentiated or undifferentiated) of various sizes are also important dispersal agents (Bowler & Rundel, 1975).

5.2.3 Diaspore transport

Some bryophytes and lichens are dispersed by animals (Bailey, 1966; van Zanten & Pócs, 1981) but the majority are wind-dispersed. The transport includes liberation of diaspores from capsules, apothecia or thallus/gametophyte, movement to a new locality and deposition on a new substrate. In bryophytes, most mechanisms tend to release spores during relatively dry conditions favourable for wind dispersal. In lichens, the release of diaspores, mostly soredia, are most efficient in moist weather (Bailey, 1966). Windstorms followed by rain are probably the optimum conditions for dispersal of pendulous forest lichens.

If dispersal is limited, more spores will establish close to the mother patch and a clumped pattern will arise. *Ptilidium pulcherrimum* has a clumped pattern over neighbourhood distances up to at least 25 m (Söderström & Jonsson, 1989) and in *Orthodontium lineare* the clumping extends for about 100 m (Hedenäs *et al.*, 1989*a*), which indicates limitation by dispersal.

A large part of the bryophyte spores are deposited within a few metres from the source (Wyatt, 1982; McQueen, 1986; Miles & Longton, 1987; Söderström & Jonsson, 1989) or even within the parent colony, as shown for *Grimmia pulvinata* (Longton & Miles, 1982). The proportion of spores dispersed over longer distances increases with decreased particle size, greater liberation height, and increased turbulence. Wind velocity may be an important variable determining spore dispersal distances, as indicated by Miles and Longton (1987). Stronger winds may catch more spores and disperse them over greater distances. Another important variable determining dispersal ability, at least on a quantitative basis, may be the spore size. It is generally assumed that small, durable diaspores can be successfully dispersed over long distances. Since bryophyte and lichen spores are usually small (average 9–25 μm in hepatics, 10–15 μm in mosses and 1–30 μm in lichens), it has often been assumed that cryptogam species have almost limitless dispersal abilities. Larger spores and most asexual propagules, which usually are larger, are supposed to disperse step-by-step and contribute more to population growth and stability than to dispersal. However, Schuster (1983) argues that large diaspore size *per se* is not enough to prohibit efficient dispersal, at least not for single occasions. Both small one-celled and large pluricelled gemmae can, theoretically, undergo long-distance dispersal, the smaller due to their size and the larger, discoid, gemmae may, due to their shape, be as effectively wind-disseminated. Larger diaspores, such as bulbils, gemmae and bryophyte fragments, have been trapped in large quantities from air at various places (e.g. Pettersson, 1940; Miller & Ambrose, 1976). An advantage of large diaspores is that they impact more efficiently against obstructions than small ones (Gregory, 1973). Even though spores may be transported over larger distances, survival may still be a problem. Van Zanten (1978) demonstrated a marked correlation between species range and survival of spores under conditions of frost and desiccation.

Many crustose lichens rely entirely on spores for dispersal since they lack soredia and isidia, but they are still widely dispersed (Bowler & Rundel, 1975). However, in many lichens the production of spores seems to be less important for dispersal than the production of the much larger soredia and isidia. Two examples illustrate this. When old-growth forests are isolated in small 'islands', as in Great Britain, some lichens become restricted to those sites (Rose, 1976). Rare, old-forest species in Britain mostly have spores, while common species have soredia or isidia, which indicates a dispersal advantage for those propagules. Of the 18 most frequent species on deciduous trees in southern Wisconsin 13 bore soredia or isidia. Of the 18 rarest species, 16 lacked any obvious vegetative diaspores (Hale, 1983).

In pendulous lichens, fragments seem to be the most important propagule. All *Alectoria* thalli from two-year-old twigs regenerated from fragments while a few thalli regenerated from soredia or spores on twigs up to 14 years old. In litter fall, small fragments of *Alectoria sarmentosa* and *Bryoria* spp. dominate, while *Usnea longissima* is present with comparatively large fragments (up to some decimetres) (Esseen, 1985). Fragments of *Usnea longissima* occur in local isolated populations while *Alectoria* and *Bryoria* species are more widely distributed (cf. Esseen *et al.*, 1989). Beyond a distance of about 400 m from the edge of a mature forest, young twigs support almost no lichens (Stevenson, 1988).

5.2.4 Establishment

Establishment depends on the number of viable diaspores deposited on a suitable substrate and their ability to germinate and compete with neighbours. In the epyxilic hepatic *Ptilidium pulcherrimum* 0·5–0·9% of all spores produced are deposited on suitable substrates (Söderström & Jonsson, 1989). The longevity of spores varies considerably, some epiphytic hepatics surviving only a few hours (Fulford, 1951) while other species may remain viable for several years and the proportion of viable spores may decrease with time. Germination of *Funaria hygrometrica* spores stored at room temperature in a capped jar fell, for example, from 100% when freshly collected to 5% after 11 years (Hoffman, 1970). When germinated, the competition with neighbours starts. There are indications that mature bryophytes suppress the growth of protonema through release of chemical substances (Watson, 1981).

5.2.5 Dispersal in time

Many plants may survive severe conditions by staying dormant, mainly in the stage of viable seeds, spores or asexually produced propagules. The strategy is best demonstrated for habitats with short duration and recurring predictably at the same place. Instead of dispersing the diaspores in space, a plant may just drop them and let them stay dormant at the same place until suitable conditions arise again. The seed bank of vascular plants has received considerable attention (e.g. Moore, 1980; Granström, 1982). B.G. Jonsson (pers. comm.) was able to grow 22 mosses and 13 hepatics from soil samples taken from a spruce forest. The disturbed patches in a forest, mainly from uprooted trees, are rapidly recolonized and the species composition differs significantly from the surrounding ground (Jonsson & Esseen, 1990). However, it is similar to the composition of the diaspore bank, which indicates that colonist species and species occurring early in the succession may survive locally by means of locally produced, dormant diaspores. Other investigations (e.g. During & ter Horst, 1983; During *et al.*, 1987; Smith, 1987) have shown that the diaspore bank consists largely of asexual propagules (tubers, protonema-gemmae, etc.) and stem fragments.

5.2.6 Regional dynamics and dispersal

The rate of new establishments and the duration of individuals or colonies determine the population size of plant species. Most bryophyte communities studied are highly dynamic (Kimmel, 1962; During & ter Horst, 1987; Herben, 1987). Preliminary results (Söderström, unpublished data) show that 17 out of 18 investigated epixylic bryophytes changed their abundance with an average of at least 80% per log (some increasing, some decreasing) during a three-year period while *Ptilidium pulcherrimum*, a seemingly stable species, changed, on average, 64% per log.

Habitat requirements (Söderström, 1988*a*) and competition (Söderström, 1990*a*) are insufficient to explain the distributions of epixylic species. Dispersal ability is postulated to be important (Söderström, 1987). Species studied showed two distinct patterns of regional distribution, some were regionally frequent and also occurred often on other (suboptimal) substrates such as boulders, stumps, litter, etc. Other species were regionally infrequent and could be further separated in two patterns with regard to their local abundances. Locally infrequent species occurred only occasionally on suboptimal substrates and locally frequent species occurred frequently on suboptimal substrates.

Hanski (1982) observed distinct patterns in species occurrence when considering only localities suitable for the species. These patterns correspond rather well with those mentioned above except that regionally infrequent but locally abundant species are lacking in his hypothesis. Following Hanski's view, Söderström (1989, 1990*b*) expanded this hypothesis to circumscribe four distribution patterns (Fig. 5.2). *Core* species (abundant on almost all available localities) may occur if the populations are stable and/or the distance between localities is less than the efficient dispersal distance, i.e. the dispersal must be efficient both over longer and shorter distances. Evidence for an efficient establishment

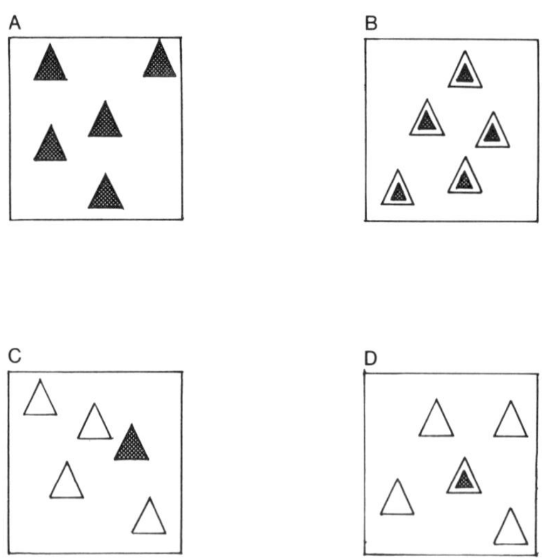

Fig. 5.2. Distribution patterns, e.g. among epixyllic mosses, related to various dispersal capabilities. ▲, locally abundant; ◬, locally sparse; △, locally absent populations; (A) core species; (B) rural species; (C) urban species; (D) satellite species.

ability is the occurrence of species on suboptimal substrates. *Urban* species (abundant but only on a few of the available localities) may exist if the environment is coarse-grained in relation to the dispersal ability. If the diaspore production is large and diaspores are easily dispersed over shorter distances, the species may maintain their populations locally. But if the diaspores are poorly dispersed over longer distances they can rarely receive diaspores from outside. Some populations will die out, even if the populations generally are large, and these localities are recolonized very slowly. *Satellite* species (rare and only on a few of the available localities) are poorly dispersed both locally and regionally. At the edge of their geographical distribution area, diaspore production and favourable establishment events may be so irregular that only a few diaspores are successful. Single successful colonizations may result in small populations which, however, are sensitive to extinction, and often are temporary. There may also be stages close to extinction, for example due to the fragmentation of habitats where the distances are so large that dispersal is prevented and the remaining pieces of habitat too small to hold a viable population. The occurrence of the fourth pattern, *rural* species (rare but occurring on almost all available localities), is dubious. If the long-distance dispersal is efficient, then the short-distance dispersal should work equally efficiently.

5.2.7 Conservation aspects of dispersal

Two important questions in nature conservation today are the consequences of habitat fragmentation and invasion of new species. In both, dispersal ability plays a major role. In common for both decreasing and increasing habitats is their patchy distribution. For many species, the new man-made habitat is unsuitable (or at least less suitable). For instance, clear-felling of natural spruce forests and transferring the land to managed timber-producing stands has a negative impact on several bryophyte species (Söderström, 1988*b*). Other species may take advantage of the increased disturbance regimes.

Fragmentation will have an impact on populations in three ways. (1) The area of each locality decreases, which will reduce the amount of temporal substrates. If the number of new habitat patches created per unit area is low, situations of substrate deficit may arise. (2) The local population size decreases, which will reduce the diaspore production. (3) The distances between the localities increase. This means that the extinction rates increase due to smaller populations and less predictable substrate availability. In addition, the diaspore dispersal becomes limited due to fewer diaspores being produced and because each diaspore has to travel longer.

In order to be able to invade, a species has to overcome a barrier and reach an environment which is open for invasion. The most suitable habitats are communities with a low average level of species saturation (e.g. sparse plant cover) and communities subject to frequent disturbance (Crawley, 1987). Invasion into a natural community is usually difficult since there are few vacant niches. Native species are also often better competitors in natural (undisturbed) habitats. Although Britain, for example, has many alien plant species, there is no evidence that any native plants have been excluded from any community by competition with aliens (Crawley, 1987).

One example of an invading species which is relatively well studied is the moss *Orthodontium lineare*, a species invading spruce plantations in southwestern Sweden. It grows on substrates also used by native species but in no case does it outcompete them (Hedenäs *et al.*, 1989*b*). *O. lineare* was introduced in England in 1911 and was first recorded in Sweden in 1969. The species now has different regional patterns in different parts of the Swedish distribution area (Hedenäs *et al.*, 1989*a*). In northern Skåne (south Sweden) the density of suitable localities is high and *O. lineare* occurs as a core species. In western Småland and in northern Halland, a little further to the north, the available localities are more scattered and *O. lineare* occurs as a satellite species. Two factors may be responsible for this, the distances between suitable localities and the possibility that invasion is not yet completed. In addition, there are some outpost localities in Västergötland, where available habitats are sparse and unevenly distributed. In some localities *O. lineare* occurs sparsely and may be a recent invader. In one such locality on the other hand, it occurs abundantly, perhaps because this locality is rather isolated—which may explain why the moss has not dispersed in the neighbourhood even though it is a rather old site.

5.3 INSECTS

5.3.1 Flight and migration in insects

Insects are generally characterized by high motility. This is due to most species being winged and able to fly, at least during a short period in their life cycle. Insects may also travel by other means. For example, young larvae of moth species may ride the wind on silken threads (ballooning, e.g. Berger, 1989). Swimming insects may be transported by streams, and terrestrial insects are sometimes carried on the water surface of lakes or the sea (Palmén, 1944). Insects may also travel moderate distances on foot. But these examples do not detract from the generalization that flight is their basic method for medium and long-range transport.

There are two types of flight movements among insects. *Migratory flights*, which are of a straightened-out and uninterrupted nature (Kennedy, 1985), enable the insect to move to new habitats for breeding or diapause during a certain phase of its life cycle. Shorter, easily interrupted, flights that aid in everyday movements within habitats, such as those between feeding stations, are called *trivial flights*. Although this distinction is useful it should be remembered that migratory and trivial flights grade into each other. Thus what starts as a typical migratory flight may end as a trivial one. For example, in aphid and butterfly species recently enclosed adults embark on straightened-out migratory flights, during which they do not respond to feeding stimuli. After flying for an hour or so they start visiting plants, the frequency of which increases concomitantly with the flight climb being less steep or flight turns becoming more common (Nielsen, 1961; Kennedy, 1985). The same insect may thus undertake both migratory and trivial flights. Walking and swimming may occasionally, and ballooning quite often, serve the function of migration, that is to take the insect to new habitats.

The study of insect migration has a strong observational and experimental basis (Glick, 1939; Williams, 1958; Johnson, 1969; Rainey, 1976; Dingle, 1978, 1984, 1985*a*; Rankin 1985; Danthanarayana, 1986). Movements of several species have been thoroughly studied in the field. It is well known that many insects regularly appear hundreds of metres above the ground. This has been demonstrated by collections made from towers (Fig. 5.3), balloons, kites and aircraft (Glick, 1939; Johnson, 1969; Farrow & Dowse, 1984) as well as from radar studies (Schaefer, 1976; Rainey, 1989). The insects collected are mainly winged adults, which are alive and in good condition. They are not accidentally displaced individuals, but are there as a result of adaptive behaviour, and they are migrating, usually downwind, to colonize new habitats. Other insects migrate at low altitudes, which gives greater opportunity for moving against the direction of the wind.

5.3.2 Two examples of migrating insects

There is great variation between insect species with regard to behaviour during migration and the tracks followed by migrants. Two examples will illustrate this.

The large pine weevil *Hylobius abietis* starts its flight by taking off

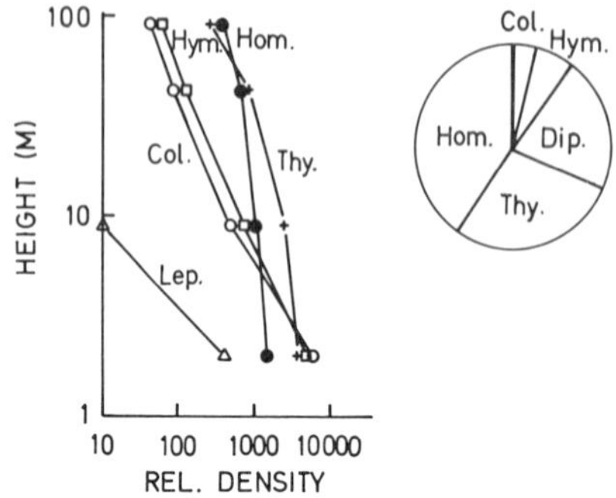

Fig. 5.3. Aerial densities of some insect orders at different heights above ground in a forested area in Värmland, Sweden, recorded by suction traps on a TV tower (2–12 June 1980). The circle shows the faunal composition of all insect groups collected at 93 m. Col., Coleoptera; Dip., Diptera; Hom., Homoptera; Hym., Hymenoptera; Lep., Lepidoptera; Thy., Thysanoptera.

against the wind. Soon, however, it turns downwind, flying upwards until well above tree-top height (Solbreck, 1980). Flying weevils have been collected more than 90 m above ground and probably also occur much higher up (Solbreck, 1985). The insect is then transported at the combined speed of its own flight and that of the wind. When encountering odours from damaged conifer trees it turns downward and against the wind, searching for suitable breeding material such as fresh conifer stumps. The flight usually does not last more than 5–30 minutes, but with the aid of wind transport this will carry the insect several kilometres. Migratory flights are repeated if the insect does not find suitable breeding material. By such repeated flights some individuals are likely to travel more than 100 km (Solbreck, 1980).

The North American monarch butterfly *Danaus plexippus* migrates on a totally different scale. In the autumn, butterflies leave breeding areas as far north as southeastern Canada to travel to overwintering sites in central Mexico, a distance of several thousand kilometres (Urquhart & Urquhart, 1978, 1979). Migration thus occurs in a specific compass direction, and does not depend on wind direction, as in the pine weevil. If winds are towards the southeast, monarchs circle on thermal updrafts (like many large migrating birds) until they have gained considerable altitudes. They then travel downwind. In crosswinds, however, they fly closer to the ground which allows them to compensate for drift. Finally, in headwinds they usually do not migrate at all (Gibo, 1986). Return flights are made by the same individuals in the next spring, but less is known about behaviour during these flights.

As these examples illustrate, flight performance varies widely between species. These differences concern number of flights, existence of return flights, flight altitude, distance, direction and degree of active control of flight track. Insects seem to be totally opportunistic with regards to the means of flight employed. There are, thus, good reasons for *not* defining migration in terms of, for example, distances, directions, orientation mechanisms or the existence of return flights (cf. Kennedy, 1985). However, migration in different species also has much in common, particularly with regard to this phenomenon as a life cycle phase, with specific behaviours and physiology.

5.3.3 Migration in the life of the individual

Migratory flights occur during a period in adult life when the insect's body functions are concentrated on maximum use of the flight apparatus. The newly eclosed adult passes through a maturation (teneral) period during which the cuticle develops and hardens. This may last from some hours to several days depending on temperature and body size. Flight muscles also develop during this period. The insect is hence not prepared for extended flight until the end of the teneral period, but before reproduction starts (Johnson, 1969; Dingle; 1985*a*, Fig. 5.4).

In most insects there is an antagonistic relationship between migration and reproduction, which is controlled by the hormonal system (Rankin *et al.*, 1986). This antagonism is obvious in species where both migratory and non-migratory forms occur. In non-migrants ovarial activity reaches a maximum immediately following the teneral period, whereas in the migratory form it is delayed and the migratory period is 'inserted' between the teneral and reproductive periods (Nielsen, 1961; Johnson, 1969; Fig. 5.4). More complex relationships also occur. Thus, in insects which diapause as adults, reproduction is typically preceded by the sequence of prediapause migration–diapause–postdiapause migration (e.g. Solbreck, 1972, 1978; Fig. 5.4). Because of the strong coupling between migration and reproduction, and because of the effects on survival, migration is also an important life history component.

Behaviourally, migration is characterized by extended rectilinear flights, and by the suppression of reproductive functions and feeding responses. At the start of the migratory flight the insect is, hence, concentrated on 'getting away'. Responses towards stimuli from the final goal tend to be suppressed during the early phases of the migratory flight

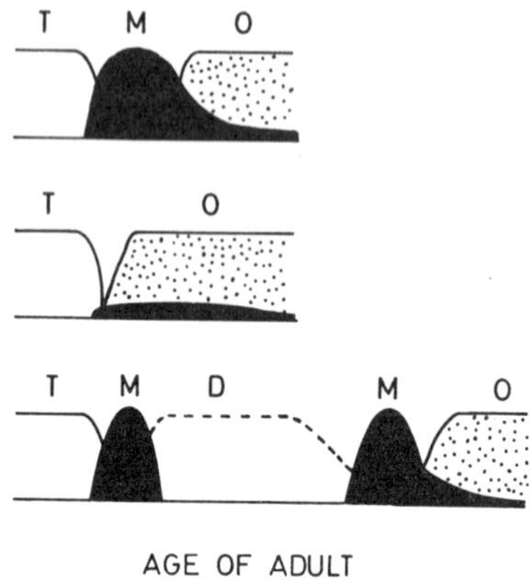

Fig. 5.4. Migratory flights in insects usually occur during a limited period (M) in the life of the adult insect. It is preceded by a maturation period (T) and usually followed by the oviposition period (O) or a dormancy period (D). A life cycle with one migratory period (I) is compared with a non-migratory life cycle (II), where flight activity is much lower and oviposition commences earlier, illustrating the common antagonism between migration and reproduction. The lowermost case (III) shows an insect with both pre- and post-dormancy migration.

(Kennedy, 1985). Migration may thus be regarded as a physiological—behavioural—life history state (Dingle, 1985*b*, 1986; Kennedy, 1985).

5.3.4 Variation in the tendency to migrate

The tendency to migrate varies widely between individuals of the same species as well as between species. This variation can be expressed in three different ways. First, only flight behaviour varies. This kind of variation can be tested in laboratory experiments, for example by measuring flight duration of insects tethered to pins or attached to the arms of flight mills, or by measuring the rate of climb in flight chambers (Dingle, 1985*a*).

A second way in which variation may be expressed concerns the internal 'flight apparatus'. Many insect species are dimorphic with regard to flight muscle development. The tendency to migrate can then be measured by dissecting insects and looking at flight muscles. Usually only two distinct conditions occur, viz. fully developed and totally reduced muscles.

Finally, variation may also be expressed in the external flight apparatus. In many species there are long-winged and short-winged (or totally wingless) forms. When wings are reduced there are usually also reductions in the size of the thorax, which harbours the flight muscles.

5.3.5 Genetic and environmental determinants of migration

The proximate causes for variation in migration, whether expressed as wing, flight muscle or behavioural polymorphism, are of several kinds. There are examples of strong genetic determinism as well as high phenotypic plasticity.

Laboratory studies of flight duration in migratory species have shown that there is considerable genetic variation for this trait. Frequently, there is also a strong response to directional selection (Caldwell & Hegmann, 1969; Dingle, 1986). Hence it is likely that strong selection may sometimes cause changes in the frequency of migrants over a few generations also under field conditions.

Many insect species respond to environmental signals such as photoperiod, temperature, food quality and quantity, and population density, and change their flight apparatus or migratory behaviour accordingly. Environmental signals affect hormonal levels, which in turn affect ontogenetic development, for example by delaying oogenesis and leaving more time for migration (cf. Fig. 5.5). In most cases, environmental signals are perceived early in life well in advance of the migratory period. In some cases, responses are even carried between generations, so-called maternal effects.

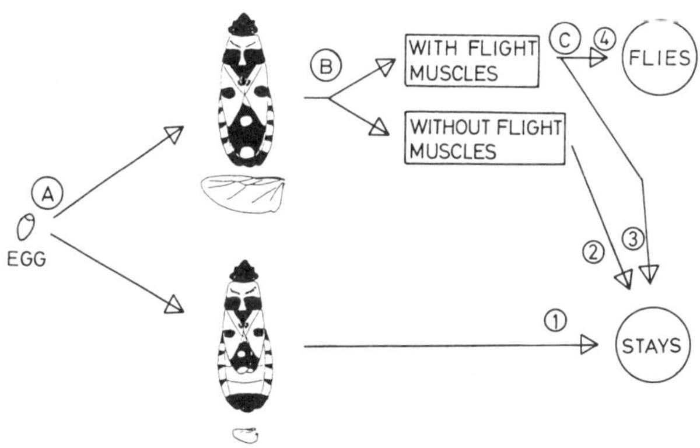

Fig. 5.5. Alternative life cycles in a lygaeid bug (*Horvathiolus gibbicolis*). An individual must develop wings, flight muscles and flight willingness to be able to perform its migratory flight. Wing length is decided (A) by the genetic constitution of an individual. Short-wingedness is determined by a dominant gene. Flight muscle development in long-winged females is environmentally controlled (B). Thus flight muscles develop when females are deprived of either males or food. Flight performance, finally, is likely to be affected by weather conditions (C). All four life cycle alternatives (1–4) will affect survival, development and reproductive patterns, but in different ways (after Solbreck *et al.*, 1990).

An example will demonstrate how both genes and environmental signals may be involved in a chain of events controlling flight polymorphism (Solbreck, 1986; Solbreck *et al.*, 1990). The lygaeid bug *Horvathiolus gibbicollis* (Fig. 5.5), which lives in dry open habitats in the Mediterranean region, feeds on various seeds on the ground. Both long- and short-winged forms occur. Wing length is controlled by a single locus two-allele system with the gene for short wings being dominant. There does not seem to be any environmental modification of this response.

Among long-winged bugs there is, further, a dimorphism among females with regard to wing muscle development. This dimorphism, however, seems to be totally controlled by environmental conditions. At least two factors are involved, namely presence of mating partner and food abundance. At an intermediate level of food abundance, long-winged females switch from laying eggs at a very low rate and maintaining flight muscles to a greatly increased egg production in combination with histolyzing wing muscles.

The above example was chosen to illustrate how complex the control of migratory potential may be. The degree of variation in migratory ability as well as the relative roles of genes and environment in controlling migration varies widely between species, and even between populations of the same species. Environmental conditions which trigger the development of migratory forms may occur at long intervals. As a result, this migratory potential is often underestimated. In certain years with, for example, high population density and favourable weather conditions, migratory rates may increase drastically compared to 'normal' years, a phenomenon which easily goes undetected in short-term studies (cf. Ehrlich & Murphy, 1987; Dempster, 1989).

5.3.6 Migration as part of a life history syndrome

The tendency to migrate is not independent of other life history traits. Migratory ability is associated with a whole complex of other traits, forming a coadapted life history syndrome (Parsons, 1983; Dingle, 1984, 1985*b*, 1986). Differences between life history syndromes of migratory and non-migratory forms are most easily demonstrated in cases with variation in wing development.

When different forms of a wing-dimorphic species are compared, short-winged forms tend to have shorter generation times and higher egg production than long-winged (Roff, 1986; Denno *et al.*, 1989). This is evidently possible as a result of less resources being needed for the flight apparatus, in particular for maintaining wing muscles (Solbreck *et al.*,

1989). There is, however, considerable variation between species concerning the trade-offs involved in changing from long-winged to short-winged morphs. For example, instead of producing more eggs, short-winged forms of some species produce larger eggs, resulting in young larvae with a higher competitive ability (Solbreck, 1986; Solbreck *et al.*, 1990).

When between-species comparisons are made, quite different patterns emerge. Since wing loss is dangerous if habitats deteriorate, it is more likely to evolve in species which have other traits that compensate for the uncertainties of being stationary. Examples of such traits are small size, polyphagy, large eggs, extended oviposition period, slow growth and high starvation tolerance. These are traits that reduce demands for food as well as consequences of starvation periods (Solbreck *et al.*, 1990).

In conclusion, a migrant is not just a good flier. Migratory capability is only one trait in a suite of coadapted characters. Migratory and sedentary species are, hence, likely to be quite different in many traits which are of fundamental importance for their population dynamics and sensitivity to habitat change, among other things. Conservation measures for a particular insect species should not consider its migratory capability in isolation but instead regard it as one of many coordinated life history attributes that affect its survival.

5.3.7 Role of migration in population dynamics

Migration is a fundamental process in many insect populations. Many species migrate to new habitats in every generation because habitats are only suitable for one generation. Examples are insects on carrion, dung, fungal fruiting bodies, dying trees, annual plants, etc. Many other species have to move to microclimatically favourable sites for hibernation and back to breeding areas in spring. Populations are therefore totally relocated twice each year. The scale of such movements varies from tens of metres in some species to thousands of kilometres in others (Johnson, 1969; Danthanarayana, 1986; Rankin, 1985).

Few insect population studies have covered both an extended period of time and an extended spatial scale. This shortage of studies concerning the basic structure and functioning of insect populations is even more pronounced for rare species in natural habitats. Because of this we have very limited information on how patchily distributed populations are connected by migration between habitat patches, and how migration interacts with survival and birth schedules in the separate habitat patches (Dempster, 1989). Often migration is undertaken as a response to changing habitat quality. Insects can thus compensate for deteriorating conditions by moving to other habitat patches. Neglecting migration or equating it with mortality or natality misses a fundamental dynamic property of the population system. Migration is a process by which mortality and natality schedules are affected (Solbreck, 1985).

The population dynamics of the lygaeid bug *Lygaeus equestris* have been studied over several years in a patchy environment in south-eastern Sweden (Solbreck & Sillén-Tullberg, 1990; Sillén-Tullberg & Solbreck, 1990). This insect uses two types of habitats during its yearly cycle. In summer it occurs in small scattered patches of its host plant *Vincetoxicum hirundinaria*, whereas during the rest of the year it stays at hibernation sites, usually crevices in southward-facing rock walls. In spring and autumn, *L. equestris* migrates by flight between these two habitat types, flights that range from tens of metres up to some kilometres (Fig. 5.6).

The percentage of host plant patches inhabited varies considerably between years, with most patches being inhabited in years of general population highs (Solbreck & Sillén-Tullberg, 1990). Large patches are more likely to be colonized than smaller ones. However, habitat quality with regard to microclimatic conditions (degree of sun exposure) is even more important than patch area.

L. equestris populations are thus structured as metapopulations, where local extinctions, migration and recolonization are important events (see Chapter 6). In one study area (comprising 24 small patches over 6 km^2), the metapopulation seemed to persist largely through interpatch movements within the study area during the first ten (1977–86) years of study (and obviously had persisted long before that). Due to a weather-induced population low in late summer 1987 this metapopulation went extinct in the winter 1987–88. No recolonization of any patch occurred in the summer of 1988 despite the fact that food and weather conditions were favourable, because possible source populations were also severely reduced due to weather effects. Recolonization by *L. equestris* succeeded in 1989 after another year of favourable weather had caused a strong increase in source populations. The correlated effects of weather over large areas had thus led to a situation where extinction coincided with a period of much lowered recolonization rates (Solbreck, 1991). Such correlations between rates of extinction and colonization strongly affects long-term survival of metapopulations (Harrison & Quinn, 1989), but few empirical studies have addressed this question.

As shown by *L. equestris* populations, unusual events, for example in weather conditions, may totally alter 'normal' patterns of population change. Periods with metapopulations persisting largely through continu-

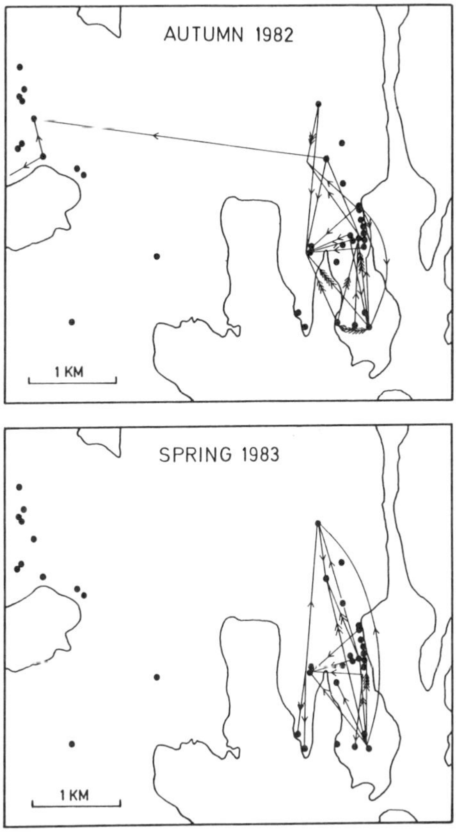

Fig. 5.6. Migrations of the lygaeid bug, *Lygaeus equestris*, before and after overwintering. Host plant patches (ranging in size from 0·5 to 170 m^2) are shown by black dots. Each arrowhead represents the movement of one marked bug. Migration links populations in different patches into an interacting network, a metapopulation (after Solbreck & Sillén-Tullberg, 1990).

ous local migration may be sporadically interrupted by periods of regional extinction when longer range migrations become critical. Unusual events, leading to population bottlenecks or extinctions, are easily overlooked in most population studies (Solbreck, 1991), but they are of particular interest for understanding the dynamics of threatened species.

The spatial scale of changes in distribution patterns varies strongly between species as well as between different time periods. Whereas the distributional changes of *L. equestris* usually involve movements of some kilometres, in other species annually recurrent movements may be on a continental scale. Thus, certain lepidoptera species such as *Vanessa atalanta*, *Cynthia cardui* and *Autographa gamma*, which do not have overwintering populations in Scandinavia, immigrate every year into this area from central or southern Europe. However, the extent of return migrations in the autumn seem less well-developed in these species than in the American monarch butterfly. It is likely that summer populations of *V. atalanta*, *C. cardui* and *A. gamma* in Scandinavia are largely 'sink' populations (cf. Pulliam, 1988).

In other cases, less frequent long-range migrations result in a brief influx or a range extension that may last for some years (Mikkola, 1967, 1971; Ryrholm & Källander, 1987). Several species recorded from Sweden are transients of this kind or represent short-duration colonizations with temporary range extensions. Kaisila (1962) summarized changes in the distribution patterns of butterflies and moths in Finland over a period of 90 years. Range extensions or retreats were often associated with short-term (decades) climatic changes.

Migration thus occurs on all spatial scales with varying effects on short-term population distribution to longer-term biogeographic patterns. A species' geographical range may expand or retreat over decades or centuries. The dependence of local populations on immigrants from outside areas (patches) also varies to a great extent. Some populations may be total 'sink' populations whereas others may be able to persist for very long periods without immigration. When evaluating conservation measures for insect species, we must consider the relative roles of processes on different temporal and spatial scales. The first priority of any population study should be to consider what scales are relevant.

5.3.8 Migration and insect fauna conservation

Migratory ability is an adaptation to survive in temporary and fluctuating environments (Southwood, 1962; Solbreck, 1978). There are, accordingly, large differences in the percentage of migratory (e.g. winged) forms between different kinds of habitats within a geographical area, as well as between different parts of the world with different histories (Darlington, 1943; Lindroth, 1949; Southwood, 1962; den Boer, 1970; Slater, 1975, 1977).

The North European fauna, for example, is a recently established one, where all species have immigrated into the area after the last glaciation. Almost no endemic insect species occur in Northern Europe, but the area seems to be characterized by a high frequency of motile and wide-ranging species. This distinguishes this fauna from that of most previously unglaciated parts of the world.

In the Scandinavian fauna peninsula effects and shortage of time since the last glaciation also seem to be important factors. Plant-feeding insects can only invade after their host plants have become established. Norway spruce, such a common tree today, is a rather recent colonist entering Sweden from the northeast (Hulthén, 1971). Its fauna seems to be impoverished in Sweden (Fig. 5.7), and faunal colonization is evidently still in progress. For example, the spruce-feeding bark beetle *Ips amitinus* is expanding northwards in Finland (Koponen, 1980) and will probably soon reach Sweden.

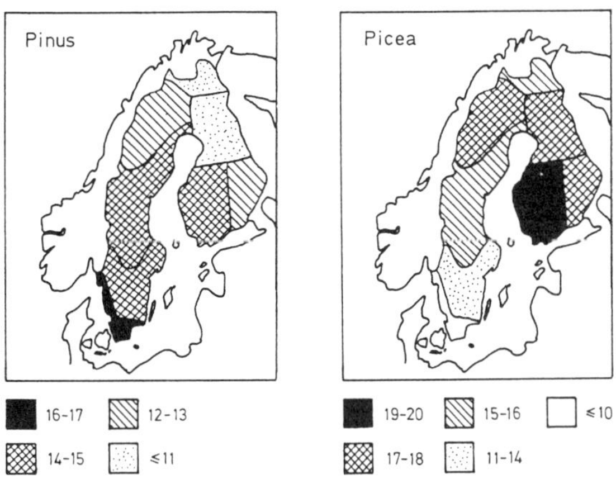

Fig. 5.7. Number of bark beetle species (Scolytidae) on Scots pine *Pinus sylvestris* and Norway spruce *Picea abies* in different parts of Finland and Sweden. The main immigration route of pine is northwards in both Finland and Sweden, whereas spruce moved from southeastern Finland northwards and then southwards through Sweden (data on scolytids from Lekander *et al.*, 1977).

The historical background of a fauna thus seems of importance when discussing conservation measures in different parts of the world. For example, the Scandinavian insect fauna, characterized by lack of endemics, continued colonization, and a high frequency of motile and wide-ranging species, will face different (and probably less severe) conservation problems than most low-latitude areas.

Both individual and collective migration (Fig. 5.1) are concepts of importance when discussing conservation problems. The view of migration as one trait in a coadapted life history syndrome points to the fact that migratory and less migratory species have different suites of life history adaptations. Hence they will respond differently to habitat change and deterioration, and are likely to exhibit considerable differences with regard to population dynamics. Individual adaptation and population dynamics interact. Conservation problems concern both migratory and non-migratory species.

Some threatened species are migratory because they live in habitats that are ephemeral on a year-to-year (or shorter but unpredictable) basis. Hence, about every new generation will have to colonize new habitats. Continued survival is dependent upon the renewal rate and distribution of suitable habitats.

In other cases, habitats may be rather stable from year to year, but seasonal variation forces the insects to move. Many insects, particularly in more open habitats, migrate seasonally between breeding and diapause habitats. Specific microclimatic demands are often the reason for these movements. It is important to know the full range of habitat requirements of these species. Conservation of breeding habitats are of little use if overwintering habitats are destroyed. Sometimes species aggregate strongly in winter habitats, which may make them more vulnerable to accidents.

The ability of the insect to move different distances is usually poorly known. Although exact values are difficult to obtain, the scale of movements can often be understood from rather limited studies (Forsse & Solbreck, 1985; Solbreck, 1980, 1985). It is also important to know if local populations studied are independent units or not. Sometimes they may exist as a result of migration from larger source populations, as in the Californian butterfly *Euphydryas editha bayensis* (Harrison *et al.*, 1988; Murphy & Weiss, 1988). In other cases, the continued existence of a species in an area is dependent upon a network of habitat patches, where no specific source populations are discernible. The spatial structure of habitat archipelagoes, the role of corridors or stepping-stone habitats, as well as temporal changes in physical factors, resources and enemies, remain poorly known for most rare or endangered insect species. Likewise, the dynamics of extinction, migration and recolonization during population bottlenecks is in great need of study (Solbreck, 1991).

5.4 SMALL MAMMALS

5.4.1 Factors releasing dispersal

Mammals are not considered to have any special morphological or physiological adaptations for dispersal. Still they may be particularly sensitive to factors releasing dispersal during certain early stages in the ontogenetic development (Holekamp & Sherman, 1989). An inherent avoidance of inbreeding is assumed to lead to sex-biased dispersal (Greenwood, 1980) but even such a behavioural adaptation for dispersal is sometimes considered dubious (e.g. Moore & Ali, 1984). Instead, dispersal at least, in small rodents is supposed to be released during normal movements by environmental, social and, especially, density-related stimuli.

The following three factors are most often discussed as causes of small mammal dispersal:

(a) Economic thresholds (Lidicker, 1962; Grant, 1978): The animals disperse when the food supply (or other resources) per individual diminishes below a certain threshold level. Density *per se* may release emigration as an indication of relative resource level.

(b) Conflicts over resources (Dobson, 1982; Moore & Ali, 1984): The resources may be food but are often mates. Females of at least herbivorous species are usually thought to compete for food and males for females. A possible conflict between parents and offspring for breeding opportunities (the Oedipus hypothesis, Liberg & von Schantz, 1985) may also be included here. Inferior (usually the young) individuals are forced to disperse, however, usually before the conflicts escalate to real battles.

(c) Avoidance of inbreeding (Dobson, 1982; Dobson & Jones, 1985): This may be a proximate as well as an ultimate factor. Sometimes individuals of only one particular sex in the litters disperse in order to be able to mate with unrelated conspecifics.

There is still no consensus regarding the relative importance of these three mechanisms or their occurrence with regard to species or habitats (see also below). However, economic thresholds have been less considered during recent years.

5.4.2 Dispersal in time and space

Observations on cyclic small rodents indicate two different types of dispersal: presaturation and saturation dispersal (Lidicker, 1975; Stenseth, 1983. Presaturation dispersal occurs during the increase phase of cyclic population developments and dispersing animals in this phase are healthy and able to settle and soon breed in new surroundings. Saturation dispersal occurs in the peak–decline phase of cyclic populations and dispersing animals at density peaks are not in very good health and obviously exhausted due to the conditions at crowding. The latter type of animals will have very low chances of reproducing anywhere. Presaturation dispersal seems more common than saturation dispersal and dispersal in cyclic populations is positively correlated with growth rates of the populations (Krebs *et al.*, 1972; Stenseth, 1983). However, most small mammal populations are not cyclic and in noncyclic populations most dispersal occurs at the seasonal end of reproduction (Garten & Smith, 1974; Hansson, 1981*a*, 1987), in south boreal–temperate areas usually in September–October. A noncyclic climax forest species, *Apodemus flavicollis*, showed early and strong dispersal at an outbreak after heavy beechmast production (Hansson, 1981*a*). Species with population outbreaks rather than cycles may have considerable dispersal at intermediate densities, similar to really cyclic populations. One experimental study (Johnson & Gaines, 1987) demonstrated high fitness of fairly low-density emigrants of the cyclic vole *Microtus ochrogaster* but the experimental design did not include survival during the critical dispersal movements. Other studies (e.g. Boyce & Boyce, 1988) have shown high survival during the whole dispersal phase.

Small mammal dispersal includes all age and sex categories but adolescent animals and males dominate in these usually polygynous animals. However, similar proportions of sexes disperse in one monogamous vole species, *M. ochrogaster*, and this has been taken as evidence that mainly competition for resources (mates) is at work (Boonstra *et al.*, 1987). However, in some vole species characteristic of secondary successions, such as *Arvicola terrestris*, *Microtus agrestis* and *Microtus oeconomus*, many adult and even pregnant females disperse between litters (Tast, 1966; Stoddart, 1970; Myllymäki, 1977). In certain Australian small marsupials *Antechinus* spp. male young disperse although their fathers, but not their mothers, consistently die shortly after reproduction (Cockburn *et al.*, 1985). This is considered as evidence for inbreeding avoidance as a cause of dispersal.

Dispersal distances are often bimodal (Howard, 1949; French *et al.*, 1968; Boonstra *et al.*, 1987). Some animals move only just outside birth places and others very large distances. This has caused speculation about environmentally (short distances) and innately (long distances) driven dispersal types (Howard, 1960). Recent reanalyses (e.g. Murray, 1967) instead interpret this pattern as due to habitat heterogeneity and varying turnover of territory holders. Dispersers have been found to differ genetically from residents (Krebs *et al.*, 1972) but this was based on serological analyses with often low accuracy. Most known dispersal distances in small mammals, based on marked animals, are often around 1 km or less (Tayor & Quy, 1978). However, Tegelström and Hansson (1987) tracked shrews *Sorex araneus* 4–5 km on a snow-covered lake.

A large number of small mammal species disperse in sheltered environments and are favoured by dispersal corridors with good cover. However, species which usually move around in open terrain in their normal home ranges, such as many mice species, also disperse through a larger variety of habitats (Hansson, 1987).

5.4.3 Dispersal and population persistence

Fairly few small mammal populations are endangered but they, and especially many rodents, may well be considered as keystone species in a large number of natural communities. Furthermore, the ubiquitous small mammals may profitably be used for 'modelling' important conservation aspects. Thus, they may be very appropriate for conservation research. Still, as models they should only be used for comparisons with species with as similar demographic or behavioural strategies as possible.

The North American mouse *Peromyscus leucopus*, a forest species, has been used in modelling the importance of connectivity between isolated woodlots (Fahrig & Merriam, 1985; Lefkovitch & Fahrig, 1985). It was found from population data gathered at the boreal distribution border of this species in Ontario that only certain combinations of woodlots, together with adjoning corridors (fence-rows), permitted persistence of either local (woodlot) populations or metapopulations, consisting of from two to numerous local populations. Furthermore, there was a need for a 'minimum dynamic area' within which the mice could disperse between the separate patches (Fig. 5.8). This model may be applicable, for example, to the endangered salt-marsh harvest mouse *Reitrodontomys raviventer*, the habitats of which in the San Francisco Bay region are becoming increasingly disjointed by urban and industrial development (Shellhammer, 1989). Similar conditions may apply to the rare boreal birch mouse *Sicista betulina*, which seems originally to have been an inhabitant of successional taiga forest patches, according to Russian observations (Hansson, 1977).

In the previously mentioned metapopulation study (Fahrig & Mer-

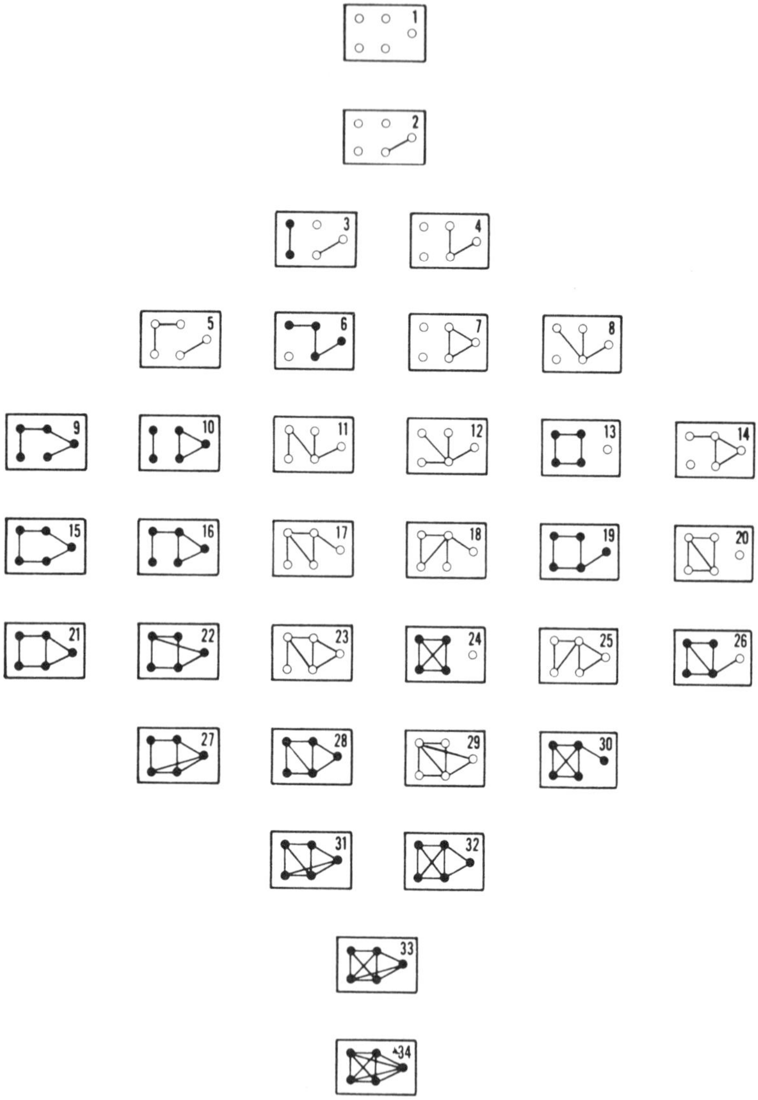

Fig. 5.8. Possible combinations of habitat patches (○) and connecting corridors (——), permitting dispersal by the forest-living mouse *Peromyscus leucopus*. Simulation models based on Canadian demographic data for this species demonstrated that only networks of extensive connecting corridors permitted population persistence in separate patches (●) or persistence of whole metapopulations. This was confirmed in field tests. (From Lefkovitch & Fahrig, 1985.)

riam, 1985), there was no source (surplus) area with a constant supply of dispersers. Such a 'mainland' to outlying habitat islands is instead assumed in the 'rescue effect' (Brown & Kodric-Brown, 1977). Many small mammal metapopulations may better agree with such a model. A good example was provided with pikas *Ochotona princeps* living in thalus slopes differing in size and distance (Smith, 1980). Many small pika populations only occurred close to larger source populations. However, various small mammal species are differently able to colonize such habitat islands; *Apodemus sylvaticus* may occur in any small woodlot in an agricultural area but *Clethrionomys glareolus* in very few (Geuse *et al.*, 1985). Thus, many species need corridors (Hansson, 1987). Furthermore, for species with low dispersal ability there is a pronounced filtering effect, so animals reaching habitat islands are younger and more fecund than in the source populations (Szacki, 1987). Such species may show especially high densities in isolated habitat islands before emigrating due to a fence effect (Krebs *et al.*, 1969). In more distant habitat islands there might be a surplus of especially wide-ranging, immigrating adult males (Gottfried, 1982), which will not sustain any vigorous populations. Thus, the recolonization potential will often not be linearly related to distance from source area. Successional species often dispersed just before and during their reproductive period (Hansson, 1988), which may cause a rapid recolonization and permit persistent populations also at very low densities (Henderson *et al.*, 1985).

Small rodents are keystone species in the boreal environment, supporting a rich predator fauna. The only small rodent predator clearly declining in numbers in Scandinavia is the barn owl *Tyto alba*, which lives in southern agricultural areas. Little is known of its feeding habits there but the increasingly homogeneous agricultural landscape of very large cultivated fields may be responsible (Shawyer, 1987), with virtually the only prey available being *Apodemus sylvaticus*, which is not easily caught (Hansson, 1981*b*). Traditional agricultural landscapes contained a multitude of habitats, supporting a diversified small mammal fauna, with common dispersal between habitat patches.

5.5 GENERAL CONCLUSIONS

A species can be rare only in certain localities (diffusive rarity) or all over its range (suffusive rarity, Schoener, 1987). Rabinowitz (1981) distinguished three variables along which a species may be rare: limited geographic ranges, narrow habitat requirements or small populations (or combinations of these). She regarded suffusive rarity as 'genuine rarity' and diffusive rarity as 'pseudo-rarity' and did not incorporate the last group in her seven forms of rarity.

For suffusively rare species the metapopulation dynamics, including recolonization of local habitats by dispersal, is crucial. Preservation of such species focuses less on local reserves and conservation may involve putting aside specific habitats over very large areas. Certainly, suffusively rare species may more often be those designated as endangered. However, the vast majority of rare species at any particular locality are common in other parts of their ranges. In the Swedish list of rare and endangered bryophytes (Ingelög *et al.*, 1987), more than 85% of the treated species have larger populations elsewhere (cf. Schoener, 1987 for comparative figures on birds). For diffusively rare species, rarity is a matter of spillover from places where the species is more common and an understanding of rarity actually involves an understanding of the dispersal power of the species, i.e. how it diffuses across the landscape. Conservation of such species involves preservation of abundance centres and ability for dispersal.

Decreases in habitat area reduce population sizes, which increases the extinction risk. Stable and non-migratory species which live in environments that have persisted for long periods of time are most sensitive. Such species are likely to possess numerous adaptations for coping with the natural vagaries of their habitats but less for finding new suitable areas. Conservation of habitats is essential for these species and knowledge of the specific habitat requirements of the individual species is important.

Some species live in habitats that are ephemeral on a year-to-year (or shorter but unpredictable) basis, and with almost every generation having to disperse and colonize new habitats. Continued survival is dependent upon the renewal rate and distribution of suitable habitats (e.g. Söderström, 1988*b*). In this case it is more important to secure the appearance of new microhabitats than to preserve old ones. For example, it is important to ensure the more or less continuous occurrence of old trees which are allowed to die a natural death and decompose. Once they have started to decay, old trees provide microhabitats and resources for mammals, birds and numerous insects, as well as for bryophytes and lichens (Ehnström & Waldén, 1986; Söderström, 1988*a*).

The ability of organisms to move different distances should be known. At the same time, the spatial structure of original and man-created habitat archipelagoes, as well as their temporal changes, should also be known. Sufficient knowledge can still only be found for a few populations, for example, the pool frog *Rana lessonae* (Sjögren, 1991).

The amount and extension of dispersal of a species should be related to the natural disturbance regime of a certain region or biotope. However, such scales may be insufficient for recolonization in productive landscapes with only few preserved remnants of original habitats. Theoretical models (e.g. Fahrig & Paloheimo, 1988) imply that the distances moved at dispersal are extremely important for the persistence of subpopulations. The problem of distance will be aggravated by a need for dispersal corridors. Fire has been the dominating disturbance agent in the boreal landscape (Zackrisson, 1977) and many boreal species, such as, for example, bark beetles, seem well adapted to find remote windfall areas or burned forests (Solbreck, 1985). Local windthrows of single trees may have been more important as disturbances in temperate forests and the dispersal ranges of a few metres per year in snails (Baur, 1986; Baker, 1988) may have permitted fairly rapid recolonization under such conditions. Such movements, and even dispersal of a few hundred metres in some small mammals, may be insufficient for permanent populations when remnant deciduous copses are separated by many kilometres, such as can be found in recent agricultural areas.

The main general principle emerging from the present survey seems to be that dispersal ability is related to the life history of a particular species. In many cases, dispersal is closely integrated with the ontogenetic development. It may be possible to partition organisms into life history types and correlate these types with various habitat types, e.g. successional stages. This has already been performed in the *r–K* gradient of demographic strategies (e.g. Southwood, 1977), where it appears, for example, that species living in habitats with short duration in relation to the generation time of the species will show especially high dispersal rates. Such a broad characterization may be of some importance for management planning but should be supplemented with further analyses of dispersal in relation to more distinctly identified strategies. There is a great need for basic ecological studies on the questions outlined above (cf. Simberloff, 1988). This would stimulate a more fruitful interplay between theory and empiricism in the future.

When considering dispersal from a conservation view point, it might thus first be observed that organisms from early successional stages generally have better dispersal abilities than those from later stages. However, this does not mean that all successional species easily (re-) colonize patches with extinct populations; i.e. animal species with concealed habits need corridors. Climax species adapted to temporary but regularly appearing resources, such as decaying logs or plants with mast seeding, may move around to quite an extent but tolerate only a limited distance between patches, in space and time, for population persistence. The main problem seems, however, to be with species with originally evenly distributed and long-lasting resources; such organisms ought to be monitored carefully at least in small nature reserves and, if necessary, may have to be reintroduced by man in order to retain community integrity.

REFERENCES

Andrewartha, H.G. & Birch, L.C. (1954). *The Distribution and Abundance of Animals*. Chicago University Press, Chicago, Illinois.

Bailey, R.H. (1966). Studies on the dispersal of lichen soredia. *Bot. J. Linn. Soc.*, **59**, 479–90.

Baker, G.H. (1988). Dispersal of *Theba pisana* (Mollusca: Helioidae). *J. Appl. Ecol.*, **25**, 879–900.

Baur, B. (1986). Patterns of dispersion, density and dispersal in alpine populations of the land snail *Arianta arbustorum* (L.) (Helicidae). *Holarct. Ecol.*, **9**, 117–25.

Berger, A. (1989). Ballooning activity of *Chilo partellus* larvae in relation to size of mother, egg batches, eggs and larvae and age of mother. *Entomol. Exp. Appl*, **50**, 125–32.

Boonstra, R., Krebs, C.J., Gaines, M.S., Johnson, M.L. & Craine, I.T.M. (1987). Natal philopatry and breeding systems in voles (*Microtus* spp.). *J. Anim. Ecol.*, **56**, 655–73.

Bowler, P.A. & Rundel, P.W. (1975). Reproductive strategies in lichens. *Bot. J. Linn. Soc.*, **70**, 325–40.

Boyce, C.C.K. & Boyce, J.L. III (1988). Population biology of *Microtus arvalis*, II. Natal and breeding dispersal of females. *J. Anim. Ecol.*, **57**, 723–36.

Brown, J.H. & Kodric-Brown, A. (1977). Turnover rates in insular biogeography: effects of immigration on extinction. *Ecology*, **58**, 445–9.

Caldwell, R.L. & Hegmann, J.P. (1969). Heritability of flight duration in the milkweed bug, *Lygaeus kalmii*. *Nature, Lond.*, **223**, 91–2.

Cockburn, A., Scott, M.P. & Scotts, D.J. (1985). Inbreeding avoidance and male-biased natal dispersal in *Antechinus* spp. (Marsupialia, Dasyuridae). *Anim. Behav.*, **33**, 908–15.

Crawley, M.J. (1987). What makes a community invasible? In *Colonization, Succession and Stability*, ed. A.J. Gray, M.J. Crawley, & P.J. Edwards. Oxford University Press, Oxford, pp. 429–53.

Danthanarayana, W. (ed.) (1986). *Insect Flight, Dispersal and Migration*. Springer-Verlag, Berlin.

Darlington, P.J. Jr (1943). Carabidae of mountains and islands: data on the evolution of isolated faunas, and on atrophy of wings. *Ecol. Monogr.*, **13**, 37–61.

Dempster, J.P. (1989). Insect introductions: natural dispersal and population persistence in insects. *Entomologist*, **108**, 5–13.
den Boer, P.J. (1970). On the significance of dispersal power for populations of Carabid beetles (Coleoptera, Carabidae). *Oecologia* (*Berl.*), **4**, 1–28.
Denno, R.F., Olmstead, K.L. & McCloud, E.S. (1989) Reproductive cost of flight capability: a comparison of life history traits in wing dimorphic planthoppers. *Ecol. Entomol.*, **14**, 31–44.
Dingle, H. (1972). Migration strategies of insects. *Science, N.Y.*, **175**, 1327–35.
Dingle, H. (1978). *Evolution of Insect Migration and Diapause*. Springer-Verlag, New York.
Dingle, H.J. (1984). Behaviour, genes, and life histories: complex adaptations in uncertain environments. In *A New Ecology. Novel Approaches to Interactive Systems*, ed. P.W. Price, C.N. Slobodchikoff & W.S. Gaud. Wiley, New York, pp. 169–94.
Dingle, H. (1985*a*). Migration. In *Comprehensive Insect Physiology, Biochemistry, and Pharmacology*, ed. G.A. Kerkut, H. & L.I. Gilbert. Pergamon Press, Oxford, pp. 375–415.
Dingle, H. (1985*b*). Migration and life histories. In *Migration: Mechanisms and Adaptive Significance*, ed. M.A. Rankin. *Contrib. Mar. Sci. Suppl.*, **27**, 27–42.
Dingle, H. (1986). Evolution and genetics of insect migration. In *Insect Flight, Dispersal and Migration*, ed. W. Danthanarayana. Springer Verlag, Berlin, pp. 11–26.
Dobson, F.S. (1982). Competition for mates and predominant juvenile dispersal in mammals. *Anim. Behav.*, **30**, 1183–92.
Dobson, F.S. & Jones, W.T. (1985). Multiple causes of dispersal. *Amer. Nat.*, **126**, 855–8.
During, H.J. & ter Horst, B.B. (1983). The diaspore bank of bryophytes and ferns in chalk grassland. *Lindbergia*, **9**, 57–64.
During, H.J. & ter Horst, B.B. (1987). Diversity and dynamics in bryophyte communities on earth banks in a Dutch forest. *Symp. Biol. Hung.*, **35**, 447–55.
During, H.J., Brugués, R.M., Cros, R.M. & Lloret, F. (1987). The diaspore bank of bryophytes and ferns in the soil in some contrasting habitat around Barcelona, Spain. *Lindbergia*, **13**, 137–49.
Ehnström, B. & Waldén, H.W. (1986). *Faunavård i Skogsbruket, Del 2. Den Lägre Faunan*. Skogsstyrelsen, Jönköping.
Ehrlich, P.R. (1983). Genetics and the extinction of butterfly populations. In *Genetics and Conservation: a Reference for Managing Wild Animal and Plant Populations*, ed. C. Schonewald-Cox, S.M. Chambers, B. MacBryde & L. Thomas. Benjamin Cummings, Menlo Park, California, pp. 152–63.
Ehrlich, P.R. & Murphy, D.D. (1987). Conservation lessons from long-term studies of checkerspot butterflies. *Conserv. Biol.*, **1**, 122–31.
Esseen, P.-A. (1985). Litter fall of epiphytic macrolichens in two old *Picea abies* forests in Sweden. *Can. J. Bot.*, **63**, 980–7.
Esseen, P.A., Ericson, L., Lindström, H. & Zackrisson, O. (1981). Occurrence and ecology of *Usnea longissima* in central Sweden. *Lichenologist*, **13**, 177–90.
Fahrig, L. & Merriam, G. (1985). Habitat patch connectivity and population survival. *Ecology*, **66**, 1762–8.

Fahrig, L. & Paloheimo, J. (1988). Determinants of local population size in patchy habitats. *Theoret. Popul. Biol.*, **34**, 194–213.

Farrow, R.A. & Dows, J.E. (1984). Method of using kites to carry tow nets in the upper air for sampling insects and its application to radar entomology. *Bull. Entomol. Res.*, **74**, 87–95.

Forsse, E. & Solbreck, C. (1985). Migration in the bark beetle *Ips typographus* L.: duration, timing and height of flight. *Z. angew. Entomol.*, **100**, 47–57.

French, N.R., Tagami, T.Y. & Hayden, P. (1968). Dispersal in a population of desert rodents. *J. Mammal.*, **49**, 272–80.

Fulford, M. (1951). Distribution patterns of the genera of Hepaticae of South America. *Evolution*, **5**, 243–64.

Garten, C.T. & Smith, M. H. (1974). Movements of oldfield mice and population regulation. *Acta Theriol.*, **19**, 513–4.

Geuse, P., Bauchau, V. & Le Boulanger, E. (1985). Distribution and population dynamics of rodents in a patchy woodland habitat in central Belgium: preliminary results. *Acta Zool. Fennica*, **173**, 65–8.

Gibo, D.L. (1986). Flight strategies of migrating monarch butterflies (*Danaus plexippus* L.) in southern Ontario. In *Insect Flight Dispersal and Migration*, ed. W. Danthanarayana, Springer Verlag, Berlin, pp. 172–84.

Glick, P. A. (1939). The distribution of insects, spiders, and mites in the air. *US Dep. Agric. Tech. Bull.*, **673**, 1–150.

Goodman, D. (1987). The demography of chance extinction. In *Viable Populations for Conservation*, ed. M. Soulé. Cambridge University Press, Cambridge, pp. 11–34.

Gottfried, B.M. (1982). A seasonal analysis of small mammal populations in woodlot islands. *Can. J. Zool.*, **60**, 1660–4.

Granström, A. (1982). Seed bank in five boreal forest stands originating between 1810 and 1963. *Can. J. Bot.*, **60**, 1815–21.

Grant, P.R. (1978). Dispersal in relation to carrying capacity. *Proc. Natn. Acad. Sci.*, **75**, 2854–8.

Greenwood, P.J. (1980). Mating systems, philopatry and dispersal in birds and mammals. *Anim. Behav.*, **28**, 1140–62.

Gregory, P.H. (1973). *Microbiology of the Air*, 2nd edn. Leonard Hill Books, London.

Hale, M.E. (1983). *The Biology of Lichens*, 3rd edn. Edward Arnold, London.

Hanski, I. (1982). Dynamics of regional distribution: the core and satellite hypothesis. *Oikos*, **38**, 210–1.

Hansson, L. (1977). Buskmus. In *Faunavård*, ed. I. Ahlén. Liber, Stockholm, p. 136.

Hansson, L. (1981*a*). Immigration to human habitations by South Swedish small mammals. *Z. angew. Zool.*, **68**, 339–55.

Hansson, L. (1981*b*). Smågnagarfaunan i beteslandskap. *Fauna Flora*, **76**, 215–22.

Hansson, L. (1987). Dispersal routes of small mammals at an abandoned field in central Sweden. *Holarct. Ecol.*, **10**, 154–9.

Hansson, L. (1988). Dispersal and patch connectivity as species-specific characteristics. In *Connectivity in Landscape Ecology*, ed. K.-F. Schreiber, Schöningh, Muenster, pp. 111–3.

Harrison, S. & Quinn, J.F. (1989). Correlated environments and the persistence of metapopulations. *Oikos*, **56**, 293–8.
Harrison, S., Murphy, D.D. & Ehrlich, P. (1988). Distribution of the bay checkerspot butterfly, *Euphydryas editha bayensis*: evidence for a metapopulation model. *Amer. Nat.*, **132**, 360–82.
Hedenäs, L., Herben, T., Rydin, H. & Söderström, L. (1989*a*). Ecology of the invading moss species *Orthodontium lineare* in Sweden: spatial distribution and population structure. *Holarct. Ecol.*, **12**, 163–72.
Hedenäs, L., Herben, T., Rydin, H. & Söderström, L. (1989*b*). Ecology of the invading moss species *Orthodontium lineare* in Sweden: substrate preference and interaction with other species. *J. Bryol.*, **15**, 565–81.
Helle, P. & Järvinen, O. (1986). Population trends of North Finnish land birds in relation to their habitat selection and changes in forest structure. *Oikos*, **46**, 107–15.
Henderson, M.T., Merriam, G. & Wegner, J. (1985). Patchy environments and species survival: chipmunks in an agricultural mosaic. *Biol. Conserv.*, **31**, 95–105.
Herben, T. (1987). The ecology of the invasion of *Orthodontium lineare* Schwaegr. in central Europe. *Symp. Biol. Hung.*, **35**, 323–33.
Hoffman, G.R. (1970). Spore viability in *Funaria hygrometrica. Bryologist*, **73**, 634–5.
Holekamp, K.E. & Sherman, P.W. (1989). Why male ground squirrels disperse. *Amer. Scient.*, **77**, 232–9.
Howard, W.E. (1949). Disperal, amount of inbreeding, and longevity in a local population of prairie deer deermice on the George Reserve, southern Michigan. *Contrib. Lab. Vertebr. Biol. Univ. Mich.*, **43**, 1–52.
Howard, W.E. (1960). Innate and environmental dispersal of individual vertebrates. *Amer. Midl. Nat.*, **63**, 152–61.
Howe, H.F. & Smallwood, J. (1982). Ecology of seed dispersal. *Ann. Rev. Ecol. Syst.*, **13**, 201–28.
Hulthén, E. (1971). *Atlas över Växternas Utbredning i Norden. Generalstabens Litografiska Anstalts Förlag*, Stockholm.
Ingelög, T., Thor, G. & Gustaffson, L. (eds) (1987). *Floravård i Skogsbruket*, 2nd edn. Skogsstyrelsen, Jönköping.
Johnson, C.G. (1969) *Migration and Dispersal of Insects by Flight*. Methuen, London.
Johnson, M.L. & Gaines, M.S. (1987). The selective basis for dispersal of the prairie vole, *Microtus ochrogaster*. *Ecology*, **68**, 684–94.
Jonsson, B.G. & Esseen, P.-A. (1990). Treefall disturbance maintains high bryophyte diversity in a boreal spruce forest. *J. Ecol.*, **78**, 924–36.
Jonsson, B.G. & Söderström, L. (1988). Growth and reproduction in the leafy hepatic *Ptilidium pulcherrimum* during a 4-year period. *J. Bryol.*, **15**, 315–25.
Kaisila, J. (1962). Immigration und Expansion der Lepidopteren in Finnland in den Jahren 1869–1960. *Acta Entomol. Fennica*, **18**, 1–452.
Kennedy, J.S. (1985). Migration, behavioural and ecological. In *Migration: Mechanisms and Adaptive Significance*, ed. M.A. Rankin, *Contrib. Mar. Sci.*, Suppl., **27**, 1–20.

Kimmel, U. (1962). Entwicklung einiger Moose und Flechten auf Dauer-Untersuchungsflächen. *Oberh. Ges. Natur- u. Heilkunde Ber. N.F. Naturw. Abt.*, **32**, 151–60.

Koponen, M. (1980). Distribution of *Ips amitinus* (Eichoff) (Coleoptera, Scolytidae) in Finland in 1974–1979. *Notulae Entomologicae*, **60**, 223–5.

Krebs, C.J., Keller, B.L. & Tamarin, R.H. (1969). *Microtus* population biology: demographic changes in fluctuation populations of *Microtus ochrogaster* and *M. pennsylvanicus* in southern Indiana. *Ecology*, **50**, 587–607.

Krebs, C.J., Gaines, M.S., Keller, B.L., Myers, J.H. & Tamarin, R.H. (1972). Population cycles in small rodents. *Science, N.Y.*, **179**, 35–41.

Kreulen, D.J.W. (1972). Spore output of moss capsules in relation to ontogeny of the archesporial tissue. *J. Bryol.*, **7**, 61–74.

Lekander, B., Beijer Petersen, B., Kangas, E. & Bakke, A. (1977). The distribution of bark beetles in the Nordic countries. *Acta Entomol. Fennica*, **32**, 1–37.

Lefkovitch, L.P. & Fahrig, L. (1985). Spatial characteristics of habitat patches and population survival. *Ecol. Modelling*, **30**, 297–308.

Liberg, O. & von Schantz, T. (1985). Sex-biased philopatry and dispersal in bird and mammals: the Oedipus hypothesis. *Amer. Nat.*, **126**, 129–35.

Lidicker, W.Z. (1962). Emigration as a possible mechanism permitting the regulation of population density below carrying capacity. *Amer. Nat.*, **96**, 29–33.

Lidicker, W.Z. (1975). The role of dispersal in the demography of small mammals. In Productivity and Dynamics of Small Mammal Populations, ed. F.B. Golley, K. Petrusewicz & L. Ryszkowski. Cambridge University Press, Cambridge, pp. 103–28.

Lindroth, C.H. (1949). *Die Fennoskandischen Carabidae. Eine Tiergeografische Studie*, III. *Allgemeiner Teil.* Göteborgs Kungl. Vetenskaps/Vitterhets-samhälles Handlingen, B,4,3, Stockholm.

Longton, R.E. (1976). Reproductive biology and evolutionary potential in bryophytes. *J. Hattari Bot. Lab.*, **41**, 205–23.

Longton, R.E. & Miles, C.J. (1982). Studies on the reproductive biology of mosses. *J. Hattori Bot. Lab.*, **52**, 219–39.

Longton, R.E. & Schuster, R.M. (1983). Reproductive biology. In *New Manual of Bryology*, ed. R.M. Schuster, Hattori Botanical Laboratory, Nichinan, pp. 386–462.

McQueen, C.B. (1986) Spatial pattern and gene flow distances in *Sphagnum subtile. Bryologist*, **88**, 333–6.

Mikkola, K. (1967). Immigrations of Lepidoptera, recorded in Finland in the years 1946–1966, in relation to air currents. *Ann. Entomol. Fennica*, **33**, 65–99.

Mikkola, K. (1971). The migratory habits of *Lymantria dispar* (Lep., Lymantridae) adults of continental Eurasia in the light of a flight to Finland. *Acta Entomol. Fennica*, **28**, 107–20.

Miles, C.J. & Longton, R.E. (1987). Life history of the moss *Atrichum undulatum* (Hedw.) P. Beauv. *Symp. Biol. Hung.*, **35**, 193–207.

Miller, N.G. & Ambrose, L.J.H. (1976). Growth in culture of wind-blown bryophyte gametophyte fragments from arctic Canada. *Bryologist*, **79**, 55–63.

Moore, J. & Ali, R. (1984). Are dispersal and inbreeding avoidance related? *Anim. Behav.*, **32**, 94–112.

Moore, P.D. (1980). Soil seed banks. *Nature, Lond.*, **284**, 123–4.

Murphy, D.D. & Weiss, S.B. (1988). Ecological studies and the conservation of the bay checkerspot butterfly, *Euphydryas editha bayensis*. *Biol. Conserv.*, **46**, 183–200.

Murray, B.G. (1967). Dispersal in vertebrates. *Ecology*, **48**, 975–8.

Myllymäki, A. (1977). Intraspecific competition and home range dynamics in the field vole *Microtus agrestis*. *Oikos*, **29**, 553–69.

Nielsen, E.T. (1961). On the habits of the migratory butterfly *Ascia monuste* L. *Biol. Medd.*, **23**, 1–81.

Palmén, E. (1944). Die anemohydrochore Ausbreitung der Insekten als zoogeographischer Faktor. *Ann. Zool. Soc. Zool. Bot. Fennicae Vanamo*, **10**(1), 1–262.

Parsons, P.A. (1983). *The Evolutionary Biology of Colonizing Species*. Cambridge University Press, Cambridge.

Petterson, B. (1940). Experimentelle Untersuchungen über die Verbreitung der Sporenpflanzen. *Acta Bot. Fennica*, **25**, 1–103.

Pulliam, H.R. (1988). Sources, sinks, and population regulation. *Amer. Nat.*, **132**, 652–61.

Rabinowitz, D. (1981). Seven forms of rarity. In *The Biological Aspects of Rare Plant Conservation*, ed. H. Synge. Wiley, Chichester, pp. 205–17.

Rainey, A.C. (1989). *Migration and Meteorology. Flight Behaviour and the Atmospheric Environment of Locusts and Other Migrant Pests*. Clarendon Press, Oxford.

Rainey, A.C. (ed.) (1976). Insect Flight. *Symp. Roy. Entomol. Soc. London*, No. 7.

Rankin, M.A. (ed.) (1985). *Migration: Mechanisms and Adaptive Significance*. *Contrib. Mar. Sci. Suppl.*, **27**.

Rankin, M.A., McAnelly, M.L. & Bodenhamer, J.E. (1986). The oogenesis flight syndrome revisited. In *Insect Flight, Dispersal and Migration*, ed. W. Danthanarayana. Springer-Verlag, Berlin, pp. 27–48.

Roff, D. (1986). The evolution of wing dimorphism in insects. *Evolution*, **40**, 1009–20.

Rose, F. (1976). Lichenological indicators of age and environmental continuity in woodlands. In *Lichenology: Progress and Problems*, ed. D.H. Brown, D.L. Hawksworth & R.H. Bailey, Academic Press, London, pp. 279–307.

Ryrholm, N. & Källander, C.F.R. (1987). Invasionen av *Autographa mandarina* (Lepidoptera, Noctuidae) i östra Sverige 1985. *Entomol. Tidskr.*, **108**, 130–4.

Schaefer, G.W. (1976). Radar observations of insect flight. *Symp. R. Entomol. Soc.*, **7**, 157–97.

Schoener, T.W. (1987). The geographical distribution of rarity. *Oecologia (Berl.)*, **74**, 161–73.

Schonewald-Cox, C., Chambers, S.M., MacBryde, B. & Thomas, L. (eds) (1983). *Genetics and Conservation. A Reference for Managing Wild Animal and Plant Populations*. Benjamin Cummings, Menlo Park, California.

Schuster, R.M. (1983). Phytogeography of the Bryophyta. In *New Manual of Bryology*, ed. R.M. Schuster. Hattori Botanical Laboratory, Nichinan. pp. 463–626.

Shaffer, M.L. (1987). Minimum viable populations: coping with uncertainty. In *Viable Populations for Conservation*, ed. M. Soulé. Cambridge University Press, Cambridge, pp. 69–86.

Shawyer, C.R. (1987). *The Barn Owl in the British Isles. Its Past, Present and Future*. The Hawk Trust, London.

Shellhammer, H.S. (1989). Salt marsh harvest mice, urban development, and rising sea levels. *Conserv. Biol.*, **3**, 59–65.

Sillén-Tullberg, B. & Solbreck, C. (1990). Population dynamics of a seed-feeding bug, *Lygaeus equestris*, 2. Temporal dynamics. *Oikos*, **58**, 210–8.

Simberloff, D. (1988). The contribution of population and community biology to conservation science. *Ann. Rev. Ecol. Syst.*, **19**, 473–511.

Sjögren, P. (1991). Extinction and isolation gradients in metapopulations: the case of the pool frog *Rana lessonae*. *Biol. J. Linn. Soc.*, **42**, 135–47.

Slater, J.A. (1975). On the biology and zoogeography of Australian Lygaeidae (Hemiptera: Heteroptera) with special reference to the Southwest fauna. *J. Austrl. Entomol. Soc.*, **14**, 47–64.

Slater, J.A. (1977). The incidence and evolutionary significance of wing polymorphism in lygaeid bugs with particular reference to those of South Africa. *Biotropica*, **9**, 217–29.

Smith, A.T. (1980). Temporal changes in insular populations of pika (*Ochotona princeps*). *Ecology*, **61**, 8–13.

Smith, R.I.L. (1987). The bryophyte propagule bank of Antarctic fellfield soils. *Symp. Biol. Hung.*, **35**, 233–45.

Söderström, L. (1987). Dispersal as a limiting factor for distribution among epixylic bryophytes. *Symp. Biol. Hung.*, **35**, 475–84.

Söderström, L. (1988*a*). Sequence of bryophytes and lichens in relation to substrate variables of decaying coniferous wood in northern Sweden. *Nord. J. Bot.*, **8**, 89–97.

Söderström, L. (1988*b*). The occurrence of epixylic bryophyte and lichen species in an old natural and a managed forest stand in northeast Sweden. *Biol. Conserv.*, **45**, 169–78.

Söderström, L. (1989). Regional distribution among bryophyte species on a temporal, patchy substrate. *Bryologist*, **92**, 349–55.

Söderström, L. (1990*a*). Niche width and niche overlap in some epixylic bryophytes. *Proc. 6th Mtg Central and East European Bryologists Working Group*, ed. T. Herben. Czechoslovak Academy of Science, Průhonice, pp. 108–19.

Söderström, L. (1990*b*). Dispersal and distribution patterns in patchy, temporal habitats. In *Spatial Processes in Plant Communities*, ed. F. Krahulec, A.D.Q. Agnew, S. Agnew & J.H. Willems. SBP Publishers, The Hague, pp. 103–13.

Söderström, L. & Jonsson, B.G. (1989). Dispersal, spatial pattern and substrate preference in the leafy hepatic *Philidium pulcherrimum*. *J. Bryol.*, **15**, 793–802.

Solbreck, C. (1972). Sexual cycle and changes in feeding activity and fat body size in relation to migration in *Lygaeus equestris* (L.) (Het., Lygaeidae). *Entomol. Scand.*, **3**, 267–74.

Solbreck, C. (1978). Migration, diapause and direct development as alternative life histories in a seed bug, *Neacoryphus bicrucis*. In *Evolution of Insect Migration and Diapause*, ed. H. Dingle. Springer-Verlag, New York, pp. 195–217.

Solbreck, C. (1980). Dispersal distances of migrating pine weevils, *Hylobius abietis*, Coleoptera: Curculionidae. *Entomol. Exp. Appl.*, **28**, 123–31.

Solbreck, C. (1985). Insect migration strategies and population dynamics. In *Migration: Mechanisms and Adaptive Significance*, ed. M.A. Rankin. *Contrib. Mar. Sci., Suppl.*, **27**, pp. 641–62.

Solbreck, C. (1986). Wing and flight muscle polymorphism in a lygaeid bud, *Horvathiolus gibbicollis*: determinants and life history consequences. *Ecol. Entomol.*, **11**, 435–44.

Solbreck, C. (1991). Unusual weather and insect population dynamics: *Lygaeus equestris* during an extinction and recovery period. *Oikos*, **60**, pp. 343–50.

Solbreck, C. & Sillén-Tullberg, B. (1990). Population dynamics of a seed feeding bug, *Lygaeus equestris*, 1. Habitat patch structure and spatial dynamics. *Oikos*, **58**, 199–209.

Solbreck, C., Anderson, D.B. & Förare, J. (1990). Migration and the coordination of life cycles as exemplified by Lygaeinae bugs. In *Insect Life Cycles: Genetics, Evolution, and Coordination*, ed. F. Gilbert. Springer Verlag, London, pp. 197–214.

Southwood, T.R.E. (1962). Migration of terrestrial arthropods in relation to habitat. *Biol. Rev.*, **37**, 171–214.

Southwood, T.R.E. (1977). Habitat, the template for ecological strategies? *J. Anim. Ecol.*, **46**, 337–66.

Stenseth, N.C. (1983). Causes and consequences of dispersal in small mammals. In *The Ecology of Animal Movements*, ed. I. Swingland & P. Greenwood. Oxford University Press, Oxford, pp. 63–101.

Stevenson, S.K. (1988). Managing second-growth forests as caribou habitat. *Proc. 5th int. Reindeer/Caribou Symp.*, Arvidsjaur.

Stoddart, D.M. (1970). Individual range, dispersion and dispersal in a population of the water vole *Arvicola terrestris* (L.). *J. Anim. Ecol.*, **39**, 403–26.

Szacki, J. (1987). Ecological corridor as a factor determining the structure and organization of a bank vole population. *Acta Theriol.*, **32**, 31–44.

Tast, J. (1966). The root vole, *Microtus oeconomus* (Pallas), as an inhabitant of seasonally flooded land. *Ann. Zool. Fennici*, **3**, 127–71.

Taylor, K.D. & Quy, R.J. (1978). Long distance movements by a common rat (*Rattus norvegicus*) revealed by radio-tracking. *Mammalia*, **42**, 63–71.

Taylor, L.R. (1986). The four kinds of migration. In *Insect Flight. Dispersal and Migration*, ed. W. Danthanarayana. Springer-Verlag, Berlin, pp. 265–80.

Taylor, L.R. & Taylor, R.A.J. (1983). Insect migration as a paradigm for survival by movement. In *The Ecology of Animal Movement*, ed. I.R. Swingland & P.J. Greenwood. Clarendon Press, Oxford, pp. 181–214.

Tegelström, H. & Hansson, L. (1987). Evidence of long-distance dispersal in the common shrew (*Sorex araneus*). *Z. Säugetierk.*, **52**, 52–4.

Urquhart, F.A. & Urquhart, N.A. (1978). Autumnal migration routes of the eastern population of the monarch butterfly (*Danaus p. plexippus* L., Danaidae: Lepidoptera) in North America to the overwintering site in the neo-volcanic plateau of Mexico. *Can. J. Zool.*, **56**, 1759–64.

Urquhart, F.A. & Urquhart, N.A. (1979). Vernal migration of the monarch butterfly (*Danaus p. plexippus*, Lepidoptera; Danaidae) in North America from the overwintering site in the neo-volcanic plateau of Mexico. *Can. Entomol.*, **111**, 15–18.

Watson, M.A. (1981). Chemically mediated interactions among mosses as possible determinants of their community structure. *J. Chem. Ecol.*, **7**, 367–76.

Williams, C.B. (1958). *Insect Migration.* Collins, London.

Wyatt, R. (1982). Population ecology of bryophytes. *J. Hattori Bot. Lab.*, **52**, 179–98.

van Zanten, B.O. (1978). Experimental studies on trans-oceanic long-range dispersal of moss spores in the Southern Hemisphere. *J. Hattori Bot. Lab.*, **44**, 455–82.

van Zanten, B.O. & Pócz, T. (1981). Distribution and dispersal of bryophytes. *Adv. Bryol.*, **1**, 479–562.

Zackrisson, O. (1977). Influence of forest fires on the North Swedish boreal forest. *Oikos*, **29**, 22–32.

Chapter 6

Archipelagoes and Theories of Insularity

STEFAN ÅS[a] JAN BENGTSSON[b] & TORBJÖRN EBENHARD[a]

[a]*Department of Zoology, Uppsala University, Box 561, S-751 22 Uppsala, Sweden*

[b]*Department of Ecology and Environmental Research, Swedish University of Agricultural Sciences, Box 7072, S-750 07 Uppsala, Sweden*

6.1 INTRODUCTION

Sweden, being part of the Scandinavian peninsula, is a country with a very long coastline and greatly influenced by its closeness to the sea. Both the Swedish and Finnish coasts along the brackish Baltic Sea and the west coast of Sweden along the Skagerak are rich in islands. This has directed much attention among Swedish and Finnish naturalists to the ecology of island life. Theories of species richness on islands, along the same lines of thinking as MacArthur and Wilson (1963, 1967) and others during the 1960s, were published as early as 1917 by Alvar Palmgren and 1921 by Olof Arrhenius.

The Swedish word 'skärgård', which can be interpreted to mean 'a fence of islands', describes the type of archipelago consisting of bedrock islands that have been exposed to glaciation. The Stockholm, Åland and Åbolands archipelagoes (Fig. 6.1) are the most island-rich of this type in the world. The enormous number of islands, the Stockholm archipelago alone consisting of approximately 30 000, their recent emergence after the latest glaciation, and the ongoing land uplift are some of the traits making this area unique. Along the Swedish coast we find regional archipelagoes in many places from the northernmost part of the Gulf of

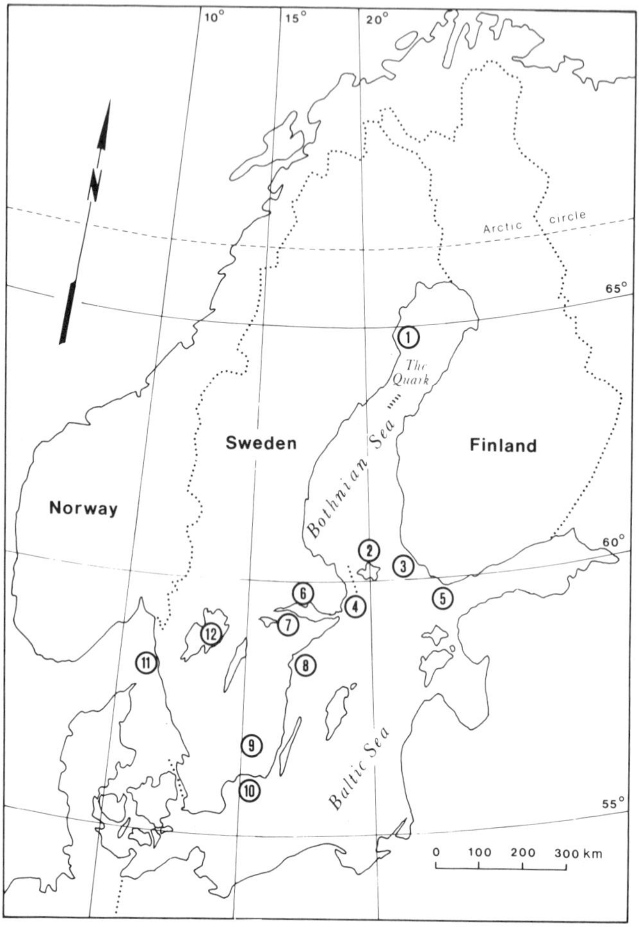

Fig. 6.1. Map of the areas discussed. 1, Luleå archipelago; 2, Åland; 3, Åbolands archipelago; 4, Stockholm archipelago; 5, Tvärminne archipelago; 6, Lake Mälaren; 7, Lake Hjälmaren; 8, Östergötland archipelago; 9, Lake Möckeln; 10, Blekinge archipelago; 11, West coast archipelago; 12, Lake Vänern.

Bothnia to the Blekinge archipelago in southern Sweden, as well as in the northern part of the Swedish west coast and in many of the larger lakes.

In this chapter we give a very brief introduction to the natural history of Swedish archipelagoes, followed by a review of some of the basic theories and hypotheses formulated and/or tested in different archipelagoes and a description of a number of investigations made in Scandinavian archipelagoes. It should be noted that this is not an exhaustive review, but rather a subjective selection among many studies. We end with some implications that can be derived from the theories and applied to conservation in general and to archipelagoes in particular.

6.1.1 Natural history of the Swedish archipelagoes

The geomorphology differs among different archipelagoes, but basically the bedrock has an age of two billion years and is what remains of the eroded Svecofennian mountain range, forming the Precambrian peneplane which in the east dips at a shallow angle into the Baltic and, with a somewhat steeper angle, into the North Sea to the west. The majority of the islands are of bedrock with only a thin layer of fertile soil, especially on the outer islands. In some archipelagoes, such as the Luleå in the north, the Blekinge in the south and most archipelagoes in lakes, the bedrock is covered with moraine or sandy deposits, giving the islands a quite different appearance.

The bedrock of the islands may be old, but the islands are still quite new and are continually changing in area and configuration due to the continual land uplift. The ice of the latest glaciation withdrew from southern Sweden approximately 15 000 BP, the Stockholm archipelago became free of ice at about 10 000 BP and the Gulf of Bothnia 2000 years later. This glaciation, in addition to the mechanical grinding effects of the rocks, is responsible for the ongoing land uplift. Its rate differs from north to south, being approximately 10 mm/year in the Luleå archipelago, 5 mm/year in the Stockholm archipelago and −1 mm/year at Falsterboudd (the southernmost peninsula of Sweden). Another important geological factor of great ecological significance, not least for vegetation and molluscs, is the distribution of limestone. In some areas this is present in the primary rock, whereas in others, as in the northern parts of Stockholm archipelago, the glacial ice has moved sedimentary limestone from the bottom of the Bothnian Sea and deposited it as lime-rich soil on the islands. The sea has influenced the presence of lime in more ways than this. Deposits, mainly consisting of shells, have occasionally formed in land uplift areas, especially on the west coast.

There is usually a gradual change in the appearance of the islands as one proceeds from the mainland to the outermost islands. This has sometimes been referred to as a zonation with an inner, a middle and an outer zone. As a consequence of the land uplift older and larger islands are mostly close to the mainland, and the distances between islands are thus also less wide. In the middle zone we find groups of islands of different sizes separated by larger distances, while the outer zone consists of widely scattered groups of small islands. This general description holds for most of the extensive archipelagoes with the exception of the most northerly. Islands close to the mainland are sheltered by other islands and hence less exposed to waves and sea spray, and there are often also climatic differences between the inner and outer zones, as well as between the archipelagoes and the adjacent mainland. Compared to the mainland the archipelagoes have a longer vegetative period, fewer nights with frost during the vegetative period, more sunshine hours and less precipitation. Although the vegetative period is prolonged, it starts and ends later as compared to the mainland. The sunnier weather, in combination with a generally coarser soil type than found on the mainland, increases the frequency of severe droughts on the islands.

The dominating forest type in Swedish archipelagoes is generally a dry Scots pine *Pinus sylvestris* forest, although deviations, for example depending on latitude and soil type, are frequent. The largest archipelago, Stockholm archipelago, and its closest neighbour to the south, the Östergötland archipelago, have an inner zone dominated by coniferous forest mixed with a relatively large amount of deciduous trees. Most of the islands are strongly affected by former, and present, agricultural activities. Alder *Alnus glutinosa* is commonly found in wet areas and along shorelines both in the inner and middle zone. The small islands in the outer zone are dominated by bare bedrock, heather moor, peat bogs and a few stunted trees. In broader parts of the archipelagoes there is a distinguishable middle zone characterized by a special type of coastal birch forest. In parts, especially in the northern Stockholm archipelago, with finer sediments rich in lime, patches of lush (i.e. southern deciduous) forest are common, often dominated by ash *Fraxinus excelsior* and with a dense understorey of guelder-rose *Viburnum xylosteum* and mountain-currant *Ribes alpinum*. But substantial parts of the middle-zone islands consist of bare rock, heather moor, and dense juniper *Juniperus communis* shrubs. On these richer soils Norway spruce *Picea abies* is the dominant conifer (Fig. 6.2).

The northermost archipelago, extending from the city of Piteå to the Finnish border, here referred to as the Luleå archipelago, is characterized by low, sandy islands covered with Norway spruce and with a border of sea buckthorn *Hippophae rhamnoides* or grey alder *Alnus incana*. The shore vegetation includes some special arctic elements such as *Primula sibirica* and *Potentilla egedii* as well as endemic species, e.g. *Artemisia botnica*.

The Blekinge archipelago is the southernmost of the Swedish archipelagoes and differs from those previously described, above all through the

Fig. 6.2. View from the island Judskär in the Ängskär island group, Stockholm archipelago. This island group is situated approximately 20 km from the coast, on the outer fringe of the middle zone. The vegetation is characterized by patches of deciduous forest, mixed with open areas covered with creeping juniper *Juniperus* shrubs, grassland or a heather moor type of vegetation. Substantial parts of the islands are bare rock, shaped by the ice during the latest glaciation, and lack higher vegetation. (Photo: Anders Forsman.)

greater prevalence of southern deciduous trees. Oak *Quercus robur*, replacing Scots pine on dry soils, beech *Fagus silvatica* and other southern deciduous trees give the landscape a more clement appearance. The southern deciduous trees also play an important role in some of the archipelagoes in the larger lakes; for example, in Lake Mälaren many islands are characterized by their great abundance, primarily of oak and small-leaved lime *Tilia cordata*. On the other hand, the archipelagoes in Lake Vänern are quite similar to the middle part of the Stockholm archipelago. Typical for the archipelago on the west coast is the scarce vegetation even on islands close to the mainland. Trees are lacking on most islands, and the shore vegetation is characterized by species adapted to more saline conditions, such as sea kale *Crambe maritima*, sea rocket *Cakile maritima* and common scurvy-grass *Cochlearia officinalis*. The forest patches that occur on large islands and in more sheltered places on smaller islands are dominated by southern deciduous trees.

6.1.2 Cultural history

Human impact on the island in the different archipelagoes started when the islands were sufficiently large to be used by man—a function of the time since an island rose above the sea. In the Luleå archipelago the mean age of the islands is approximately 1000 years and the maximum about 3000 years, thus human impact is a relatively recent phenomenon. Two of the larger islands in the Stockholm archipelago, Yxlan and Blidö, probably had enough arable land to sustain a human settlement in 1000–1500 BP. The first documentation of people living on these islands is a tax roll from the early 16th century, but man has probably been permanently present there from the Viking era or early medieval times. Up to the beginning of this century agricultural activities were intense on most islands supporting enough vegetation. As farmers normally combined fishing and farming, most families had fishing cottages in the outer parts of the archipelagoes. In summer cattle and/or sheep were released to graze on many of the islands. On others grasses and herbs from the shore meadows and foliage of deciduous trees were harvested and stored to feed the livestock during winter. Hunting, primarily for birds such as eider *Somateria mollissima* and velvet scooter *Melanitta fusca* and for grey seal *Halichoerus grypus*, harbour seal *Phoca vitulina* or ringed seal *Phoca hispida*, was conducted in the outer parts of the archipelago.

Nowadays only a few active farms remain in the archipelagoes, on some of the largest islands closest to the mainland. The small cottages on the islands in the outer parts of the archipelago are used for recreation during a few weeks in summer only. On the other hand, a new group of people has become much more abundant, namely holiday-makers yachting among the islands.

Summer visitors began to favour the archipelagoes as early as in the 18th century. Influenced by the romantic trends of the time, and of the smell in the cities, wealthy citizens fled from them during the summer. Because of difficulties of transportation only the closest islands were favoured by these vacationists. In the 1840s the first regular steamship routes from Stockholm were inaugurated, marking the start of a new era. In 1930 the number of people leaving Stockholm to spend the summer in the archipelago had increased to 30 000–40 000, and similar developments took place in the other archipelagoes. In the 1940s and 1950s the movement of residents from the islands to the urban areas accelerated, while the number of summer visitors continued to increase. With a holiday for all stipulated by law and a continually increasing standard of living, the number of summer cottages in the inner parts of the archipelago, and on the adjacent mainland, grew rapidly as did the number of yachts. This pattern is even more pronounced today, especially in the number of yachts cruising in the archipelago.

6.1.3 Why are archipelagoes important for nature conservation?

A prime reason for preserving archipelagoes is that the islands often present types of natural habitat not found elsewhere. As noted earlier, the unique combination of climatic, geological and topographical factors creates very special environmental conditions. This leads to the formation of unique species assemblages and, eventually, to the evolution of varieties, subspecies and even species. For instance, the vegetation of the outer parts of the archipelagoes resembles that of high alpine areas more than anything else, but still possesses many species and varieties special to the coastal areas. Three endemic plant species, *Deschampsia bottnica*, *Euphrasia bottnica* and *Artemisia bottnica*, occur in the northern most archipelagoes, and a number of endemic subspecies of other species have been described from various archipelagoes. Another example is the ongoing vegetation succession, caused by new land being continually made available as a result of the land rise. Both this and the ice disturbance of the shore vegetation are favourable to species like the sea buckthorn. The closeness of the sea also creates other special biotic conditions for terrestrial organisms, since material from the aquatic ecosystem is perpetually being transferred to the terrestrial, and vice versa. The archipelagoes also have a special avifauna with species such as white-tailed eagle *Haliaeetus albicilla*, eagle owl *Bubo bubo*, turnstone *Arenaria interpres*, arctic tern *Sterna paradisaea*, Caspian tern *Sterna caspia* and the rock pipit *Anthus spinoletta littoralis*.

Effects of insularity as such can change the prerequisites for different ecological processes; thus, for example, a lower number of predators and competitors enables species to exist which elsewhere might be unable to survive. As an example it has been suggested that reduced competition from other anuran species may allow the natterjack toad *Bufo calamita* to persist on a number of islands in the archipelago off the west coast (Andrén & Nilsson, 1985). Even though no endemic vertebrate species have evolved in any Swedish archipelago, examples can easily be found where natural selection has favoured different traits in the archipelago than on the mainland. Thus, for example, the mean weight of an adult field vole *Microtus agrestis* in the Stockholm archipelago is approximately 50% greater than that of an average mainland adult, and this difference is genetically determined (Ebenhard, 1990; Ebenhard & Forsman, in prep.). Mainland and island populations of the most important predator of field voles in the Baltic archipelagoes, the adder *Vipera berus*, exhibit genetically determined differences. In mainland adder populations only a few percent of the individuals are melanistic but figures up to 40% are not exceptional among island populations (Forsman & Ås, 1987). Island adders also have larger heads in relation to their body length compared to those from a mainland population. It seems probable that the morphology of the adder head is related to the size distribution of prey available, and the variation in relative head size between different island populations is indeed correlated with the mean body size of their main prey, the field vole (Forsman, 1991).

When comparing to what extent habitat variables affected the composition of carabid beetle species assemblages it was found that environmental descriptors explained substantially less of the variation in species composition in both archipelagoes investigated than they did in adjacent mainland areas. This indicates that different ecological processes vary in importance in an archipelago compared to the mainland (Ås, 1985).

Remnants of habitats which have largely disappeared on the mainland may still occur in archipelagoes, especially in lakes and close to the mainland, since islands have not been subjected to habitat changes to the same extent as have most mainland areas in the past hundred years. This is true both for old cultural and for natural habitats. An example can be taken from a comparison between forests in the archipelagoes and those on the adjacent mainland made by the Swedish Forestry Department (Skogsvårdsstyrelsen). The forests in the Stockholm archipelago and in the archipelagoes in Lake Mälaren have more old trees than found on the mainland and, despite a somewhat lower productivity, more trees with a stem diameter exceeding 25 cm for all tree species compared (Kihlblom, 1990).

Another important conservation biology function of archipelagoes is that they can serve as model systems in which knowledge about the effects of landscape fragmentation can be gained. Most of the hypotheses used to predict changes in fauna and flora due to fragmentation were originally developed for islands.

6.2 THEORIES OF INSULARITY

6.2.1 A static versus a dynamic viewpoint

Formerly, nature was looked upon as a static system where the distribution of a species solely depended on whether its habitat was present or not. If a species was missing from a suitable area, an historical explanation was sought; for example, the species was thought still to be expanding its range and not yet to have reached the area. This view was an inevitable consequence of the descriptive approach to biogeography up to the late 1950s. The autecology of each species was studied as a separate unit (e.g. Boycott, 1934; Lindroth, 1945, 1949). Since it is never possible to look at a habitat with the animals' eyes, the absence of a species from an apparently suitable area, although it was present nearby, was assumed to be due to the lack of some essential property of the habitat, not yet understood. But the number of unexplainable absences accumulated, and it gradually became apparent that species were missing from areas that they were perfectly able to inhabit and ought to have colonized a long time ago (see, for example, Carter & Prince, 1988).

The empirical evidence of relationships between species number and island area dates back to the beginning of this century. These early works mostly described specific situations (Brenner, 1921), or addressed questions such as whether power or logarithmic functions best fitted the relationships observed (Arrhenius, 1921; Gleason, 1922). The Finnish plant ecologist Alvar Palmgren even suggested an explanation for the relationships between the number of plant species and island areas as early as 1917 (see Haila & Järvinen, 1982). These early investigators invariably attributed such relationships either to habitat differences or phenomena related to the sampling hypothesis described below.

In 1948 E.G. Munroe, in his doctoral thesis on the distribution of butterflies in the West Indies, formulated a theory where the number of species was determined by an equilibrium between immigration and extinction rates. In the scientific climate at that time, his theory passed virtually unnoticed. In the 1960s, when MacArthur and Wilson (1963, 1967) and Preston (1962) independently formulated theories similar to those of Munroe, the scientific community was prepared. The success of a mathematical approach in formulating population and genetic models, eventually leading to the synthetic evolutionary ecology, and a growing desire to leave the descriptive mode of science, had produced a soil in which these novel ideas could grow and release an immense amount of research. Thus theories were formulated in which a species' presence in an isolated area was thought to depend not only on the availability of suitable habitat, but also on how often individuals might reach it, and their probability of building up a population and then persisting for a certain time span (see discussion in Haila & Järvinen, 1982; Brown & Lomolino, 1989).

Many of the theories advanced to explain species richness and community composition are not mutually exclusive and their predicted effects are probably not detectable in all situations. Their practical usefulness is that they have provided us with a conceptual framework within which we can evaluate the importance of certain mechanisms, and they have directed our attention to the importance of dynamic processes, such as immigrations and extinctions, as opposed to static patterns.

6.2.2 Species richness theories

As mentioned above, ecologists have for a long time known that the number of species in an area tends to increase with its size. This holds both for islands of different sizes and for an increasing number of samples examined within a community. In many cases, this relationship is linear when log number of species is plotted against log area (e.g. Arrhenius, 1921; Williamson, 1981), although the plot of species number against log area sometimes gives a better fit than the log–log plot (e.g. Gleason, 1922; Connor & McCoy, 1979). When written in the log–log form, the regression equation for species number (S) against area (A) is:

$$\log S = c + z \log A$$

where the coefficient z is usually between 0·15 and 0·40 (Williamson, 1981). Several theories have been put forward to explain the species–area relation. Although they are not mutually exclusive, they may be divided for convenience into two groups: (1) those acting through an increase in the total number of individuals with increasing area, and therefore more species, including the equilibrium theory of MacArthur and Wilson (1967); and (2) those acting through an increase with area in habitat diversity and/or number of niches.

More individuals, more species: area per se

All else being equal, larger areas can support more individuals of a given taxon, e.g. carabid beetles or vascular plants. Since there usually is a positive relationship between the number of individuals and the number of species, it follows that a large island with many individuals will hold more species than a small island with few individuals. This explanation for species–area curves assumes that such relations are a result of a sampling phenomenon where larger areas receive a larger sample of individuals than smaller ones. The sample should ideally be drawn from the propagule pool of dispersers (e.g. Connor & McCoy, 1979), but this pool is usually approximated to either a population consisting of all individuals in an archipelago (Coleman *et al.*, 1982) or a population in a larger geographical region (Haila, 1983). This view denies the importance of population processes and habitat diversity as explanations for species–area curves.

Under the sampling hypotheses, the total amount of resources is assumed to increase as island size increases, and therefore more individuals will be found on larger islands. The number of individuals on an island may also depend on productivity per unit area, more productive areas holding more individuals and hence more species. If productivity increases with island area, this would also contribute to the positive relation between species number and area, and increase the slope of the log S–log A-regression above that found according to the other hypotheses discussed in this section.

The sampling hypotheses together with the equilibrium theory (MacArthur & Wilson, 1967) may be viewed as area *per se* hypotheses for species–area relations when contrasted to the theories for species richness acting through increased diversity (below). However, the equilibrium theory of MacArthur and Wilson (1967) is different from the sampling hypotheses in several respects. These authors suggested that the number of species on an island is determined by a balance between immigration of new species and extinction of species already present. As the number of species on an island increases, the immigration rate (i.e. the number of new species arriving per time unit) decreases as a greater proportion of individuals arriving on the island will belong to species already present, and when an island has all the species in the pool, immigration rate is 0 (Fig. 6.3). Extinction rate (i.e. the number of extinctions per time unit), on the other hand, was suggested to increase with the number of species on the island. Assuming that the per-species extinction probabilities are equal in all species, the more species there are, the more extinctions will occur (the non-interactive extinction rate in Fig. 6.3). If, on the other hand, the species on an island interact in ways that increase the probabilities of each species going extinct if more species are present, such an interactive extinction rate will curve upwards compared to the non-interactive extinction rate (Fig. 6.3). Ecological processes giving rise to such increased per-species extinction probabilities when more species are present include interspecific competition (e.g. Bengtsson, 1989) and perhaps also various types of predation (see below under 'Changes in species composition with fragmentation').

MacArthur and Wilson (1967) argued that the probability of a species going extinct on an island was dependent on its population size, small populations being more prone to go extinct, while large ones could persist for a very long time. Since, on average, population sizes are likely to be

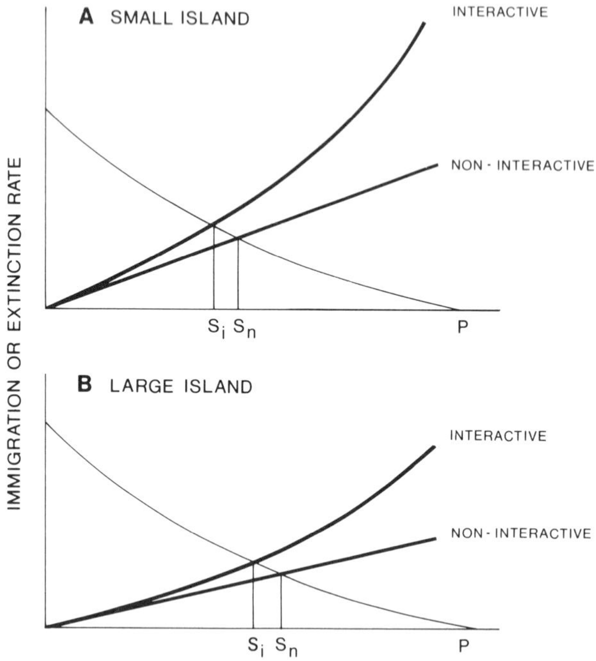

Fig. 6.3. The MacArthur and Wilson (1967) equilibrium theory of island biogeography. *x*-axis, number of species on an island; *y*-axis, immigration or extinction rates (in number of immigrants or extinctions in a given time unit); ——, immigration rate; ——, extinction rates (interactive and non-interactive). The intersections of the immigration and extinction rates give the equilibrium number of species on the island, S_i and S_n for interactive and non-interactive species, respectively. *P* is the number of species in the species pool (usually the mainland). The number of species is shown for a small island with high extinction rates (A) and a larger island with lower extinction rates (B). Note that the equilibrium species number is higher on the larger island, and thus the equilibrium theory is one of several that can explain the species–area relation. See text for further explanations.

larger on larger islands than on smaller ones, extinction rates are likely to be lower on the former. This is illustrated by comparing the upper and lower parts of Fig. 6.3. It is through this mechanism that the equilibrium model can generate the species–area relation, since the equilibrium species number becomes lower with higher extinction rates.

The equilibrium theory also suggests that the number of species should be lower on isolated islands than on islands of similar size near the mainland or in archipelagoes, since immigration rates are lower the more isolated the islands are. Since its publication, this theory has been refined in various ways, for example by incorporating an effect of isolation on extinction rates (Brown & Kodric-Brown, 1977). This 'rescue-effect' leads to decreased extinction rates on islands close to each other, since populations dwindling towards extinction can receive immigrants from the nearby islands, thus enhancing the probability of survival.

Several predictions result from the equilibrium theory, for example, that while there is a turnover of species on an island, its species number will remain fairly constant. The equilibrium theory, as well as the sampling hypotheses and that of habitat diversity, predicts that there will be a positive relation between species number and area. As will be seen below, the equilibrium theory need not be in conflict with some versions of the sampling or habitat diversity hypotheses, and it may actually often be difficult to distinguish between them as causes for a particular species–area relationship.

The MacArthur and Wilson (1967) theory has been criticized in various ways (see, for example Williamson, 1981 or a discussion of its pros and cons). For example, its predictions may be equivocal, and evolutionary processes were ignored in the original versions. Nonetheless, the theory was immediately a huge success, and has been tremendously important in several areas of ecology, such as community ecology, life history theory, and conservation biology. This was not because it could be used to explain species–area relationships; it has rarely been shown to be the cause for such a relationship. Rather, it offered a conceptual innovation and focused the attention of researchers on important and interesting ecological processes on islands and in patchy environments in general, i.e. dispersal, immigration, colonization and extinction (Haila & Järvinen, 1982).

Increased habitat diversity

For a given area, the number of species is often higher when habitat diversity or the diversity of resourses increases. As more niches 'become available', more species may co-exist. Similarly, as larger areas are investigated, more habitats will be included and the number of species will consequently increase with area (Williams, 1964). If environmental heterogeneity increases with increasing island area, as has been suggested in many cases, then large islands will contain more species, regardless of sampling phenomena and the population and community processes discussed by the equilibrium theory. Habitat diversity has often been suggested to be *the* cause of species–area relationships (Lack, 1976), although the circumstances when this hypothesis can be stringently evaluated against the area *per se* hypotheses are rather special (see below).

There are other factors than habitat diversity that may give rise to differences in the amount of niche space and hence in species richness within an area—for example, predation and disturbance (Paine, 1966; Harper, 1969; Huston, 1979). Many have been largely overlooked in relation to species–area relationships, and their effect on species richness is often far from clear. The influence of some of these factors on species richness may be dependent on habitat area. For example, disturbance may be more severe on smaller habitat patches (McGuinness, 1984). Large predators may be absent in small patches, and the impact of predation on community composition may thus be different on differently sized islands (cf. Oksanen *et al.*, 1981; Fretwell, 1987; Oksanen, 1988).

Thus there are many factors that may give rise to differences in species richness between habitat patches. If the effects of a factor depend on patch or island size, it may generate a relation between species number and the size of patches or islands. While predictions from the several sampling hypotheses, the equilibrium hypothesis and the habitat diversity hypothesis for patterns of species–area relationships can be formulated, at least in principle, the effects of productivity, predation, disturbance, and presumably many other factors, on this relationship are still largely unexplored. However, many of these factors may act jointly to determine local species richness.

6.2.3 Theories of metapopulation dynamics

The equilibrium theory of MacArthur and Wilson (1967) dealt primarily with extinction and immigration rates as determinants of species richness on islands, i.e. a community level phenomenon. It soon became clear, however, that the theory could be reduced and applied to individual species and their distributions, dynamics and persistence in archipelagoes and patchy environments in continental areas. Levins (1969, 1970) coined the term 'metapopulation' for a regional population consisting of semi-isolated local populations which are established by colonists, persist for a shorter or longer time, and eventually go extinct. Models of metapopulations dynamics and regional persistence of species have become increasingly important in conservation biology, but empirical evaluations of the concept have been scarce. Archipelagoes, being naturally patchy environments, provide excellent opportunities for testing the various metapopulation models.

The basic metapopulation model is simple. Local dynamics within patches are ignored, and instead the number of local populations in a region in relation to extinction and immigration rates in modelled (see Hanski, 1987 for a review). The basic model discussed by Levins (1969) can be used as an example of this type of model.

In this model, the change in the proportion of habitat patches occupied by a local population of a species is given by

$$\frac{\mathrm{d}p}{\mathrm{d}t} = mp(1-p) - ep$$

where m is a colonization parameter, e is an extinction parameter, and $p = N/T$, where N is number of occupied patches and T is the total number of available patches in the region. At equilibrium,

$$\frac{\mathrm{d}p}{\mathrm{d}t} = 0$$

and thus the equilibrium proportion of occupied patches $\hat{p}$ is

$$\hat{p} = 1 - \frac{e}{m}$$

The key parameters in this model are m and e. The colonization parameter m measures the number of patches receiving colonizers from one particular patch in a given time unit (including already inhabited patches). In empirical studies this can usually be calculated as the probability that an empty patch is colonized divided by the proportion of occupied patches. Likewise, the extinction parameter e can usually be calculated as the probability of a local population going extinct in a given time interval (see, for example, Bengtsson, 1991). Thus the proportion of occupied patches depends on extinction and colonization parameters in the system. In this deterministic model, a species persists if the equilibrium number of occupied patches, $\hat{N}$ is larger than 1, which may be achieved by increasing T or m, or by decreasing e. It makes one point of special importance for conservation biology. If, because of fragmentation and other human activities, the number of available habitat patches decreases, the distance between them increases, and/or their size decreases (i.e. decreasing T and m, and increasing e, respectively), then a species formerly common may become regionally rare, i.e the proportion of available patches that are occupied will be reduced (Hanski, 1982). If the fragmentation processes are allowed to continue, a regional population may suddenly disappear despite there still being a number of inhabitable patches present.

The basic model can be extended in various directions. Extinction and colonization rates can be influenced by interspecific interactions such as competition (e.g. Slatkin, 1974; Hanski, 1983) or predation, and degree of isolation. Differences in population size and density between local populations can also be taken into account. An extreme version is the core-satellite model (Boorman & Levitt, 1973), in which the local subpopulations can be classified into a few that are very large and highly persistent, and several that are small and unstable. In such a model the regional persistence is strictly dependent on the large patches. In other models, such as the source-sink model (Pulliam, 1988), demographic processes may differ between patches, so that certain patches (sources) support subpopulations producing large numbers of emigrants, while other patches (sinks) are mainly recipients of emigrating animals from the source patches (Harrison, 1991).

A stochastic version of the Levins model (Nisbet & Gurney, 1982) makes one particularly important prediction. In the deterministic model, a species persists as long as $N \geqslant 1$, but when demographic stochasticity is introduced, the equilibrium number of occupied patches $\hat{N}$ must be larger than approximately $3\sqrt{T}$ for a species to persist in the long term. This means that the number of available patches, T, in most cases must be considerably larger than 10.

All the extensions of the model have in common that they focus on several key factors in metapopulation systems, i.e. dispersal, colonization, persistence, and extinction. Metapopulations may be described as systems in which patterns of distribution and dynamics in an entire region are important for processes at the scale of local populations, and vice versa. Recent Scandinavian studies of organisms in various kinds of archipelagoes, such as voles on islands, frogs in coastal ponds, and waterfleas in rockpools on islands, will be used to illustrate the utility of metapopulation models as a conceptual framework for populations living in patchy environments (see 'Scandinavian case studies').

6.2.4 What are the features of a successful island species?

r- and K-selection and island populations

A question with a long history in ecology is: 'What are the features and traits that characterize colonizers and island species?'. It has often been assumed that pioneers and colonizing species are generally *r*-selected species, and that long-established island species are K-selected (Baker & Stebbins, 1965; MacArthur & Wilson, 1967; Pianka, 1978; Safriel & Ritte, 1983). r- and K-selected species were also assumed to be characterized by a suite of correlated demographic and morphological traits. A typical r-selected species was described as a small, short-lived, semelparous species with rapid sexual maturation, large litters and low competitive ability, while a K-selected species would have the opposite traits. Empirical patterns found in island species sometimes fit these descriptions, but never completely, and often not at all. Furthermore, the theoretical basis for such descriptions of island species is weak. Theoretically, r- and K-selection may occur under certain conditions (Boyce, 1984), but those conditions are not necessarily met during island colonizations and persistence.

The rise and fall of an island population

The dynamic island biogeography theory presented by MacArthur and Wilson (1967) describes an island population as finite both in time and space. This means that it has a defined time to extinction, and that 'empty' islands have a certain probability of being colonized by a given species. The life of an island population starts when dispersing individuals, the propagule, reach the empty island. Then follows an increase in numbers during the colonization phase, which ends when some kind of equilibrium population size has been reached. At any time during the colonization, or subsequently, the population may go extinct.

The colonization phase is usually much shorter that the ensuing period of persistence, and different factors determine the extinction probability during the two phases. Hence it is practical to describe colonization success as $P_s(K, m)$, the probability of population survival during the increase in numbers from a propagule with m individuals to an equilibrium population size (K, 'carrying capacity') and to express persistence as T_K, the mean time to extinction of a population which has reached K.

Dispersal. Dispersal from a source area such as a mainland or neighbouring islands is a prerequisite for a colonization to take place. Two components determine the size and frequency of propagules, viz. the degree of isolation on the island and the ability of the individuals to move through or over the intervening 'sea'. Any trait which increases this ability will increase the probability of a propagule reaching the island. In voles, for example, large body size which minimizes heat loss and maximizes endurability during swimming may be such a trait (Lomolino,

1984). In other groups, the opposite may be true; for terrestrial snails small body size has been assumed to increase dispersal ability (Vagvölgyi, 1975). Wings may be thought of as good adaptations for dispersal, but this depends on whether wings or some other trait, such as floating ability in carabid beetles (Ås, 1984), are the most efficient means of dispersing.

Colonization. In a simple analytical model employing demographic stochasticity MacArthur and Wilson (1967) estimated colonization success as

$$P_s(K, m) = 1 - \left(\frac{\mu}{\lambda}\right)^m$$

where μ and λ are instantaneous rates of mortality and natality, respectively, and m is propagule size. Colonization success in this model is hence dependent on propagule size, as well as on the ratio of natality to mortality, as opposed to the difference ($r = \lambda - \mu$) between them. The kind of demographic stochasticity used by MacArthur and Wilson may be described as homogeneous, which means that this model is applicable to organisms in which all individuals are demographically similar, i.e. have identical probabilities of dying or giving birth.

In a simulation model Ebenhard (in prep.) analysed the effects on colonization success of heterogeneous demographic stochasticity, in this case sexual reproduction with two sexes, age-structure, time-lags due to pregnancy, and non-random family sizes. The ratio λ/μ and propagule size were found to remain important determinators, but this model generally predicts lower colonization success than does MacArthur and Wilson's, due to the larger stochastic fluctuation in population size caused by the heterogeneous demography.

Leigh (1981) analysed the effects of a varying environment on mean time to extinction of a propagule, through a model with environmental stochasticity. He found r, the intrinsic rate of increase in population growth ($r = \lambda - \mu$), to be the most important parameter. The main difference between λ/μ (proportional to R_0) and r is the lack of a time dimension in the ratio. Depending on the time-scale of environmental variation during the colonization, λ/μ or r may be the most important demographic parameter. If environmental disturbances are infrequent, there is a probability that most colonization phases, which are relatively short, will not be affected by any disturbance. In this case a high ratio λ/μ is important. If, on the other hand, disturbances are frequent, increasing the population size as rapidly as possible, i.e. a high r, is an important feature reducing the risk of extinction.

Good colonizers have often been described as 'r-selected' species (*sensu* Pianka, 1970), but this is a simplification that can be misleading, for several reasons. First, a high value of r is not always advantageous; depending on the frequency and duration of environmental disturbances, a high ratio λ/μ may be more important. This may be achieved even though r is very small. Second, in the classical 'r-selection' syndrome of demographic traits, a high r is thought to be achieved through a high reproductive rate (λ), whereas a good colonizer may achieve a high ratio λ/μ or r through a low mortality (μ), and hence rather resemble a classical 'K-selected' species. Third, a good colonizer should have a low demographic heterogeneity, that is, early sexual maturity and continuous asexual reproduction with a short gravidity time. Such traits have also been attributed to 'r-selected' species, since they generally increase the reproductive output. That effect may be important to an island colonizer, but decreased fluctuations in population size due to demographic stochasticity, granted by low demographic heterogeneity, may be equally important (note, however, that a demographically heterogeneous population facing environmental stochasticity may show less fluctuation than a homogeneous population, due to the buffering effect of heterogeneity (Lomnicki, 1988)). Fourth, 'r-selected' species are often described as small, using allometrical relationships obtained in comparisons between a wide range of species, from bacteria to elephants. Within taxonomically more restricted groups body size is a less good predictor of r. It is, as an example, impossible to use body weight as a single criterion for colonization success when comparing different vole species. It is, however, generally thought that species with wide ecological niches, i.e. generalists, are better colonizers, since they are able to realize a high λ and a low μ in a wider range of habitats. This may be expressed as a greater phenotypic plasticity.

Persistence. In persistence models with only homogeneous demographic stochasticity (MacArthur & Wilson, 1967; Richter-Dyn & Goel, 1972), the time to extinction of a population at equilibrium, T_K, is a function of e^K, which makes extinction highly improbable in moderately large populations. More realistic models, often called minimum viable population (MVP) models, include environmental stochasticity as well (e.g. Shaffer, 1983). Such models predict T_K to be directly proportional to K, but also to be affected by temporal variations in K caused by the varying environment. The risk incurred during occasional periods of worse conditions can be minimized through two different strategies. If adverse conditions mainly affect reproduction, a low adult mortality is advantageous. Otherwise a high *r* is important to guarantee a rapid recover from a population low (Goodman, 1987).

Dispersal to an island may continue during the persistence phase, allowing population increase without reproduction (rescue effect), which should decrease the risk of extinction when the population is small (Brown & Kodric-Brown, 1977).

In these models, the size of the island—and hence of the population—and the pattern of temporal variations in resources are the most important determinators of population persistence on islands. The demographic properties of the organisms are comparatively less important. Genetic processes in small populations may also be of importance for long-term persistence (see chapter 4).

6.3 SCANDINAVIAN CASE-STUDIES

6.3.1 Island size and species richness. Area *per se*, habitat diversity or sampling?

One mistake often made is to assume that different theories for the species–area relationship should be mutually exclusive. It is as wrong to reject an area *per se* hypothesis when a connection between habitat diversity and species diversity is found, as it is to reject a habitat diversity hypothesis if species number is found to vary with island area. A related problem in studies of the effect of island area on species diversity is the normally high covariation between factors which all might influence an observed species–area relation. The only way to handle this problem of covariation is to find a situation or arrage an experiment, where island area and habitat diversity are uncorrelated.

A different type of problem is to distinguish between predictions from the sampling hypothesis and other area *per se* hypotheses, respectively. A simple null-model predicts a species' chance of reaching an island to be directly proportional to its abundance in the surroundings. More realistic models include assumptions about dispersal abilities, often based upon the distribution of the species in the archipelago under investigation (Simberloff, 1978*b*); however, the use of such null-models may reduce the possibility of detecting differences between the hypotheses.

The importance of habitat

In order to examine the relative importance of the area *per se* hypotheses and the habitat diversity hypothesis respectively, species richness and total densities of carabid beetles, land snails and woody plants on 17 islands in Lake Mälaren, central Sweden, were investigated. Since a detailed study of the vegetation on the islands had been made, it was possible to assess habitat diversity and calculate a number of estimates related to habitat heterogeneity as well as measures of vegetation structure, soil productivity and soil wetness (Nilsson *et al.*, 1988).

A prediction of the habitat diversity hypothesis (in the sense that the *reason* for a species–area relation is an increased number of habitats with area) is that area should not correlate with species number, unless number of habitats correlates with area. The islands investigated in Lake Mälaren did not show any relation between area and number of habitats or between area and habitat heterogeneity. Yet, species number of all three organism groups examined significantly increased with area (Fig. 6.4), and only land snails revealed a significant relation with any measure of habitat variability in a stepwise regression analysis. Measures of habitat quality were, however, significantly correlated with species number of both carabid beetles and land snails. A couple of years earlier the number of breeding land bird species had been counted on the same islands (Ahlén & Nilsson, 1982). The number of bird species showed a relation with area, as did carabid beetles, land snails and woody plants (Table 6.1). The conclusion to be drawn from this is that habitat-related factors can influence species richness but that area *per se* has a relation to species richness of its own.

Between 1882 and 1886 the water level of Lake Hjälmaren was lowered approximately 130 cm. As a consequence, a large number of new islands arose. Lists of plant species from these newly formed islands are available from 1886, 1892, 1903–04, and 1927–28. In 1984–85 a new inventory was made and the relations between area, habitat numbers, island shape and location on the one hand and species richness of vascular plants on the other were examined (Rydin & Borgegård, 1988). The number of species on each island had increased with time, indicating an ongoing succession, but nevertheless significant species–area relations were found, with *z*-values which were within one standard deviation of the expected, thus indicating no obvious relation between this value and equilibrium or succession. The logarithm of island area was the best predictor of species richness but for individual islands deviations from species number predicted from island area were best explained by habitat quality.

Nilsson and Nilsson (1978, 1982, 1983, 1985) studied the distribution and turnover of vascular plants on small islands (0·03–2·19 ha) in Lake Möckeln, south Sweden. Most of the islands were formed when the lake

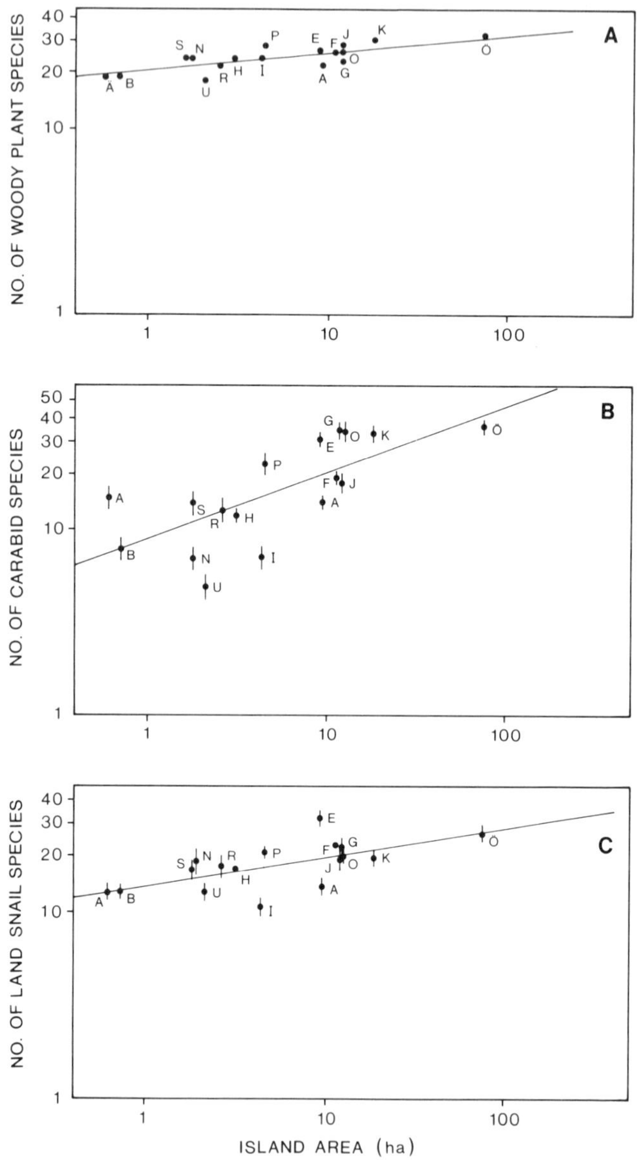

Fig. 6.4. Species richness of (A) woody plants, (B) carabid beetles and (C) land snails on islands in Lake Mälaren in relation to island area. Logarithmic scale on all axes. Species richness for carabid beetles and land snails are jack-knife estimates of the total number of species inhabiting the island. Bars show one standard deviation of the jack-knife estimate. See also Table 6.1. From Nilsson *et al.*, 1988.

Table 6.1
The relation between species number and island area in different organism groups on islands in Lake Mälaren—Results of linear regression between log species number and log island area (log S = z*log A + C) (n = 17). (from Nilsson *et al.*, 1988)

Organism group[a]	*z-values (slope)* (±95%*CL*)	*t-test*[b] *(slope)*	*Intercept* (±95% *CL*)	*t-test*[b] *intercept*	F	$P<$	R^2
Woody plants	0·101(±0·043)	A	1·315(±0·038)	A	20·8	0·0004	0·581
Land snails	0·156(±0·091)	AB	1·144(±0·080)	B	11·4	0·0042	0·431
Carabid beetles	0·361(±0·180)	B	0·948(±0·158)	C	15·6	0·0013	0·509
Forest birds	0·616(±0·109)	C	0·604(±0·096)	D	122·1	0·0001	0·890

[a] Jack-knife estimates of species number were used for land snails and carabid beetles. Species number for forest birds are from Ahlén & Nilsson (1982).
[b] *z*-values (or intercepts) with the same letter do not differ significantly on the 0·05 probability level (t-test; d.f. adjusted because of unequal variances, but always larger than 17·8).

was lowered about 1 m in the 1850s. There were no significant differences in habitat diversity between large and small islands, but there was still a significant log species–log area relation with slope (z-value) of 0·30. In addition to area, some variables such as the amount of stony ground and canopy density had an influence on species richness on these islands, in both cases negatively. Data on turnover suggest that the species–area relation on these islands can be explained, at least partly, by the equilibrium theory (Nilsson & Nilsson, 1982; see below).

The sampling hypotheses

In its simplest form the sampling hypothesis can be considered as a 'null model', invoking no biological processes to explain the species–area relationship (Connor & McCoy, 1979). It was tested by Niemelä (1988) and Niemelä *et al.* (1985) using data on the distribution of carabid beetles from the Åland archipelago. By randomly drawing the same number of individuals caught on an island, or habitat patch, from a pool of species where each species occurred in the same proportion as in the mainland sample, a random species abundance distribution was created. Repeating this a sufficient number of times makes it possible to calculate expected properties of a community as well as the probability of finding a particular species in a smaller sample from an island. This so-called rarefaction technique (Simberloff, 1978*a*; James & Rathbun, 1981) is regularly used to evaluate deviations from a random sample.

The rarefaction technique was applied to the species abundance patterns of carabid beetles in the Åland archipelago both on different islands and different patches on the main island (considered as mainland) (Niemelä, 1988), and as comparisons between specific habitat types on main Åland and different islands (Niemelä *et al.*, 1985).

Looking at patches of lush deciduous forest on main Åland and on small islands in the archipelago, a simple sampling hypothesis was found inadequate to explain the patterns found. No difference in vegetation structure between island and mainland patches could be detected, but species assemblages of carabid beetles were clearly different (Fig. 6.5). The number of individuals caught on main Åland was more than four times that caught on the islands, with the same sampling effort. Since the rarefaction curves calculated from the pooled material from mainland and island areas, respectively, did not differ significantly, the increased number of species found in main Åland locations was explicable by the differences in sample size. But seven of the 13 most abundant carabid species deviated significantly on the islands from the expected abundance values, based on mainland species abundances (Fig. 6.6).

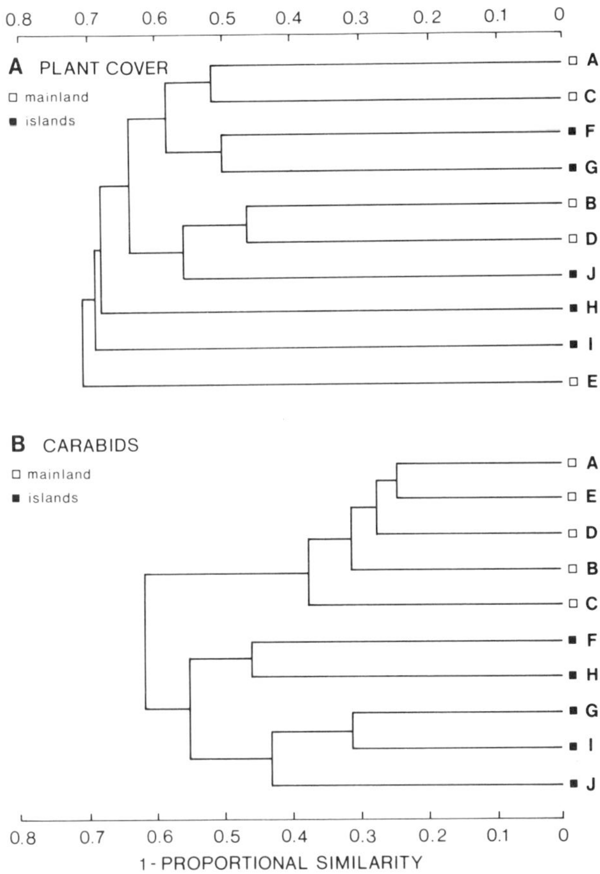

Fig. 6.5. Carabid beetle communities in different sampling sites on islands in the Åland archipelago (sites F–J) and sampling sites on main Åland (sites A–E). The sampling sites are grouped with a cluster analysis (proportional similarity); each site is connected to the most similar site, or group of sites, according to their vegetation (A, above) and carabid communities (B, below). The position of the connections between the horizontal lines along the *x*-axis reflects the degree of similarity. From Niemelä *et al.*, 1985.

A similar pattern was also found in data from several habitat types on islands and mainland. Species were found to use roughly the same habitats on the islands as on the mainland. Within lush deciduous forest patches the distribution of the species abundances was more even on islands than on main Åland, and species composition was also more similar between habitats on islands than on the mainland. The explana-

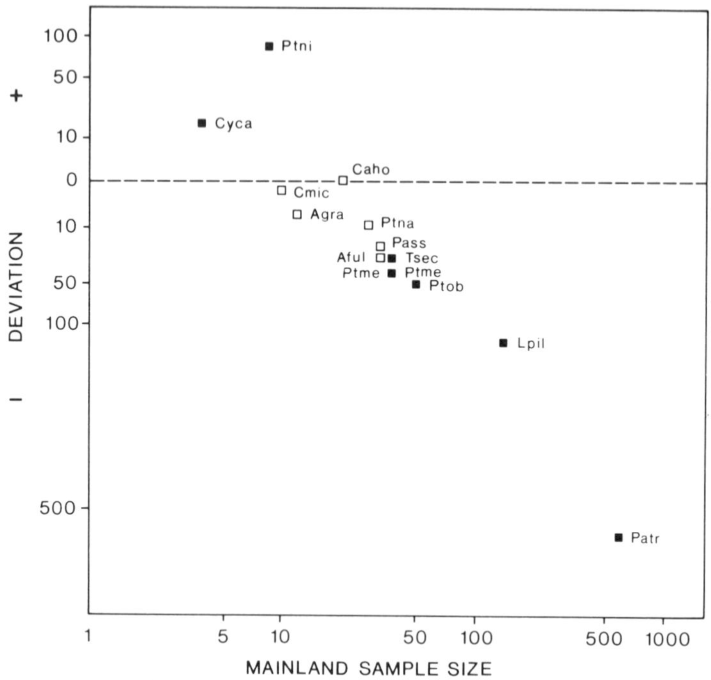

Fig. 6.6. The deviation between island and mainland abundances for 13 carabid species from the Åland archipelago, as a function of their mainland abundance (■, statistically significant difference from equal abundance on mainland and islands). For name codes, see original reference. From Niemelä *et al.*, 1985.

tions suggested for the deviations from the passive sampling hypothesis were differences in habitats between islands and Åland (the size of the habitat patches might be important), differences in dispersal ability, and interspecific interactions. Even clearer deviations from a passive sampling model, in terms of species numbers and species composition, were found in a comparison between the carabid fauna on the Dutch mainland and recently formed polders (Ranta & Ås, 1982).

Thus, there exist species–area relations in Nordic insular environments which cannot be explained by a habitat diversity–area relation, as on the Lake Mälaren islands and others, for which a passive sampling model is not enough, for example, islands in the Åland archipelago. It is imperative that the habitat diversity and the different area *per se* hypotheses are tested against each other, using the same empirical data set, which has not so far been done.

The situation for land birds breeding on islands in the Åland archipelago is quite different from that of ground-dwelling beetles. Most species are migratory and 'go extinct' once a year and recolonize next spring. Most land birds can also move between islands. Another important aspect for migratory birds is that they may be limited by factors operating elsewhere than at their breeding grounds.

The number of land bird species on any of the islands examined in the Åland archipelago was lower than predicted from a straightforward sampling hypothesis (Haila, 1983). If the sampling hypothesis is correct, each island should possess a random sample of the entire avifauna of the Åland archipelago. This, however, is likely to overestimate the expected number of land bird species on these islands, because pooling all islands generates a habitat diversity value higher than existing on any single island. However, taking habitat differences into account by dividing the islands examined into habitat groups, and by applying the rarefaction technique to each group, the observed species numbers were in accordance with those predicted from this modified sampling hypothesis (Haila, 1983).

The prevalence function derived by Haila *et al.* (1983) is defined as the ratio between the observed number of individuals of a species on an island and the expected number derived from its abundance in mainland areas. Prevalence functions have been calculated for breeding land bird species on islands in the Åland archipelago, using main Åland as mainland. The expected numbers were calculated as the density of the species in a particular habitat on main Åland times the area of the habitat on the island. The majority of species did not deviate from expected abundances, and Haila *et al.* (1983) suggested that deviations, with few exceptions, were caused by microhabitat differences, more edges on the islands and a mosaic structure of small patches on small islands. Rare birds did not occur less often than expected on islands in the archipelago or in habitats on a large island. The conclusions drawn from these land bird studies on islands in the Åland archipelago were that the land bird fauna on poorly isolated islands in northern environments is best viewed as samples from a species pool, where each species' probability of being represented depends on its autecological requirements. The only special feature of an island species was that habitat generalists were probably better off in insular as compared to continental situations.

By contrast, Nilsson (1986) found that there were differences in species composition of forest birds between small islands, large islands, and mainland areas in and around Lake Möckeln. The four most abundant species were over-represented on the small islands, and the proportion of hole-nesting birds decreased from the mainland to small islands. Nilsson (1977) suggested that relaxed interspecific competition allowed the common bird species to be more numerous on small islands—so-called 'density compensation'—although this conclusion was disputed by Haila *et al.* (1983). Alternative explanations for the density compensation on the small islands were later examined by Nilsson and co-workers. Nilsson *et al.* (1985*a*) found that rates of nest predation differ between small and large islands and mainland areas, but that the difference was too small to explain the observed differences in density. The food supply on the mainland and different-sized islands was also compared (Nilsson & Ebenman, 1981; Nilsson *et al.*, 1985*b*). In several instances, food density appeared to be lower on smaller islands. The authors concluded that interspecific competition was indeed the explanation for the increased densities of the common birds on small islands. Once again, the results pointed in the direction that habitat generalists will be favoured by fragmentation, while specialists such as hole-nesters are negatively affected by small habitat size.

6.3.2 Colonizations, extinctions, and metapopulation dynamics

Island colonizations and features of colonizing species

As noted earlier, the process of colonization must be preceded by a dispersal event. This has been investigated for plants on islands in two lakes, for field voles in the Stockholm archipelago and for carabid beetles in the Stockholm and Åland archipelagoes. The second phase, the establishment of a population, has to some extent been examined for carabid beetles in different archipelagoes. Several suggestions have been made on what features characterize a successful colonizing species, apart from good dispersal ability. To study this life history traits of land snails that have successfully colonized different islands have been examined, and for bank voles an experiment on the importance of propagule size and demographic parameters on colonization success is reported.

A major problem of interpretation is that observed patterns in species number or species presence/absence on islands are usually regarded as reflecting colonizing ability, when they are in fact accumulated effects of abilities to disperse, colonize, and persist on the island. In some of the case studies described below attempts have been made to control for at least one of these processes (e.g. Ebenhard, 1987*a*,*b*, 1989; Niemälä *et al.*, 1988).

Dispersal and colonization in plants. The data on plant species composition on the islands in Lake Hjälmaren formed after the lowering of the lake surface in the 1880s were used by Rydin and Borgegård (1991) to examine differences in a number of life-history traits between species having their greatest abundances on different successional stages, i.e. pioneers, early, mid and late successional species. They found that the pioneers formed a distinct group, differing from the other groups in a number of characteristics. Pioneers were usually annuals, predominantly autogamous with self-fertilization, and lacking vegetative reproduction. Many pioneers were also shore species. Among the other groups, no clear differences could be found in life history traits. Rydin and Borgegård also noted that some pioneer species later (after about 100 years) acted as vegetation formers, e.g. *Epilobium angustifolium* and *Populus tremula.* These appear to be a kind of 'superspecies', being both superior colonizers and good competitors late in the succession (contrary to the classical *r–K* selection theory; e.g. MacArthur & Wilson, 1967; Pianka, 1978).

Many of the plant species colonizing the Lake Hjälmaren islands appeared to be dispersed by floating on water, either as seeds or as vegetative parts. Wind dispersal by itself was probably less important, but the combination of wind and water dispersal, by which seeds were carried by wind to the lake and then by the water to the islands, was suggested to be frequent. Such 'anemohydrochorous' dispersal has previously been suggested to be common among insects in the Baltic (e.g. Palmén, 1944). On the Lake Hjälmaren islands, plant species which are dispersed by animals—mainly birds—arrived late in the succession.

Nilsson and Nilsson (1978) reported similar results for plants on islands in Lake Möckeln. Water-dispersed species and some wind-dispersed species had colonized a large number of the islands, while species dispersed by animals were found in lower frequency on the islands than on the mainland.

The conclusion that water dispersal is important does of course, hold only for islands surrounded by water. In fragmented mainland biotopes, wind dispersal is likely to be more important, and perhaps also dispersal by birds. As far as we know, few data exist examining this question.

Carabids in the Stockholm and Åland archipelagoes. The importance of a good dispersal ability for a successful island colonizer seems obvious in most cases. But is flight necessarily the best way of travelling to, or between, islands? Ås (1984) analysed differences in the proportion of ground-living carabid beetle species between those that always possess long hind wings and those always possessing rudimentary wings, between islands in the Stockholm archipelago. He showed that fewer long-winged species had colonized the islands than would be expected from a hypothesis of equal dispersal ability among the two types of species. Models of passive wind dispersal (exponential decline with distance) and rafting or floating on the sea surface (normal decline with distance) showed that the proportion of short-winged species should change with distance (see Fig. 6.7). The proportion of short-winged species predicted from the model, supplied with values for flight duration, survival time in brackish water and speed of small objects dispersed by waves (references in Ås, 1984), was very similar to the observed proportions on islands in the Stockholm archipelago. Actually, data suggested that floating on the water surface is a better means of travelling than is flying at medium

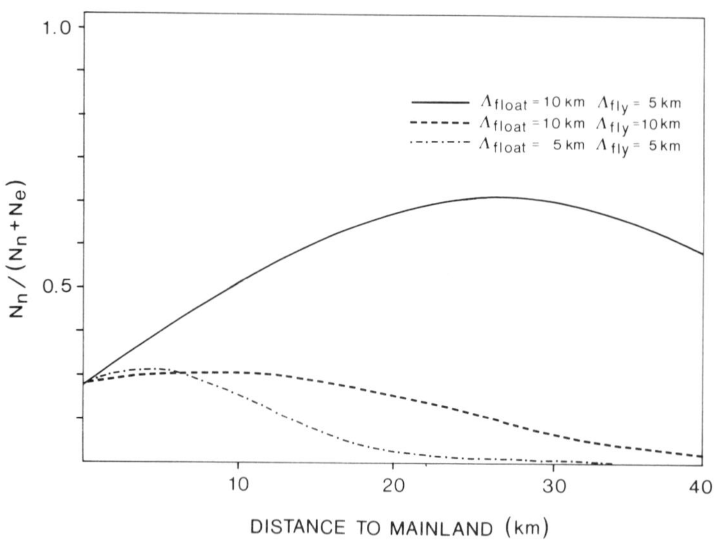

Fig. 6.7. The expected proportion of water-dispersed animals ($N_n/(N_n+N_e)$) at different distances from the mainland, calculated for three different combinations of mean distance travelling by sea, i.e. the mean distance of dispersal to islands for short-winged species (Λ float) and by air, i.e. the mean flight distance for long-winged species capable of flying (Λ fly). Λ float is dependent on survival time in water and Λ fly on flight duration and ability. From Ås, 1984.

distances under these conditions. The results underline that dispersal ability is important for carabid beetles habitually colonizing new islands and stress the importance of identifying the characteristics of a good disperser.

Niemelä *et al.* (1988) found clear differences in the colonization ability of carabid species, when analysing the differences in species composition between areas on main Åland, islands close to main Åland and two isolated islands (10 and 20 km off main Åland, respectively). But the good colonizers seemed to share only a few traits. They were found in the habitats that are most common on the distant islands, but the stochasticity of succeeding in long-distance dispersal over water seemed also to be important. One trait that appeared to feature prominently among most of the good colonizers on Åland as well as in the Stockholm archipelago is that they are habitat generalists. In the Stockholm archipelago, the mean habitat niche breadth for the species caught on islands in 1980 increased with distance to mainland (Ås, unpublished) (Fig. 6.8).

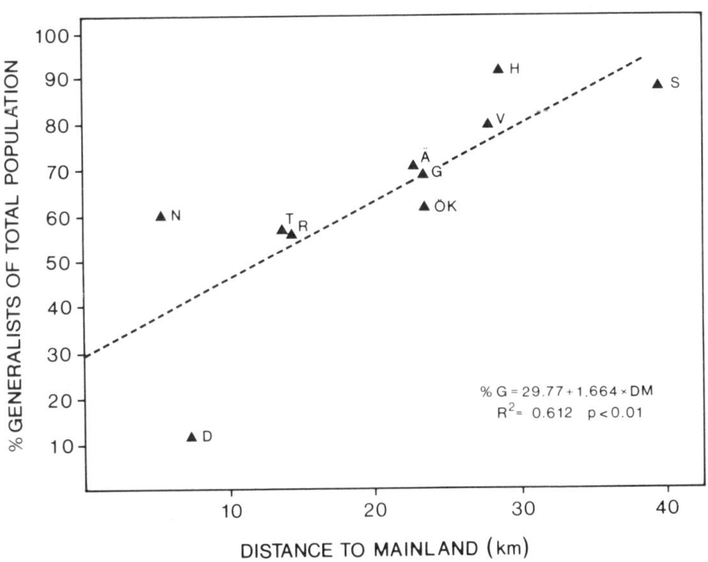

Fig. 6.8. The relation between the proportion of carabid beetle species with large habitat niche breadth ($B > 3{\cdot}0$; $G\%$) on island and distance to mainland (DM). Niche breadth was calculated with reciprocal Simpson index ($1/\Sigma\, p_i^2$, p_i being the proportion of occurrence of the species in the ith habitat) (Ås, unpublished data).

The parameter of niche breadth used in this investigation may be criticized, since it was calculated from the species distributions on the islands, but it is supported by autecological data in Lindroth (1945).

When analysing carabid beetle species assemblages on islands in comparison with areas on the adjacent mainland, a larger proportion of the variation in composition could be explained by habitat composition on the mainland than on the islands. This was observed both in a comparison between islands in Lake Mälaren and surrounding mainland and in one between islands in the Stockholm archipelago and the mainland close to the coast. In both mainland areas habitat accounted for approximately 60% of the variation in species composition while the corresponding figure for the islands was approximately 40% both in Lake Mälaren and in the Stockholm archipelago (Ås, 1985). This result is to be expected if stochasticity in the dispersal phase is important, for example, if an early immigrant species can exclude later arrivals and the order of arrival is a matter of chance. It might also be expected if dispersal ability is of greater importance than the habitat composition, and also if species found on islands on average occupy more habitats than do mainland species, i.e. island species are more generalistic.

Colonizing ability of land snails. Land snails have often been regarded as being relatively bad colonizers. Baur and Bengtsson (1987) and Bengtsson and Baur (unpublished data) examined whether good colonizers among land snails possessed any particular life history traits associated with *r*-selection, or any other recurrent characteristics. Data on land snail distribution on islands of different ages in two land-uplift archipelagoes in Finland, viz. Tvärminne and Quarken, have been published by Valovirta (1977, 1979). Baur and Bengtsson used these data and information on life history traits, abundance in different habitats, and mating systems. Three different measures of colonizing ability were used. In a first attempt, life history traits were correlated with (a) the observed order of colonization of different species, and (b) a colonization order that took into account the habitat requirements of the species (Baur & Bengtsson, 1987). A problem in many studies of life history evolution, however, is that many traits scale allometrically with body size, which may confound interpretations of life-history differences between organisms of different size. To overcome this problem, Bengtsson and Baur in a later study examined whether pioneer species possessed *r*-selected traits relative to non-pioneers, using an analysis of the residuals from the trait–body size regressions, rather than the actual values for the traits.

As regards the presumed *r*-selected properties of colonizing species, the results from these studies are clear. No correlations could be found between the life history traits usually associated with *r*-selection, e.g. body size, age at first reproduction, clutch size, egg size, lifetime fecundity, or simple measures of reproductive effort, and the two measurements of colonization ability (Baur & Bengtsson, 1987). When body size was taken into account and pioneers and non-pioneers were compared, the results again were inconclusive, showing no clear evidence for pioneers having particular life history traits associated with *r*-selection (Bengtsson & Baur, unpublished data). The conclusion from these studies, as well as from other studies of colonizing ability among land snails species, is that pioneers and colonizing species cannot be regarded as *r*-selected, but rather that several different combinations of life–history traits can be found among such species, these combinations having little in common with the *r*–*K*-selection concept.

If good colonizers among land snails are not *r*-selected, what kind of species are they? Habitat requirements clearly play a role in which species are found on islands of different age, but this factor is not the only one determining the orders of colonization, since many species are apparently absent despite the presence of a suitable habitat (Baur & Bengtsson, 1987). The data suggest that the good colonizers are mainly found among species that are habitat generalists rather than specialized on one or a few habitats (Fig. 6.9), and they are also more abundant in the source

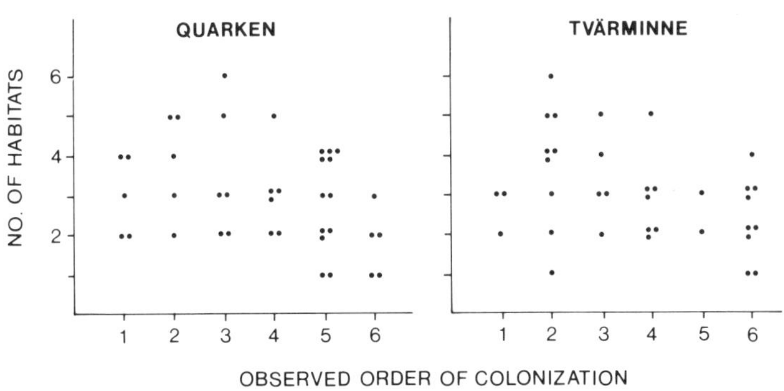

Fig. 6.9. Habitat niche breadth (the number of macro-habitat types in which a species is fairly common or abundant) of land snail species in relation to a measurement of colonizing ability: the order of colonization (1, early successional species; 6, species found only on the nearby mainland), in the Quarken ($r_s = -0{\cdot}31$, $p = 0{\cdot}052$)and Tvärminne (($r_s = -0{\cdot}34$, $p = 0{\cdot}043$) archipelagoes. From Baur and Bengtsson, 1987.

area, i.e. the nearby mainland. The limited data available also suggested that self-fertilization is more frequent among good colonizers, but this point should be treated with caution, since information about the mating system of many land snail species is highly inadequate (Baur & Bengtsson, 1987).

Thus, once again, these studies suggest that if anything characterizes good colonizers, it is that such species are habitat generalists rather than specialists.

Dispersal in field voles Microtus agrestis. As further described below, the field vole *Microtus agrestis* has a wide distribution in the Stockholm archipelago. Its local populations go extinct and recolonize islands in a pattern characterized by a high turnover rate. According to Lomolino (1984) there should be a selective advantage for large voles during the dispersal stage of such colonizations. Among the islands with recurrent vole colonizations more isolated islands should hence support larger voles. Ebenhard and Forsman (Ebenhard, 1988; Ebenhard & Forsman, in prep.) analysed voles from six groups of islands, along a gradient of isolation, and from two mainland localities, with the intention of testing Lomolino's hypothesis.

Insular field voles were found to be considerably larger (45–50% heavier as adults), and a principal component summarizing eight body and skull characters in a general size parameter was found to be positively correlated with both distance to mainland, maximum water-gap which had to be crossed during dispersal from the mainland, and degree of fragmentation within island groups. These three variables were also inter-correlated. Island area had no effect on the principal component. The results thus support the hypothesis that a trait that will maximize dispersal ability, and hence propagule size and/or rate of arrival, will be selected for in a patchy environment with population turnover.

Colonization experiments with bank voles Clethrionomys glareolus. The basic island colonization theory presented by MacArthur and Wilson (1967), relating the probability of colonization to the ratio of natality to mortality (λ/μ) and propagule size (m), has been tested by Ebenhard (1987*a*,*b*, 1989) in the Stockholm archipelago. A total of 40 propagules, consisting of bank voles *Clethrionomys glareolus* bred in a laboratory, were released on small islands in order to study the colonization process. Twenty propagules consisted of bank voles descended from voles caught at Ammarnäs, Lappland, at 66°N (N), and the remaining were voles from Skåne, at 56°N (S). Reproductive differences between the two provenances have been established in the laboratory (e.g. Gustafsson *et al.*, 1983), voles from the north being better reproducers through a larger litter size. But λ and μ are not wholly genetically determined population-specific constants; on the islands both natality and mortality also depend on the environment. During the field experiment, northern voles achieved a higher reproductive rate (λ) on the islands (Table 6.2),

Table 6.2

Natality (λ) and mortality (μ) rates and colonization success for different propagule sizes (m), for Bank Voles *Clethrionomys glareolus* from Northern (N) and Southern (S) Sweden in an introduction experiment on islands in Stockholm Archipelago

Provenance	λ	μ	λ/μ	*Colonization probability*			
				$m=2$	$m=4$	$m=10$	$m=20$
N	0·019	0·032	0·803	0·00	0·20	0·80	0·20
S	0·016	0·015	1·124	0·20	0·20	0·60	0·80

but their adult mortality was much higher than among the southern voles. This difference was probably caused by the greater amount of time the northern bank voles spent outside the protective runways. As a result, the ratio λ/μ achieved on the islands was larger for the southern voles. Furthermore, in this experiment the propagule size was varied ($m=2$, 4, 10, 20). The general prediction then was that large propagules should have a higher probability of colonization, and the southern voles should succeed more often. Counting absence of voles on an island after three months and the presence of voles of only one sex as events of extinctions during colonization, 85% of the small propagules (two and four individuals) went extinct, while only 40% of the large ones succumbed. Among northern voles (N) 70% of propagules failed to colonize, and among southern (S) 55% failed (Table 6.2).

It may be surprising that any northern propagules at all succeeded in colonizing an island, considering their ratio of natality to mortality (Table 6.2)—but this was calculated as a mean over 20 islands, and on a few islands it may have been significantly higher. These results confirm the paramount importance of demographic traits for colonization success.

Studies of extinction on islands and habitat patches

One of the most important research topics in conservation biology is the study of extinctions. What are their causes, and what makes certain species more extinction-prone than others? The MacArthur and Wilson equilibrium theory (1967) placed weight on one important determinant of extinction, namely population size, which in most cases is related to area. Thus extinction rates in their model increased with decreasing area. Other factors causing extinction, e.g. interspecific interactions (predation and competition), were also discussed by MacArthur and Wilson. Although they were not necessary for an equilibrium to occur, it was suggested that interspecific interactions increase the per-species extinction rates as the number of species on an island increase. These pioneering ideas have been elaborated and extended during the last decade. The basic metapopulation model for one species (Levins, 1969) has been extended to situations where interspecific competition influences extinction and colonization rates by, for example, Slatkin (1974), Hanski (1983) and Hanski and Ranta (1983) (see Bengtsson, 1991 for a review). It has been suggested that the important factor causing natural extinctions may be variability in population size, but not necessarily small population size *per se* (e.g. Williamson, 1983; Wright & Hubbell, 1983), although these factors may often act in combination. Recently a start has been made on analysing the effects of environmental variability and of patchiness on population extinction rates (e.g. Forney and Gilpin, 1989). Several studies also suggest that differences exist in extinction-proneness between species in different trophic levels and with different body sizes (e.g. Soulé *et al.*, 1979; Dickerson & Robinson, 1986; Hanski, 1986; Belovsky, 1987; Shaffer, 1987).

Experiments with rockpool Daphnia. Bengtsson (1988, 1989), using three common waterflea species from rockpools (see below), performed a four-year field experiment in artificial rockpools of different sizes and with different species combinations, to examine the effects of interspecific competition, habitat size (volume), population density and population variability on extinction rates. The results are summarized in Table 6.3. It was evident that extinction rates were higher in smaller volumes, and also higher in three-species experiments than in two-species experiments. No extinctions occurred in the one-species controls. Data on population dynamics and reproduction indicated that the species used similar resources and that interspecific competition was intense during large parts of the year (Bengtsson, 1988). The populations that subsequently

Table 6.3

Extinction rates in one-, two- and three-species experiments with *Daphnia* in artificial rockpools of different volumes, and average mean densities and average coefficients of variation in density in persisting and extinct populations. n = initial number of populations (from Bengtsson, 1989)

Volume (litres)	*Extinction probability (per population per year) ± SD* — *One-species experiments*	*Two-species experiments*	*Three-species experiments*
4	0(—)($n=6$)	0·21(±0·078)($n=28$)	0·28(±0·11)($n=18$)
12	0(—)($n=3$)	0·029(±0·019)($n=22$)	0·20(±0·10)($n=12$)
50	0(—)($n=4$)	0(—)($n=26$)	0·10(±0·047)($n=15$)
300	—	0(—)($n=2$)	0·030(±0·029)($n=9$)

	Average mean population density (individuals per litre) ± SD		*Average coefficient of variation in density ± SD*	
Volume (litres)	*Persisting populations*	*Extinct populations*	*Persisting populations*	*Extinct populations*
4	—	—	—	—
12	125(±97·9)($n=22$)	5·4(±3·66)($n=5$)	131(±38·3)($n=22$)	122(±40·8)($n=5$)
50	54·1(±30·9)($n=26$)	10·6(±10·4)($n=4$)	143(±37·1)($n=26$)	188(±64·7)($n=4$)
300	32·3(±16·6)($n=10$)	5·6(—)($n=1$)	159(±38·5)($n=10$)	137(—)($n=1$)

went extinct had lower mean population densities, but were not more variable (Table 6.3). Thus, in this experiment, both interspecific competition and small population size made populations vulnerable to extinction.

The effect of habitat size on extinction rates observed in this experiment is similar to that found in many other organisms, e.g. spiders on Bahama islands (Schoener & Spiller, 1987), anthropods on *Spartina* islands in Florida (Rey, 1981), and other examples discussed below. An increase in extinction rates with species number (i.e. interspecific competition) was implicit in the MacArthur and Wilson (1967) theory, but has seldom been demonstrated in nature. Moulton and Pimm (1986) found such an effect among birds introduced to Hawaiian islands, but the *Daphnia* study was the first experimental demonstration.

Plants on islands in Lake Möckeln. Nilsson and Nilsson (1982, 1983, 1985) studied the turnover of vascular plant species on 41 forested islands in Lake Möckeln. They found extinction rates to be negatively correlated to area, and argued that this was mainly due to an increased proportion of one-individual populations as island size decreased. These results are in accordance with the equilibrium theory. The same workers also pointed out one important problem with studies of extinctions in the field, namely that of sampling. While presence can be assessed with confidence this is not true for absence. In many cases, field estimates of extinction or turnover rates may be much higher than the real values, due to sampling errors. Thus a large part of the turnover rates published may in fact be such pseudo-turnover events rather than real extinction or immigration. In plants, the existence of a seed bank may also confound estimates of rates of extinction and immigration, and the problem is similar in animals with resting stages (e.g. ephippia in *Daphnia*).

Shrews on islands in Finnish lakes. Hanski (1986) and Hanski and Kuitunen (1986) used epigenetic cranial traits in the common shrew *Sorex araneus* to study population persistence on small islands in lakes. They found that populations on larger islands, but not on smaller ones, had differentiated in comparison with the mainland population, implying a relatively longer persistence time on larger islands than on smaller. This demonstrates how area may affect extinction probability. The pattern of island occupancy for three different shrew species was also studied. The common shrew occurred on most islands larger than 2 ha, but was absent from most smaller islands, while the smaller *S. caecutiens* and *S. minutus* showed a much weaker area-dependence. This pattern implies that extinctions are area-dependent in the common shrew, but not in the smaller species. Models of population extinction have suggested that time to extinction should have a stronger area-dependence under pure demographic stochasticity than under environmental stochasticity. Hence Hanski's conclusion was that populations of the larger *S. araneus* were sensitive mainly to demographic stochasticity, and the two smaller species to environmental stochasticity.

Studies of metapopulations

Surprisingly, the theory of metapopulation dynamics outlined above has not been applied to field situations except in a few cases. In part, this may be because long-term studies in the field are required to evaluate the concept. Studies of various organisms in Scandinavian archipelagoes represent an important part of all existing field studies of metapopulation dynamics.

The field vole Microtus agrestis *on islands in the Baltic.* Only 240 km due east of the Stockholm archipelago, across the Baltic Sea, the dynamics of field voles *Microtus agrestis* on small islands in archipelagoes have been closely studied by Pokki (1981). His study covered 71 islands of the Tvärminne archipelago, and lasted from 1972 to 1977. Sixty-nine islands were inhabited in at least one year, but the mean yearly occupancy was only 38·8 islands ($\bar{N}$, SD 8·2) and the proportion $\hat{p}$ was 0·546. The total turnover rate was found to be 0·48 colonizations and extinctions per year and island (Table 6.4). This system can be modelled as a

Table 6.4

Data on colonizations and extinctions of field voles *Microtus agrestis* on islands in Tvärminne Archipelago, Southern Finland (adapted from Pokki, 1981)

Island area (ha)	*n*	$\bar{N}$	$\bar{N}/T$ %	*Extinctions*	*Extinctions/ occupied island and year*	*Colonizations*	*Colonizations/ unoccupied island and year*
0 –1·0	40	15·2	38	54	0·71	50	0·40
1·1–5·0	18	13·5	75	16	0·24	21	0·93
75·0	13	10·2	78	11	0·22	13	0·92
Total	71	38·8	55	81	0·42	84	0·52

metapopulation system, and certain predictions about turnover can be tested. The simple equations given by Levins (1969, see above) should predict equilibrium number of islands occupied, and according to MacArthur and Wilson (1967) extinction should be a function of island area, and colonization a function of isolation. Table 6.4 gives the data necessary for these tests.

During the five seasons m (the rate at which each island is colonized) was calculated as 0·951 per year, while e (the extinction rate per occupied island) was 0·418 per year. The yearly rate of change in occupied patches (Δp) was only 0·008, meaning that this system shows no trend, only temporal variations in number of occupied patches. $\hat{p}$ was estimated as 0·563, which is very close to the observed mean of 0·546. As $\hat{N} > 3\sqrt{T}$; (38·8 > 25·3), this system will persist for a long time.

The smallest islands, which were also generally the most isolated, had higher extinction and lower colonization rates. Thus extinction rates on these islands may have been exacerbated also through lack of immigration.

The field vole in the Stockholm archipelago, although not as closely studied as the Finnish populations, showed similar turnover rates, and a metapopulation structure is highly probable. Selection for demographic traits which increase colonizing ability should be important in insular field voles, due to frequent colonization attempts. A comparative study was therefore performed by Ebenhard (1988, in press) to test theoretical models of demographic aspects of colonization success.

Two colonies of voles, one started with insular field voles and one with animals from the mainland, were kept in a laboratory for about two years. A total of 463 litters were produced by 100 females. Insular field voles were found to be larger at birth and as adults, and to grow faster. Insular females made a larger reproductive effort per litter, in relation to their weight, and in addition produced larger litters (5·0 versus 4·1). The age of sexual maturation was similar in the two vole provenances. The natality rate (λ) was therefore considerably larger among insular field voles, and mortality (μ), at least of juveniles, may be lower, since there is known to exist a negative correlation between neonate body weight and mortality. The large body size in insular field voles, which improves their dispersal ability, is therefore probably also advantageous during the colonization phase, because it permits a higher reproductive rate. This result is unexpected, considering classic allometric relationships and r–K-selection theory (*sensu* Pianka, 1970), but is wholly consistent with several demographic colonization models (e.g. MacArthur and Wilson, 1967).

The pool frog Rana lessonae *in isolated coastal ponds.* The Swedish population of the pool frog *Rana lessonae* is restricted to about 60 localities along the coast of northern Uppland and on islands in the Baltic. Its metapopulation structure has been closely studied by Sjögren (1988, 1989, 1991) and earlier by Forselius (1962). In the distribution area about 140 suitable waters exist, but only about half that number were occupied at any time. The turnover rate is slower than in the field vole system; extinction rate is only 2% per occupied pond and year, but in the period from 1960 to 1988 26 local extinctions and 39 colonizations were recorded. More isolated ponds, which do not necessarily support smaller populations, have had more than their share of extinctions, which Sjögren interprets as a result of low immigration rates (want of a rescue effect). The frog system is destabilized, as yet to an unknown degree, by a high sensitivity to environmental stochasticity, namely particularly cool summers, which once or twice per decade cause total reproductive failure in most ponds. Adult mortality is, however, low, and unless complete failures of reproduction occur in several consecutive years, the expected time to regional extinction is considerable.

Daphnia *in rockpools—the importance of interspecific competition.* Rockpools, i.e., small bedrock depressions containing fresh or brackish water, are common in coastal areas around Scandinavia. Many small rockpools dry up in summer, and most freeze solid every winter. Among the most common animals in rockpools are three plankton species of the genus *Daphnia* (Cladocera), *D. magna*, *D. pulex* and *D. longispina*, which co-occur regionally along the coasts of Finland and Sweden (Hanski & Ranta, 1983; Pajunen, 1986; Bengtsson, 1988). Their niches show high overlaps along several niche axes, including that of food. For *Daphnia* the rockpool habitat may be viewed as islands on islands, each pool on an island being a local population, while each island has a small-scale metapopulation, and the metapopulations on adjacent islands (an island group) make up the regional population.

Hanski and Ranta (1983) formulated a metapopulation model for the regional distribution and dynamics of the three *Daphnia* species. They suggested that interspecific competition increased extinction rates and decreased colonization rates, but neither experimental nor field evidence was provided. The model has been evaluated in the field by Pajunen (1986) and Bengtsson (1988, 1989), working on islands in the Tvärminne archipelago, in southwest Finland, and three areas on the east and west coast of Sweden, respectively.

Both Pajunen and Bengtsson found that extinctions and colonizations were frequent among rockpool *Daphnia*. On average, approximately 10% of the populations went extinct each year, and colonization rates measured as the proportion of newly established populations were of similar magnitude. Thus a metapopulation model is applicable to this system.

As in the experiment in artificial rockpools discussed above, natural extinction rates in rockpools were higher in pools with more species present. No three-species pools were found, but the per-population extinction rate in two-species pools on average was 22%, while in one-species pools 11% of the populations went extinct per year (Table 6.5). Thus interspecific competition could be demonstrated to affect extinction rates in this system, as conjectured by Hanski and Ranta (1983). However, interspecific competition did not appear to influence colonization rates, since these were similar in one- and two-species pools (Bengtsson, 1988, 1991).

Table 6.5
Extinction rates of *Daphnia* in one- and two-species natural rockpools in four areas in Scandinavia
Data from Bengtsson (1988, 1989) and Pajunen (1986; Tvärminne)

Extinction rates (probability of extinction per population and year) (±SD) (n)						
Site	*1 species present*			*2 species present*		
Flatholmen	0·13	(±0·037)	(82)	0·15	(±0·046)	(58)
Mönster	0·12	(±0·038	(74)	0·42	(±0·14)	(12)
Ängskär	0·097	(±0·025)	(143)	0·17	(±0·051)	(54)
Tvärminne	0·11	(±0·028)	(123)	0·16	(±0·052)	(50)

Although interspecific competition was important, it was not the only factor causing extinctions among rockpool *Daphnia* populations. Population turnovers in the field appeared to be more frequent in pools with higher and more variable salinities (Bengtsson, 1988), suggesting that environmental variability, too, may play a role. In addition, predators such as newts and backswimmers *Notonecta* spp. may drive *Daphnia* populations extinct. In many cases, these factors may interact with competition in causing an extinction event.

These results suggest that interspecific competition may limit species richness of cladoceran zooplankton species assemblages in the rockpool metapopulation system. Two *Daphnia* species are able to coexist in single pools for a longer time period, despite little or no niche differentiation (Bengtsson, 1988; Table 6.3). Three species actually coexist regionally in several areas in Scandinavia, which is consistent with the finding from immigration–extinction models that similar competitors may coexist regionally in patchy habitats (e.g. Slatkin, 1974; Hanski and Ranta, 1983; Hanski, 1987). However, the increased per-population extinction rates with increasing species number demonstrated in the field experiment as well as in natural rockpools indicate that regional coexistence of a large number of cladoceran species in this system is unlikely.

The *Daphnia* study suggests that species interactions can be important in determining distribution patterns and regional dynamics of species. This indicates that metapopulation models of single species may at times less accurately predict long-term persistence of these species. Perhaps 'metacommunity dynamics' will become a keyword for future research. In any case, conservation biologists working with the perservation of single species should pay attention to potential interactions with competitors, predators, and prey.

6.4 IMPLICATIONS FOR CONSERVATION

6.4.1 General recommendations

Conservation is mostly concerned with rare and threatened species. However, it is not justifiable to perform the type of experimental studies discussed above, nor any studies in which animals inevitably are killed, using such species. Consequently, in order to improve our understanding of how populations and communities are affected by fragmentation processes, it is necessary to use more numerous and widespread species in naturally patchy biotopes such as islands in an archipelago. The results still have considerable relevance for conservation in mainland habitats also. Below we discuss some of the recommendations for conservation that have emerged from recent studies of archipelagoes of our own, as well as of other authors.

6.4.2 Metapopulations

The general results of metapopulations research indicate that a species consisting of metapopulations will often be absent from some of the patches/islands where potentially it should flourish. Only a proportion of the available patches may be inhabited at any single point in time. If the total number of patches where the species can be present (T) is low (less than approximately 10), a species is likely to go regionally extinct even if it occupies a high proportion of the patches. On the other hand, if T

increases to, say, between 30 and 50, the species may persist in the long run, even if less than 50% of the patches are inhabited at any single time. Thus it is important to preserve not only the places where a rare species is present, but also currently unoccupied sites which it can reach and where it potentially can survive and reproduce. Detailed studies of the habitat requirements and other details of the autecology of such species may be necessary in order to identify such uninhabited patches.

The proportion of patches inhabited by a metapopulation will increase if dispersal and migration between patches is made easier. Thus, for a metapopulation to persist, it may be important to increase migration rates artificially, for example, by creating dispersal corridors and improving survival possibilities in the area between suitable patches (i.e. in the so-called matrix). Note, however, that it may still be desirable that patches to some extent are isolated from each other because of other factors such as contagious diseases (e.g. Soulé & Simberloff, 1986).

In general, it has been shown that small populations or small subpopulations in a metapopulation have higher rates of extinction (cf. above). Therefore small isolated areas (reserves, islands, etc.) should be avoided. This follows from ecological-demographic processes in small populations, regardless of the possible genetic effects, which may or may not play an additional role (see chapter 4). A larger number of patches for a metapopulation should never be created through subdivision of larger patches.

6.4.3 Changes in species composition with fragmentation

The studies of archipelagoes also allow some predictions about which species will be favoured by, or, rather, be less susceptible to, fragmentation processes, and how communities may change as large continuous areas are broken up into more or less isolated patches. Since small populations are more extinction-prone than larger ones, it follows that species with lower densities are the first to disappear in the process of fragmentation. This can have several consequences.

The trophic structure of islands often differs from that of the mainland. Often predators, with larger body sizes and larger area requirements, are under-represented or absent on islands (e.g. Heaney, 1984; Angerbjörn, 1985). As a consequence, other processes than predation, such as competition, may regulate populations at the next lower trophic level (cf., for example, Oksanen *et al.*, 1981; Fretwell, 1987; Oksanen, 1988). Also, different kinds of predators may be favoured. Thus, although little theoretical work has been done in this field, we expect that fragmentation will cause communities and ecosystems to be differently structured and regulated as compared to unfragmented areas. As the factors regulating populations may change rather suddenly with decreasing habitat size, fragmentation may cause unexpected changes in species composition with only minor changes in the degree of fragmentation.

Studies of species composition of carabids, land snails, and migrant birds on archipelagoes (Haila *et al.*, 1983; Nilsson, 1986; Baur & Bengtsson, 1987; Ås, unpublished) all suggest it to be a general rule that generalist species are less susceptible to habitat fragmentation than habitat specialists. Such species are more likely to find a suitable habitat in a given patch. When the intervening matrix between habitat islands is not completely hostile, e.g. in mainland areas, habitat generalists may also have a higher survival than specialists in the matrix. Species with good dispersal abilities may also be among those favoured in the process of fragmentation. Since generalist species and species with good dispersal abilities are seldom considered rare and threatened, fragmentation will usually have undesirable effects from the point of view of conservation.

6.4.4 Conservation in an archipelago

In many cases, general recommendations can easily be adapted to the specific conditions found in an archipelago. Yet it is obvious that some recommendations are not applicable to archipelagoes; thus, for example, it is hard to do much about the hostility of the matrix (the sea) for a terrestrial organism. The main conclusion from both species richness and metapopulation theories is that populations in an archipelago are most susceptible to extinction. Generally, smaller population sizes will increase the risk of extinction on islands compared to the mainland, hence making island populations, on average, more sensitive to disturbances. On the other hand, there is a possibility that species living in such a patchy habitat are in fact those which are adapted to these conditions (generalists, good dispersal ability, etc.). Fragmentation of a continuous mainland area might thus cause even greater damage than that to be expected on the basis of studies in archipelagoes. Nevertheless, the combination of small populations and dispersal obstacles calls for extreme caution in this type of ecosystem.

Many populations in archipelagoes are probably metapopulations consisting of subpopulations spread over many islands in one or several island groups. A consequence of this is that when preserving islands in order to protect the species, all islands included in the metapopulation system must be preserved, including those that do not at the time support a subpopulation. A general recommendation is to protect groups of islands rather than to single out one island within a group.

Finally it has to be realized that, compared to an ordinary coast, an archipelago has a tremendously long edge shore line towards the sea. Many of the organisms are truly dependent on both systems for their survival. Therefore one of the most important tasks is to protect the surrounding sea.

ACKNOWLEDGEMENTS

We thank Sven G. Nilsson, Håkan Rydin, and above all Staffan Ulfstrand for discussions and comments on earlier versions. Our research reported in this chapter has been financed by grants from SNV to Sven G. Nilsson, Torbjörn Ebenhard and Stefan Ås, from NFR to Sven G. Nilsson, Jan Bengtsson and Staffan Ulfstrand, from Sven and Lilly Lawski's Foundation for Natural Science Research to Jan Bengtsson, and, extremely generously, from Stiftelsen Olle Engqvist, byggmästare.

REFERENCES

Ahlén, I. & Nilsson, S.G. (1982). Species richness and area requirements of forest bird species on islands with natural forests in Lake Mälaren and Hjälmaren. *Vår Fågelvärld*, **41**, 161–84 (in Swedish with English summary).

Andrén, C. & Nilsson, G. (1985). Breeding pool characteristics and reproduction in an island population of natterjack toads *Bufo calamita* Laur., at the Swedish west coast. *Amphibia-Reptilia*, **6**, 137–42.

Angerbjörn, A. (1985). The evolution of body size in mammals on islands: some comments. *Amer. Nat.*, **125**, 304–9.

Arrhenius, O. (1921). Species and area. *J. Ecol.*, **9**, 95–9.

Ås, S. (1984). To fly or not to fly? Colonization of Baltic islands by winged and wingless carabid beetles. *J. Biogeogr.*, **11**, 413–26.

Ås, S. (1985). Biological community patterns in insular environments. *Acta Universitatis Upsaliensis, Abstracts of Uppsala Dissertations from the Faculty of Science*, No. 792, Uppsala.

Baker, H.G. & Stebbins, G.L. (1965). *The Genetics of Colonizing Species*. Academic Press, New York.

Baur, B. & Bengtsson, J. (1987). Colonizing ability in land snails on Baltic uplift archipelagos. *J. Biogeogr.*, **14**, 329–41.

Belovsky, G.E. (1987). Extinction models and mammalian persistence. In *Viable Populations for Conservation*, ed. M.E. Soulé. Cambridge University Press, Cambridge, pp. 35–57.

Bengtsson, J. (1988). Life histories, interspecific competition and regional distribution of three rockpool *Daphnia* species. *Acta Universitatis Upsaliensis, Comprehensive Summaries of Uppsala Dissertations from the Faculty of Science*, No. 142.

Bengtsson, J. (1989). Interspecific competition increases local extinction rate in a metapopulation system. *Nature*, (*Lond.*), **340**, 713–5.

Bengtsson, J. (1991). Interspecific competition in metapopulations. *Biol. J. Linn. Soc.*, **42**, 219–37.

Boorman, S.A. & Levitt, P.R. (1973). Group selection on the boundary of a stable population. *Theoret. Popul. Biol.*, **4**, 85–128.

Boyce, M.S. (1984). Restitution of *r*- and *K*-selection as a model of density dependent natural selection. *Ann. Rev. Ecol. Syst.*, **15**, 427–47.

Boycott, A.E. (1934). The habitats of land mollusca in Britain. *J. Ecol.*, **22**, 1–38.

Brenner, W. (1921). Växtgeografiska studier i Barösund skärgård. *Acta Soc. Fauna Flora Fenn*, **49**, 1–151.

Brown, J.H. & Kodric-Brown, A. (1977). Turnover rates in insular biogeography: effect of immigration on extinction. *Ecology*, **58**, 445–9.

Brown, J.H. & Lomolino, M.V. (1989). Independent discovery of the equilibrium theory of island biogeography. *Ecology*, **70**, 1954–7.

Carter, R.N. & Prince, S.D. (1988). Distribution limits from a demographic viewpoint. In *Plant Population Biology*, ed. A.J. Davy, M.J. Hutchings & A.R. Watkinson. Blackwell, Oxford, pp. 165–84.

Coleman, B.D., Mares, M.A., Willig, M.R. & Hsieh, Y.-H. (1982). Randomness, area, and species richness. *Ecology*, **63**, 1121–33.

Connor, E.F. & McCoy, E.D. (1979). The statistics and biology of the species–area relationship. *Amer. Nat.*, **113**, 791–833.

Dickerson, J.E., Jr & Robinson, J.V. (1986). The controlled assembly of microcosmic communities: the selective extinction hypothesis. *Oecologia*, (*Berl.*), **71**, 12–17.

Ebenhard, T. (1987*a*). An experimental test of the island colonization survival model: bank vole (*Clethrionomys glareolus*) populations with different demographic parameter values. *J. Biogeogr.*, **14**, 213–23.

Ebenhard, T. (1987*b*). High activity in bank voles in relation to colonization ability. *Oikos*, **49**, 297–302.

Ebenhard, T. (1988). Demography and island colonization of experimentally introduced and natural vole populations. *Acta Universitatis Upsaliensis, Comprehensive Summaries of Uppsala Dissertations from the Faculty of Science*, No. 141.

Ebenhard, T. (1989). Bank vole (*Clethrionomys glareolus* (Schreber, 1780)) propagules of different sizes and island colonization. *J. Biogeogr.*, **16**, 173–80.

Ebenhard, T. (1990). Reproductive traits and body size as adaptations to a colonization strategy in field voles (*Microtus agrestis*). *Ecology*, **71**, 1833–48.

Forney, K.A. & Gilpin, M.E. (1989). Spatial structure and population extinction: a study with *Drosophila* flies. *Conserv. Biol.*, **3**, 45–51.

Forselius, S. (1962). Distribution and reproductive behaviour of *Rana esculenta* L., (*sic*) in the coastal area of N Uppland, C Sweden. *Zoologiska Bidrag från Uppsala*, **35**, 517–28.

Forsman, A. & Ås, S. (1987). Maintenance of colour polymorphism in adder, *Vipera berus*, populations: a test of a popular hypothesis. *Oikos*, **50**, 13–16.

Forsman, A. (1991). Adaptive variation in head size in *Vipera berus* L. populations. *Biol. J. Linn. Soc.*, **43**, 281–96.

Fretwell, S.D. (1987). Food chain dynamics: the central theory of population ecology? *Oikos*, **50**, 291–301.

Gleason, H.A. (1922). On the relation between species and area. *Ecology*, **3**, 158–62.

Goodman, D. (1987). The demography of chance extinction. In *Viable Populations for Conservation*, ed. M.E. Soulé. Cambridge University Press, Cambridge, pp. 11–34.

Gustafsson, T., Andersson, B. & Westlin, L. (1983). Reproduction in laboratory colonies of bank vole, *Clethrionomys glareolus*, originating from populations with different degrees of cyclicity. *Oikos*, **40**, 182–8.

Haila, Y. (1983). Land birds on northern islands: a sampling metaphor for insular colonization. *Oikos*, **41**, 334–49.

Haila, Y., Järvinen, O. & Kuusela, S. (1983). Colonization of islands by land birds: prevalence functions in a Finnish archipelago. *J. Biogeogr.*, **10**, 499–531.

Haila, Y. & Järvinen, O. (1982). The role of theoretical concepts in understanding the ecological theatre: a case study on island biogeography. In *Conceptual Issues in Ecology*, ed. E. Saarinen, D. Reidel, Dordrecht, pp. 261–78.

Hanski, I. (1982). Dynamics of regional distribution: the core and satellite species hypothesis. *Oikos*, **38**, 210–21.

Hanski, I. (1983). Coexistence of competitors in patchy environments. *Ecology*, **64**, 493–500.

Hanski, I. (1986). Population dynamics of shrews on small islands accord with the equilibrium model. *Biol. J. Linn. Soc.*, **28**, 23–36.

Hanski, I. (1987). Colonization of ephemeral habitats. In *Colonization, Succession and Stability*, ed. A.J. Gray, M.J. Crawley & P.J. Edwards. Blackwell Scientific Publications, Oxford, pp. 155–85.

Hanski, I. & Kuitunen, J. (1986). Shrews on small islands: epigenetic variation elucidates population stability. *Holoarct. Ecol.*, **9**, 193–204.

Hanski, I. & Ranta, E. (1983). Coexistence in a patchy environment: three species of *Daphnia* in rock pools. *J. Anim. Ecol.*, **52**, 263–79.

Harper, J.L. (1969). The role of predation in vegetational diversity. In *Diversity and Stability in Ecological Systems, Brookhaven Symposium in Biology*, No. 22, pp. 48–61.

Harrison, S. (1991). Local extinction in a metapopulation context: An empirical evaluation. *Biol. J. Linn. Soc.*, **42**, 73–88.

Heaney, L.R. (1984). Mammalian species richness on islands on the Sunda Shelf, Southeast Asia. *Oecologia* (*Berl.*), **61**, 11–17.

Huston, M. (1979). A general hypothesis of species diversity. *Amer. Nat.*, **113**, 81–101.

James, F.C. & Rathbun, S. (1981). Rarefaction, relative abundance, and diversity of avian communities. *Auk*, **98**, 785–800.

Kihlblom, D. (1990). *Skärgårdsskog–Ekologi och Skötsel*. Skogsstyrelsen, Jönköping (in Swedish).

Lack, D. (1976). *Island Biology, Illustrated by the Land Birds of Jamaica*. Blackwell, Oxford.

Leigh, E.G. Jr (1981). The average lifetime of a population in a varying environment. *J. Theoret. Biol.*, **90**, 213–39.
Levins, R. (1969). Some demographic consequences of environmental heterogeneity for biological control. *Bull. Entomol. Soc. Amer.*, **15**, 237–40.
Levins, R. (1970). Extinction. In *Lectures on Mathematics in the Life Sciences. Vol. 2*, ed. M.L. Gerstenhaber. American Mathematical Society, Providence, RI, pp. 77–107.
Lindroth, C.H. (1945). Die Fennoskandischen Carabidae, I. Spezieller Teil. Die Karten, *Kungl. Vetensk. Vitterh. Samh. Handl. Sjätte Följden, B* **4**(1,2).
Lindroth, C.H. (1949). Die Fennoskandischen Carabidae, III. Allgemeiner Teil. *Kungl. Vetensk. Vitterh. Samh. Handl. Sjätte Följden, B*, **4**(3).
Lomnicki, A. (1988). *Population Ecology of Individuals*. Princeton University Press, Princeton, New Jersey.
Lomolino, M.V. (1984). Immigrant selection, predation, and the distributions of *Microtus pennsylvanicus* and *Blarina brevicauda* on islands. *Amer. Nat.*, **123**, 468–83.
MacArthur, R.H. & Wilson, E.O. (1963). An equilibrium theory of insular zoogeography. *Evolution*, **17**, 373–87.
MacArthur, R.H. & Wilson, E.O. (1967). *The Theory of Island Biogeography*. Princeton Universtiy Press, Princeton, New Jersey.
McGuinness, K.A. (1984). Equations and explanations in the study of species–area curves. *Biol. Rev.*, **59**, 423–40.
Moulton, M.P. & Pimm, S.L. (1986). The extent of competition in shaping an introduced avifauna. In *Community Ecology*, ed. J. Diamond & T.J. Case. Harper & Row, New York, pp. 80–97.
Munroe, E.G. (1948). The geographical distribution of butterflies in the West Indies. PhD thesis, Cornell University, Ithaca, New York.
Niemelä, J. (1988). Habitat occupancy of carabid beetles on small islands and the adjacent Åland mainland, SW Finland. *Ann. Zool. Fennici.*, **25**, 121–31.
Niemelä, J., Ranta, E. & Haila, Y. (1985). Carabid beetles in lush forest patches on the Åland Islands, south-west Finland: an island–mainland comparison. *J. Biogeogr.*, **12**, 109–20.
Niemelä, J., Haila, Y. & Halme, E. (1988). Carabid beetles on isolated Baltic Islands and on the adjacent Åland mainland: variation in colonization success. *Ann. Zool. Fennici.*, **25**, 133–43.
Nilsson, I.N. & Nilsson, S.G. (1982). Turnover of vascular plant species on small islands in Lake Möckeln, South Sweden, 1976–1980. *Oecologia* (*Berl.*), **53**, 128–33.
Nilsson, S.G. (1977). Density compensation and competition among birds on small islands in a South Swedish lake. *Oikos*, **28**, 170–6.
Nilsson, S.G. (1986). Are bird communities in small biotope patches random samples from communities in large patches? *Biol. Conserv.*, **38**, 179–204.
Nilsson, S.G. & Ebenman, B. (1981). Density changes and niche differences in island and mainland willow warblers *Phylloscopus trochilus* at a lake in southern Sweden. *Ornis Scand.*, **12**, 62–7.
Nilsson, S.G. & Nilsson, I.N. (1978). Species richness and dispersal of vascular plants to islands in Lake Möckeln, southern Sweden. *Ecology*, **59**, 473–80.

Nilsson, S.G. & Nilsson, I.N. (1983). Are estimated species turnover rates on islands largely sampling errors? *Amer. Nat.*, **121**, 595–7.

Nilsson, S.G. & Nilsson, I.N. (1985). Experimental estimates of census efficiency and pseudoturnover on islands: error trend and between-observer variation when recording vascular plants. *J. Ecol.*, **73**, 65–70.

Nilsson, S.G., Björkman, C., Forslund, P. & Höglund, J. (1985*a*). Egg predation in forest bird communities on islands and mainland. *Oecologia. (Berl.)*., **66**, 511–5.

Nilsson, S.G., Björkman, C., Forslund, P. & Höglund, J. (1985*b*). Nesting holes and food supply in relation to forest bird densities on islands and mainland. *Oecologia. (Berl.)*, **66**, 516–21.

Nilsson, S.G., Bengtsson, J. & Ås, S. (1988). Habitat diversity or area *per se*? Species richness of woody plants, carabid beetles and land snails on islands. *J. Anim. Ecol.*, **57**, 685–704.

Nisbet, R.M. & Gurney, W.S.C. (1982). *Modelling Fluctuating Populations.* Wiley, New York.

Paine, R.T. (1966). Food web complexity and species diversity. *Amer. Nat.*, **100**, 65–76.

Oksanen, L. (1988). Ecosystem organization: mutualism and cybernetics or plain Darwinian struggle for existence? *Amer. Nat.*, **131**, 424–44.

Oksanen, L. & Fretwell, S.D., Arruda, J. & Niemelä, P. (1981). Exploitation ecosystems in gradients of primary productivity. *Amer. Nat.*, **118**, 240–61.

Pajunen, V.I. (1986). Distributional dynamics of *Daphnia* species in a rockpool environment. *Ann. Zool. Fennici*, **23**, 131–40.

Palmén, E. (1944). Die anemohydrochore Ausbreitung der Insekten als zoogeographischer Faktor. *Ann. Zool. Soc. Zool. Bot. Fennicae. Vanamo* **10**(1), 1–262.

Palmgren, A. (1917). Studier öfver löfängsområdena på Åland, III. Statistisk behandling af floran. *Acta Soc. Fauna Flora Fennica*, **42**, 475–634.

Pianka, E.R. (1970). On '*r*' and '*K*'-selection. *Amer. Nat.*, **104**, 592–7.

Pianka, E.R. (1978). *Evolutionary Ecology*, 2nd edn. Harper & Row, New York.

Pokki, J. (1981). Distribution, demography and dispersal of the field vole, *Microtus agrestis* (L.), in the Tvärminne archipelago, Finland. *Acta Zool. Fennica*, **164**, 1–48.

Preston, F.W. (1962). The canonical distribution of commonness and rarity. *Ecology*, **43**, 185–215, 410–32.

Pulliam, H.R. (1988). Sources, sinks, and population regulation. *Amer. Nat.*, **132**, 652–61.

Ranta, E. & Ås, S. (1982). Non-random colonization of habitat islands by carabid beetles. *Ann. Zool. Fennici*, **19**, 175–81.

Rey, J.R. (1981). Ecological biogeography of anthropods on *Spartina alterniflora* islands in Northwest Florida. *Ecol. Monogr.*, **51**, 237–65.

Richter-Dyn, N. & Goel, N.S. (1972). On the extinction of a colonizing species. *Theoret. Popul. Biol.*, **3**, 406–33.

Rydin, H. & Borgegård, S.-O. (1988). Plant species richness on islands over a century of primary succession: Lake Hjälmaren. *Ecology*, **69**, 916–27.

Rydin, H. & Borgegård, S.-O. (1991). Plant characteristics over a century of primary succession on islands: Lake Hjälmaren. *Ecology*, **72**, 1089–101.

Safriel, U.N. & Ritte, U. (1983). Universal correlates of colonizing ability. In *The*

Ecology of Animal Movement, ed. I.R. Swingland & P.J. Greenwood. Clarendon Press, Oxford, pp. 215–39.
Schoener, T.W. & Spiller, D.A. (1987). High population persistence in a system with high turnover. *Nature*, (*Lond.*), **330**, 474–7.
Shaffer, M.L. (1983). Determining minimum viable population sizes for the grizzly bear. *International Conference of Bear Research and Management*, **5**, 133–9.
Shaffer, M.L. (1987). Minimum viable populations: coping with uncertainty. In *Viable Populations for Conservation*, ed., M.E. Soulé. Cambridge University Press, Cambridge, pp. 69–86.
Simberloff, D.S. (1978*a*). Use of rarefaction and related methods in ecology. In *Biological Data in Water Pollution Assessment: Quantitative and Statistical Analyses*, ed., K.L. Dickson, J. Garins & R.J. Livingston. *ASTM STP*, **652**, pp. 150–65.
Simberloff, D.S. (1978*b*). Using island biogeographic distributions to determine if colonization is stochastic. *Amer. Nat.*, **112**, 713–26.
Sjögren, P. (1988). Metapopulation biology of *Rana lessonae* Camerano on the northern periphery of its range. *Acta Universitatis Upsaliensis. Comprehensive Summaries of Uppsala Dissertations from the Faculty of Science*, No. 157.
Sjögren, P. (1989). Extinction patterns and metapopulation dynamics of Swedish pool frogs *Rana lessonae* and other taxa. *Swedish Nature Environment Board*, MP 3686 (in Swedish with English summary).
Sjögren, P. (1991). Extinction and isolation gradients in metapopulations: the case of the pool frog *Rana lessonae*. *Biol. J. Linn. Soc.*, **42**, 135–47.
Slatkin, M. (1974). Competition and regional coexistence. *Ecology*, **55**, 128–34.
Soulé, M.E., Wilcox, B.A. & Holtby, C. (1979). Benign neglect: a model of faunal collapse in the game reserves of East Africa. *Biol. Conserv.*, **15**, 259–72.
Soulé, M.E. & Simberloff, D. (1986). What do genetics and ecology tell us about the design of nature reserves? *Biol. Conserv.*, **35**, 19–40.
Vagvölgyi, J. (1975). Body size, areal dispersal and origin of the pacific land snail fauna. *Syst. Zool.*, **24**, 465–88.
Valovirta, I. (1977). The Baltic island survey, a malacological cooperation between Scotland and Finland, 1974. *Malacologia*, **16**, 271–7.
Valovirta, I. (1979). Primary succession of land molluscs in an uplift archipelago of the Baltic. *Malacologia*, **18**, 169–76.
Williams, C.B. (1964). *Patterns in the Balance of Nature*. Academic Press, London.
Williamson, M. (1981). *Island Populations*. Oxford University Press, Oxford.
Williamson, M. (1983). Island biogeography: variation in population density and extinction. *Nature*, (*Lond.*), **303**, 201.
Wright, S.J. & Hubbell, S.P. (1983). Stochastic extinction and reserve size: a focal species approach. *Oikos*, **41**, 466–76.

Chapter 7

Boreal Forests—The Focal Habitats of Fennoscandia

PER-ANDERS ESSEEN[a], BENGT EHNSTRÖM[b], LARS ERICSON[a] & KJELL SJÖBERG[c]

[a] *Department of Ecological Botany, University of Umeå, S-901 87 Umeå, Sweden*

[b] *Department of Plant and Forest Protection, Swedish University of Agricultural Sciences, S-750 07 Uppsala, Sweden*

[c] *Department of Wildlife Ecology, Swedish University of Agricultural Sciences, S-901 83 Umeå, Sweden*

7.1 INTRODUCTION

This chapter deals with the question of maintaining biological diversity in the boreal forests of Fennoscandia, i.e. the Scandinavian Peninsula and Finland. The boreal coniferous forest, or taiga, is the dominant biome, with a latitudinal extension from 56°N to 69°N (Fig. 7.1). The total area of the Fennoscandian forest amounts to more than 50 million ha, of which 13% occurs in Norway, 40% in Finland and 47% in Sweden (Nordic Statistical Secretariat, 1990). Man has utilized the boreal forest, for various purposes, during a very long period of time (Tenow, 1974). However, the most intense utilization has occurred during the last 300 years in connection with the development of the forest industry. At present, Fennoscandian forestry is amongst the most mechanized and efficient in the world. The result is that almost all forest land is now used for production of saw-timber and wood pulp. This has a tremendous impact on the structure and function of the boreal ecosystem. Cutting of old-growth forest and other consequences of forestry have a destructive

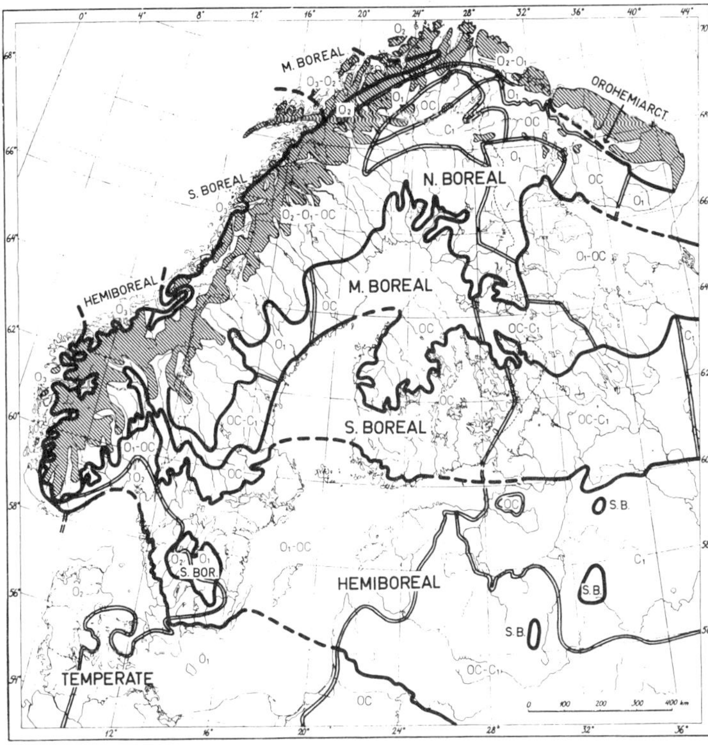

Fig. 7.1. Vegetation zones (—, – – –) and their sections, (=) in north-western Europe. O_3, highly oceanic; O_2, suboceanic; O_1, slightly oceanic; OC, indifferent. Hatched areas are dominated by orohemiarctic zones. From Ahti *et al.* (1968).

impact on a large number of boreal organisms, despite the fact that some species are favoured by forest management. The result is decreasing populations for many hundreds of plants and animals, as reflected in recent Red Data lists (Andersson *et al.*, 1987; Ahlén & Tjernberg, 1988; Databanken för hotarer och Naturvårdsverket, 1991).

In this chapter we review some significant ecological patterns and processes in boreal forests. First, we give a brief description of the plant and animal communities occurring in boreal Fennoscandia, with particular emphasis on the mechanisms that determine habitat quality in the natural forest. Then we discuss man's past and present impact, with special reference to the effects of forest management on biodiversity.

Finally, we propose a strategy for conservation and management of boreal organisms and ecosystems.

We place special emphasis on small-sized organisms, such as bryophytes, fungi, lichens and invertebrates, (1) because the vast majority of endangered species in Fennoscandia belongs to these groups of organisms (Ehnström & Waldén, 1986; Ingelög *et al.*, 1987) and (2) because these species have been largely neglected by conservation biologists (Synge, 1981; Wilson, 1987). A thorough review of all major aspects of boreal ecology is beyond the scope of this chapter. Interested readers may consult, for example, Ahti *et al.* (1968); Korchagin and Karpov (1974); Tamm (1976); Larsen (1980, 1982); Persson (1980); Wein *et al.* (1983); Van Cleve *et al.* (1986); Elliott-Fisk (1988) and Bonan and Shugart (1989), for additional information.

7.2 BIOTIC ZONATION

7.2.1 Vegetation zones

Most of Fennoscandia falls within the boreal zone, with the exception of the south-west, which belongs to the temperate zone, and the far north in the hemiarctic zone (Fig. 7.1). The subdivision (biotic zonation) of this area has attributed much interest and comprises a large number of works with different terminology, even if the subdivisions usually show only minor deviations (e.g. Du Rietz, 1925; Hustich, 1960; Sjörs, 1963; Ahti *et al.*, 1968; Nordiska Ministerrådet, 1984). The most widely accepted scheme is that by Ahti *et al.* (1968), which identified the hemiboreal, southern, middle and northern boreal zones (Fig. 7.1), based mainly on large-scale differences in tree species composition. For example, the hemiboreal zone is characterized by the presence of temperate broad-leaved tree species, such as ash *Fraxinus excelsior*, elm *Ulmus glabra*, lime *Tilia cordata* and oak *Quercus robur*, which in Sweden are termed 'noble', mainly on better soils, and conifers on poor soils. The south boreal zone is characterized by scattered occurrences of the same broad-leaved trees and a dominance of conifers. Both the middle and northern boreal zones have a conifer dominance, with birch *Betula* spp. as the main broad-leaved tree, and are separated on floristic criteria, the latter zone being characterized by the presence of a large number of northern plants, often with an Amphi-Atlantic distribution (Dahl, 1989). The zonal differentiation has been particularly intensively studied in Finland (see e.g. Kalela, 1961; Kalliola, 1973 and Oksanen, 1983, for forest vegetation). One reason is that regional variations in the vegetation are most easily observed in Finland, where small changes in altitude result in a more uniform regional (south–north) shift in the thermic and hygric climate. For example, Lahti *et al.* (1988) have shown that species richness of vascular plants slowly decreases from south to north in Finland. In Sweden, the situation is more complicated as there is a more striking increase in altitude along the east–west gradient, from the Baltic Sea to the Scandinavian mountain chain. Northern Sweden can geographically, from east to west, be roughly divided into a coastal plain, an interior forest region and a mountain region. From northwest to southeast, rivers divide the landscape into river valleys and elevated areas between the rivers. The situation is even more complicated in the highly undulating Norwegian terrain.

Beside the latitudinal zonal pattern, mainly caused by an increasingly more chilly climate and hence a decreasing evaporation, there is an azonal pattern from west to east, from prevailing oceanic in the west to continental conditions in the east (Tuhkanen, 1980). The identification of these patterns caused Ahti *et al.* (1968) to identify a number of subzones (Fig. 7.1). This is also observed on a global scale, with Fennoscandia standing out as under oceanic influence compared with the Eurasian part of the boreal zone. Among characteristic features are thick snow cover during the winter and frequent shifts between cold and warm periods. In the vegetation, the oceanic influence is above all seen in the presence of the mountain birch *Betula pubescens* ssp. *tortuosa* forests in the Scandes and in the large number of plant species which extend much further towards the north in Fennoscandia compared to eastern Russia and Siberia (e.g. Ahti *et al.*, 1968).

7.2.2 Vegetation classification

Classification of forest vegetation has been the focus of a large number of studies in Fennoscandia, with the object of formulating forest site type schemes for silvicultural purposes, e.g. Cajander (1909), Arnborg (1943), Malmström (1949), Ebeling (1978) and Hägglund and Lundmark (1982). This work was initiated by Cajander (1909), who recognized that the composition of ground vegetation largely depended on the fertility of the soil, i.e. local moisture and nutrient conditions. However, as pointed out by Ebeling (1978), an intense fire frequency may cause a drastic reduction in the nutrient content of the soil, so called heath-degeneration, with the result that a lichen-dominated (*Cladonia* spp.) ground flora, otherwise characteristic for nutrient-poor and very dry sites, could equally well develop on mesic sites. Another weakness is that the focus on conifers and forest management has resulted in the various stages of post-fire succession being largely neglected in the vegetation classification. In Finland, several workers, e.g. Kalela (1961); Hämet-Ahti (1963); Oksanen (1983), have continued the work of Cajander and elaborated detailed schemes of zonal parallel-types describing changes in species composition and dominance due to zonal position. These zonal differences may often be rather small and hard to detect. However, there are some general trends which are of significance (Kalliola, 1973; Ericson & Oksanen, 1987). First, there is a gradient from south to north showing that a large number of herbs face an increasingly narrowed habitat amplitude towards the north, thus becoming more restricted to productive vegetation types. Second, there is a marked decrease in evergreen herbs towards the north, and a corresponding increase of evergreen dwarf-shrubs. Beside this, the north boreal zone deviates from the middle boreal zone by the addition of a large number of northern plant species as well as an abundance of willow *Salix* spp. thickets and various tall-herb communities and the absence of low-herb spruce forests and *Vaccinium vitis-idaea* forests (Dahl, 1989).

Hemiboreal forests are generally characterized by the dominance of a large number of herbs. Easily observed species which these forests have in common with south boreal forests are the grass *Calamagrostis arundinacea* and bracken *Pteridium aquilinum.* The main change in the south boreal forest is a marked decrease among herbs and shrubs, although evergreen herbs still are important (e.g. *Hepatica nobilis* and *Veronica officinalis*). The middle boreal forest is generally characterized by the absence of evergreen herbs; however, evergreen rosette graminoids such as *Luzula pilosa* may still be common. Herbs are mainly restricted to nutrient-rich sites. This situation is still more accentuated in the north boreal zone, where bryophytes and lichens also become increasingly abundant.

7.2.3 Biotic zonation of animals

According to Järvinen and Väisänen (1973) the zonation of the avifauna corresponds closely to the vegetation zones of Ahti *et al.* (1968). They also found that the diversity of birds decreased towards the north, which is in agreement with general diversity patterns in bird communities (MacArthur, 1965; Tramer, 1974). Bird density also decreased northwards, but only in pine forest, not in spruce. A plausible explanation is that the higher proportion of birch in spruce forest in northern areas increases foliage diversity. However, as to numbers of bird species, spruce forest shows a northward decrease. This may be due to adverse climatic conditions rather than variation in habitat structure (Virkkala, 1987).

The bird fauna on mires of both Sweden and Finland seems to be an exception from the general trend of decreasing diversity towards the north (Järvinen & Sammalisto, 1976; Boström & Nilsson, 1983: Järvinen *et al.*, 1987).

7.3 THE NATURAL FOREST

7.3.1 Community structure

Plant community structure

The structure of the Fennoscandian boreal forest is relatively homogeneous due to the overall dominance of only two conifer species, Scots pine *Pinus sylvestris* and Norway spruce *Picea abies*, especially compared with the diversity of conifers in the North American boreal forests (Elliott-Fisk, 1988). This is furthermore accentuated by forestry which aims at suppressing broad-leaved trees, e.g. *Betula pubescens*, *B. pendula*, *Populus tremula* and *Salix caprea*, and shrub species. Both conifers have extremely broad habitat amplitudes and may grow from very dry (e.g. rocky outcrops, sharp alluvial heaths) to wet habitats (e.g. wet forests, mires). Spruce generally prevails on mesic-moist soils, in areas with a more oceanic climate and on sites with a low fire frequency. Pine, on the other hand, prevails on drier soils, in areas with a more continental climate and a high fire frequency. It should be mentioned, however, that spruce is a rather recent colonizer in Sweden; expansion started in the northeast, about 3100–3400 years ago (Tallantire, 1972), and the species is still spreading in the southernmost part.

There is considerable variation in the size and age structure of boreal forest stands due to local variations in disturbance history, climate and soil conditions. Pine forests with a high fire frequency are often characterized by bimodal or multiple modal diameter distributions. This is due to a peak in tree regeneration following fire and because the largest pine trees often survive one or more fires. Old-growth spruce forests, on the contrary, normally have strongly skewed diameter distributions with large numbers of small-sized, suppressed trees (Arnborg, 1943; Sirén, 1955). Another typical feature of old-growth spruce forests is the frequent occurrence of snow-breaks (Hesselman, 1912). The production of viable conifer seeds at high elevations is strongly dependent on warm summers, and hence very unpredictable (Hagner, 1957). Years with abundant seed production and favourable germination conditions may cause peaks in the age structure of the forest (Linder, 1984; Hytteborn *et al.*, 1987; Steijlen & Zackrisson, 1987; Ågren & Zackrisson, 1990; but see Kullman, 1987). As a consequence, many high-altitude forests have age structures in the form of a damped wave (Steijlen & Zackrisson, 1987).

Another feature of the Fennoscandian boreal forest is the low significance played by shrubs compared to other boreal areas. If present, the shrub layer is mostly comprised of tree saplings and suppressed individuals of the existing tree species. Most of the large number of shrub species characteristic of hemiboreal forest are absent, and those that occur are often reminiscent of earlier successional stages, for example, a few willow species such as *Salix starkeana* and *S. xerophila* and juniper *Juniperus communis*. However, on mesic sites, saplings of rowan *Sorbus aucuparia* and goat willow *Salix caprea* constitute a regular and characteristic feature. As these species are heavily browsed during winter by moose *Alces alces* and mountain hare *Lepus timidus*, they nowadays rarely attain tree size. More extensive thickets of deciduous shrubs are mainly found on moister soils, e.g. at mire edges, river banks, ravines and on the rising (land up-lift) Bothnian sea-shore.

The vascular plant flora is dominated by perennials. Dwarf shrubs are usually the most significant species in the field layer, with *Calluna vulgaris* and *Empetrum hermaphroditum* on the driest sites and *Vaccinium vitis-idaea* and *V. myrtillus* on mesic and moist sites. Due to the decreased evaporation towards the north, species such as *Vaccinium uliginosum* and *Ledum palustre*, which are otherwise known as mire plants, may become dominant even on drier heaths. The number of herbs and ferns is rather low in the Fennoscandian boreal forest and herbs will not reach any abundance except in mesic or moist sites with a mobile groundwater.

The forest floor vegetation usually contains considerably more bryophyte and lichen species than vascular plants. For example, in six pine stands of different age in a dry, middle boreal lichen heath, only nine species of vascular plants were observed compared to 38 lichens and 20 bryophytes (Esseen, 1983). Higher numbers of vascular plants and bryophytes occur in mesic spruce forests. For example, seven spruce stands of different age located on the border between the southern and middle boreal zones contained 57 vascular plants, 77 bryophytes and 27 lichens (Ericson *et al.*, unpublished data).

Bryophytes usually dominate the bottom layer in conifer forests on mesic and moist sites (Cajander, 1909; Kujala, 1926*a*,*b*; Larsen, 1980). The bryophyte carpet can suppress other plants, mainly by preventing seeds from reaching the soil, but also by overgrowing tree seedlings (Sirén, 1955). The common feather mosses *Hylocomium splendens* and *Pleurozium schreberi* usually reach high abundance on mesic soils and are typical for spruce forests of low to medium productivity.

Lichens also contribute significantly to plant diversity in Fennoscandian forests, especially towards the north. They are particularly abundant in the ground vegetation of heath forests, on rocks and as tree epiphytes (Ahlner, 1948; Ahti, 1977; Esseen, 1983; Ahti & Oksanen, 1990). Lichens in the genus *Cladonia*, occasionally in the genus *Stereocaulon*, dominate the bottom layer in pine forests on dry, sandy ground (Oksanen & Ahti, 1982). Lichen-rich forests are the most important winter habitats for domesticated reindeer *Rangifer tarandus tarandus* and grazing intensity is an important factor influencing lichen abundance and diversity (Richardson & Young, 1977; Helle & Aspi, 1983; Ahti & Oksanen, 1990).

Structure of the animal community

The large carnivores which are able to catch large-sized prey, for example, the wolverine *Gulo gulo*, and particularly the wolf *Canis lupus*, are at present very rare in Sweden, mainly as a result of intense hunting in the past. The wolverine is limited to about 100 animals in the mountain region and the Swedish wolf population consists of a small pack of about 10 individuals in the woodland of the southwestern part of the country (Anon., 1987). Even the brown bear *Ursus arctos* and the lynx *Lynx lynx* are heavily regulated by hunting, and cannot, except very locally, influence the population development of browsing mammals. The Swedish populations are estimated at 500–700 animals for brown bear (Anon., 1987) and about 200 animals for lynx.

The low density of large predators is an important factor responsible for the increase in the populations of large-sized browsing mammals during the present century. However, decreased cattle grazing in forests and an increased food supply, caused by clear-cutting, have also contributed. The number of moose killed by hunting has increased rapidly during the 20th century (Cederlund & Markgren, 1987). It reached a peak during the 1980s, when about 170 000 animals were killed annually in Sweden (Anon., 1987). The current winter population of moose is about

350 000 animals. The population of roe deer *Capreolus capreolus* has also increased markedly and expanded its range over most of the country.

The geographic distribution of birds is well demonstrated along the north–south gradient. Tiainen (1980) compared with the bird fauna of pine forests—a habitat with similar structure—from Poland, through Central Finland to Finnish Lapland. The bird community in Poland and Central Finland showed a greater similarity than that of Central Finland and Lapland. In addition, the number of species present or absent in either Polish or Finnish pine forests was smaller than the difference between Lapland and other areas. Thus, the results seem to support the interpretation by Järvinen and Väisänen (1973) that the North Finnish forests have an impoverished bird community.

Virkkala (1987) studied the regional variation in bird communities in a gradient over mid-boreal, north-boreal and hemiarctic zones in northern Finland. He identified a northern group of taiga species which prefer virgin, old forests, i.e. *Bombycilla garrulus*, *Parus cinctus*, *Perisoreus infaustus*, *Picoides tridactylus* and *Pinicola enucleator*, and a group of southern species, i.e. *Bonasia bonasia*, *Carduelis spinus*, *Certhia familiaris*, *Erithacus rubecula*, *Fringilla coelebs*, *Parus cristatus*, *P. major*, *Phylloscopus collybita*, *P. sibilatrix*, *Prunella modularis*, *Pyrrhula pyrrhula*, *Regulus regulus*, *Tringa ochropus*, and *Turdus philomelos*, which prefer spruce-dominated forests.

More species have colonized Denmark, Norway, Sweden and Finland from 1850 to 1970, compared with the number that has disappeared—2·8 and 0·6 species per decade and country, respectively (Järvinen & Ulfstrand, 1980). More nonpasserines have disappeared than passerines. Over 50 of the 88 'winners' have been favoured by man's activities. However, it should be stressed that the 'winners' are generalists with broad ecological amplitude, while the 'losers' are specialized species, for example, species confined to old-growth forest.

Habitat changes such as eutrophication, reduced cattle grazing and conifer plantations stand out as major factors behind the changes in population levels (Järvinen & Ulfstrand, 1980). In Finland, of 86 species which in 1945 had a population size greater than 20 000 pairs, 43% increased by 1% or more per year, 28% were stable and 29% decreased by at least 1% per year in 30 years (Järvinen & Väisänen, 1977). Because many 'winners' have invaded habitats that have been available for a long time, North European birds communities are probably often unsaturated. For general views on changes in the bird fauna see von Haartman (1973, 1978), Järvinen and Väisänen (1977), Järvinen and Ulfstrand (1980).

7.3.2 Community dynamics

Successional transitions occur in the boreal forest at several levels of scale and time. They range from the large-scale primary succession after the latest glaciation, which ended *c.* 10 000 years ago, through secondary forest succession to recolonization of small, locally disturbed patches (West *et al.*, 1981; Urban *et al.*, 1987). Forest fires and storms are the most important natural disturbances initiating forest succession in Fennoscandia. Other disturbances such as snow avalanches, landslides, erosion (White, 1979) and damage by ice (Zackrisson, 1978) have only local impact.

Fire regimes

The large-scale impact of fire perturbations on the boreal landscape and its importance in governing the structure and dynamics of the forest is well documented (e.g. Högbom, 1934; Tirén, 1937; Uggla, 1958; Rowe & Scotter, 1973; Zackrisson, 1977; Heinselman, 1981; Wein & MacLean, 1983; Engelmark, 1984). For example, it has been estimated that up to 200 fire catastrophes have occurred at many north boreal sites in Sweden during the Holocene (Zackrisson, 1985). Fire frequency varies considerably, both within and between regions. Important factors are tree species composition, stand structure, soil conditions, exposure, topography and climate (Zackrisson, 1977; Engelmark, 1987). Although most fires are caused by lightning (Högbom, 1934; Kinnman, 1936), man also increased fire frequency. Local fire chronologies show that it is highest in the interior due to a higher frequency of thunder-storms. Evidently, there is also a south–north gradient with a generally higher fire frequency towards the south boreal zone (cf. Payette *et al.*, 1989). In southern Norrland, the middle part of Sweden, the highest fire frequencies occur in stands currently dominated by broad-leaved trees. A mean fire interval of about 30 years has been found in such stands, which suggest that conifers have little opportunity to become dominant (Sturesson, 1983; Zackrisson, 1985).

Pine forest burns more frequently than spruce. Along the Vindel River valley, in the middle and north boreal Sweden, pine forests of the lichen–*Calluna* type on flat, sandy ground burnt at an average interval of 46 years, while mixed coniferous forests of *Vaccinium myrtillus* type on north-facing hill-slopes burned with an interval of 122 years (Zackrisson, 1977). In Muddus National Park, north boreal Sweden, pine forests burned with an interval of 81–90 years while spruce forests had a very low fire frequency (Engelmark, 1984). Fire-free areas (fire refugia) probably occur over most of Fennoscandia but seem to be most common in the high-altitude conifer forest. Fire refugia usually have spruce as the dominant tree species and are mainly found on north-facing hill-slopes, in wet forests, at mire edges as well as on forested islands in large mire complexes.

The very large-scale impact of fire is reflected in that, prior to 1876, about 0·6–1·9% of the forests burned annually along the Vindel River valley (Zackrisson, 1977; Fig. 7.2). Since the end of the 19th century, fires have been passively and actively suppressed, resulting in a drastic decrease in both fire frequency and areal extent of fires (Fig. 7.2). During the period 1971–79, only 0·001–0·02%, 300 to 5 000 ha, of all forest land in Sweden was burnt each year (Anon., 1981).

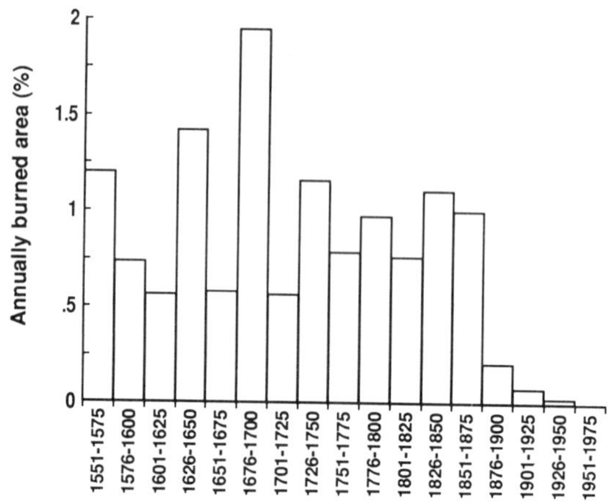

Fig. 7.2. Percentage of forest land burned annually from 1551 to 1975 along the Vindel River valley, North Sweden. Modified after Zackrisson (1977).

Plant succession following fire

The change in ground vegetation during post-fire succession on lichen-rich, heath forests is rather well-known and is similar in almost all boreal areas (Ahti, 1977). In early successional stands on pine heaths, crustose lichens in the genera *Placynthiella* and *Trapelia* as well as cup- and needle-like *Cladonia*-species dominate (Ahti, 1977; Oksanen & Ahti, 1982; Oksanen, 1983). In mid-successional stands, fruticose reindeer lichens, i.e. *Cladonia arbuscula*, *C. mitis* and *C. rangiferina*, take over the dominant position. In the final stages of post-fire succession *Cladonia stellaris* usually becomes the most abundant species (Ahti, 1977; Kershaw, 1977).

Only a few studies of the change in plant diversity during post-fire succession have been made in Fennoscandia. However, studies in both pine (Bråkenhielm & Persson, 1980; Oksanen & Ahti, 1982; Esseen, 1983) and spruce forests (Lindholm & Vasander, 1987) support the view that plant diversity peaks early in post-fire succession and thereafter slowly declines with increased stand age (Loucks, 1970; Shafi & Yarranton, 1973; Connell, 1978; Schoonmaker & McKee, 1988). The mechanism is due to a lag in plant colonization after disturbance, to changes in competitive interactions and in habitat heterogeneity as succession proceeds. However, there is evidence that the early peak is absent in some spruce stands, mainly due to a high diversity of wood-inhabiting bryophytes in late successional stands (Ericson *et al.*, unpublished data).

Storm-felling and gap dynamics

Storm-felling normally occurs on a much smaller scale compared to forest fire and ranges from a single tree to multiple tree-fall gaps covering several hectares (Sernander, 1936; Pickett & Thompson, 1978; White, 1979; Runkle, 1985; Shugart, 1987; Hytteborn *et al.*, 1987; Foster, 1988). Storm damage is especially important in the dynamics of old-growth stands. Gaps in the forest canopy caused by the death of living trees and tree falls radically change the environmental conditions for the forest floor community (Troedsson & Lyford, 1973; Falinski, 1978; Putz *et al.*, 1983; Beatty & Stone, 1986). Gap formation initiates vegetation changes by altering the light regime and by creating colonizable space (Poulsen & Platt, 1989). It can lead to changes in the demographic patterns and diversity of forest herbs (Falinski, 1978; Thompson, 1980; Beatty, 1984; Collins *et al.*, 1985; Collins & Pickett, 1987). Gaps also promote seedling establishment in birch and spruce (Arnborg, 1942). The effect of soil disturbance, caused by uprooting (Fig. 7.3), on bryophyte diversity in a north boreal spruce forest has been studied by Jonsson and Esseen (1990). The number of bryophyte species was twice as high in disturbed patches compared to undisturbed forest floor, 112 and 52 species, respectively. In total, 11% of all Swedish bryophytes were found in disturbed patches, which shows that uprooting creates local centres of high bryophyte diversity.

Cyclic animal populations

Regular population fluctuations, usually with a 3–4 year cycle, in multispecies communities of microtine rodents are characteristic in boreal

Fig. 7.3. A recently uprooted spruce in the old-growth spruce forest at Kirjesålandet, North boreal Sweden. Pits, mounds and logs create a diversity of microhabitats for a large number of bryophytes, fungi, lichens and invertebrates. Photograph P.-A. Esseen.

Fennoscandia. Rodent population cycles have a geographic trend and occur from the south to north boreal zones, but in temperate-hemiboreal areas, their populations are usually non-cyclic (Angelstam *et al.*, 1985; Hansson & Henttonen, 1985; Lindén, 1988). Other vertebrates, e.g. black grouse *Tetrao tetrix*, mountain hare *Lepus timidus* and red fox *Vulpes vulpes*, are more or less synchronized with rodent fluctuations (Hagen, 1969; Hörnfeldt, 1978, 1991; Angelstam *et al.*, 1985; Danell & Hörnfeldt, 1987; but see Lindén, 1989). Both food plant quantity/quality and predation have been proposed as the synchronizing agents (Angelstam *et al.*, 1985). In the non-cyclic rodent populations of south Sweden, 80% of the individuals are taken by generalist predators such as foxes, buzzards *Buteo* spp. and domestic cats (Erlinge *et al.*, 1983). However, in northern areas specialist predators may be more important in limiting cyclic rodent populations (Henttonen *et al.*, 1987). Curiously enough, rodent cycles are less pronounced in middle boreal Finland compared to middle boreal Sweden (Lindén, 1988), which may be an indication of the occurrence of generalist predators in more productive agricultural areas.

Järvinen (1979) suggests a relation between the vole cycle fluctuation and predators in north Fennoscandia as a possible explanation of the instability of the northern communities, especially the predator switching theory, e.g. vole predators may switch to small passerines after the vole populations have crashed. Data on most bird populations are not yet available, but the phenomenon is well-known for, for example, tetraonids and the mountain hare (Hagen, 1969; Linkola & Myllymäki, 1969; Hörnfeldt, 1978, and is further analysed by Angelstam *et al.*, 1985, Lindén, 1988, 1989 and Korpimäki *et al.*, 1990). No single theory, e.g. regulation by intrinsic factors, seems as yet been able to explain rodent population cycles. It appears that they are likely to depend on several interacting factors such as pathogens, parasites, predators, snow-cover and habitat heterogeneity (Hansson & Henttonen, 1988).

During rodent peak years, bark-feeding of willows and other preferred plants constitutes a characteristic feature in boreal areas (Ericson, 1977; Danell *et al.*, 1981). It is known that winter grazing may result in increased mortality in willows (Elmqvist *et al.*, 1988) and that grazed patches and runways may provide regeneration sites for both forest and meadow plants (Ericson, 1977). Ericson and Oksanen (1987) have claimed that regional plant distribution patterns, above all the decreased habitat amplitude of preferred plants, to more productive sites in northern areas is due to rodent grazing.

7.3.3 Features of special importance

The fauna and flora of Fennoscandia is well-known compared to many other parts of the world. It is therefore quite surprising that there exists no estimate of total plant and animal diversity for any boreal habitat in this region. However, despite a general low biodiversity compared to temperate and tropical forests, it is evident that many habitats have high plant and animal diversity, as reflected by numerous studies dealing with specific taxa. Muhr (1986), for example, recorded as many as 353 lichen species in a 10–30 m wide and 1 km long canyon, in Värmland, western Sweden. Several patterns and processes contribute to the biological and structural diversity occurring in natural boreal forests (Table 7.1). Three factors are especially important, i.e. forest fire, a long undisturbed forest continuity and the occurrence of coarse woody debris.

Impact of forest fire

Many boreal organisms are more or less associated with fire perturbations. A number have evidently evolved adaptations in response to different fire regimes (Rowe, 1983; Zackrisson, 1985), for example, by having a persistent seed bank (Granström, 1986). Two major groups of species can be separated: (1) those for which fire is necessary for

Table 7.1
Structural components, spatial patterns and processes of particular importance for plant and animal diversity in natural Fennoscandian boreal forests (features lacking or altered in managed stands)

Structural components
Very old pine and spruce trees
Old broad-leaved trees, particularly of *Populus tremula* and *Salix caprea*
Trees with heavy load of epiphytic lichens
Broken top, stag-headed, and leaning trees
Trees with holes and cavities
Dead standing trees (snags)
Fire-scarred trees, snags and stumps
Stumps with uneven surfaces
Large-sized logs in various stage of decomposition
Spatial patterns
A developed understorey of tree saplings and shrubs
Mixed stands, with both conifers and broad-leaved trees
Patchy distribution of tree species and tree stems
Uneven-aged stand structure
Multilayered tree canopies
Processes
Post-fire succession
Succession with tree-species replacement
Self-thinning
Gap formation
Snag and log formation

colonization and establishment; and (2) those which are largely confined to post-fire successional stands. The first group, 'fire-dependent' species, include a number of vascular plants, fungi, lichens and invertebrates (Table 7.2). The second group, which we may term 'fire-related' species, includes many species from almost all plant and animal groups. A group of a more intermediate character includes clonal, rhizomatous vascular plants which predominantly establish after fire, e.g. *Calamagrostis epigeios*, *Convallaria majalis*, *Lycopodium annotinum*, *L. clavatum*, *L. complanatum*, *Maianthemum bifolium* and *Pteridium aquilinum* (Oinonen, 1967, 1969).

Fire-dependent species. Some vascular plants, lichens and fungi are more or less confined to fire-perturbated stands (Table 7.2). For example, a few crustose lichens, i.e. some *Hypocenomyce* spp., are exclusively found on fire-scarred wood (Timdal, 1984). These lichens can even occur in very

Table 7.2
Examples of fire-dependent organisms in Fennoscandia

Species	*Red data category*[a]	*Species*	*Red data category*[a]
Vascular plants[b]		**Invertebrates**[e]	
Anemone patens	2	*Agonum bogemanni*	0
Anemone vernalis	4	*Agonum quadripunctatum*	—
Chimaphila umbellata	4	*Aradus angularis*	1
Geranium bohemicum	3	*Aradus anisotomus*	—
Geranium lanuginosum	2	*Aradus aterrimus*	1
		Aradus laeviusculus	2
Fungi[c]		*Aradus lugubris*	—
Coprinus angulatus	—	*Aradus signaticornis*	2
Coprinus lagopides	—	*Corticaria planula*	—
Daldinia concentrica	—	*Denticollis borealis*	—
Geopyxis carbonaria	—	*Laemophloeus muticus*	—
Morchella elata	—	*Melanophila acuminata*	—
Naucoria amarescens	—	*Melanophila cyanea*	—
Naucoria pseudoamarescens	—	*Paranopleta inhabilis*	—
Peziza echinospora	—	*Platyrhinus resinosus*	—
Peziza ostracoderma	—	*Pterostichus quadrifoveolatus*	—
Peziza praetervisa	—	*Scoloposcelidea phryganophila*	—
Peziza violacea	—	*Scoloposcelis obscurella*	3
Pholiota highlandensis	—	*Sphaeriestes stockmanni*	—
Rhizina undulata	—	*Stephanopachys linearis*	—
		Stephanopachys substriatus	—
Lichens[d]			
Hypocenomyce anthracophila	—		
Hypocenomyce castaneocinerea	—		
Hypocenomyce friesii	—		
Hypocenomyce scalaris	—		
Lecidea botryosa	—		

[a]Red data categories: 0, extinct; 1, endangered; 2, vulnerable; 3, rare; 4, at risk.
[b]Ingelög *et al.* (1987).
[c]Ryman & Holmåsen (1984).
[d]Santesson (1984); Timdal (1984).
[e]Heliövaara & Väisänen (1984); Lundberg (1984); Ehnström & Waldén (1986).

late post-fire successional stands. They have been observed on burnt pine snags in a forest stand which burned over 290 years ago, and may survive there because of the slow decay rate of fire-scarred pine wood. Fire-dependent fungi mainly include pyrophilous discomycetes (Ahlgren, 1974)—for example, *Peziza violacea* (Table 7.2).

Many insects associated with dead trees are favoured by forest fires or even require fire-scarred wood (Palm, 1951; Heilövaara & Väisänen, 1984; Lundberg, 1984; Ehnström & Waldén, 1986). On the one hand, recently burned stands can provide an abundant supply of nutrients and breeding material for many eurotopic bark- and wood-inhabiting species. On the other hand, some insects are adapted to the specific climatic conditions occurring in fire-perturbated stands (Lundberg, 1984). Insects specializing on fire-damaged trees include some beetles living under bark, those associated with forest-floor fungi and several mycophagous flat bugs (Heliövaara & Väisänen, 1983). Most fire-dependent insects seem to have good dispersal abilities (see chapter 5) and some beetles can even track fires at several kilometres distance (Evans, 1966).

Fire-related species. Boreal post-fire successions contain a broad range of fire-related plants and animals. For example, a small group of bryophytes, e.g. *Ceratodon purpureus*, *Funaria hygrometrica*, *Polytrichum juniperinum*, *P. piliferum* and *Marchantia polymorpha*, can become extremely abundant on fire-scarred ground. A characteristic feature of early to intermediate post-fire successional stands on mesic, often productive, forest ground is the high proportion of pioneer broad-leaved tree species, such as aspen and birch (Fig. 7.4). This successional phase, the deciduous stage, evidently occupied substantial areas of the boreal forest prior to fire elimination, and may have been more or less continuous in some central Swedish forests characterized by very high fire frequency (Zackrisson, 1985).

Broad-leaved trees usually dominate the stand for about 80–150 years following a fire, when conifers take over. However, single large trees, e.g. *Salix caprea*, may be present even in 200–300 year old post-fire successions. Existing data suggest that many herb-rich vegetation types, at least in boreal Sweden, are very late post-fire successional stages. The deciduous stage is probably among the most species-rich forest types in boreal Fennoscandia. Examples of species largely confined to the deciduous stage include perennial herbs, e.g. *Pyrola media*, and disturbance-related lichens such as some *Peltigera* spp. The deciduous stage is also the primary habitat for a number of plants and animals associated with aspen, for example some epiphytic bryophytes and lichens, including crustose, foliose (e.g. *Collema* spp.) and fruticose (*e.g. Ramalina sinensis*) species. Evidently, a large number of micro- and macrofungi are also largely confined to the deciduous stage. Examples of fungi that predominantly occur on broad-leaved trees in late post-fire successions include *Haploporus odorus* on *Salix caprea*, *Gloiodon strigosus* on *Populus tremula*

Fig. 7.4. Natural post-fire successional stand in middle boreal Sweden dominated by broad-leaved trees, mainly *Betula pubescens* and *Populus tremula*. The stand originated after a fire in 1888. Note the invading spruce *Picea abies*. Only some small and scattered stands of this type remain in Fennoscandia as a result of efficient fire suppression. Photograph P.-A. Esseen.

and *Salix caprea*, and *Daedaleopsis septentrionalis* on *Betula pubescens*.

The deciduous stage is the primary habitat for a large number of insects (e.g. Ehnström & Waldén, 1986). In a study of broad-leaved dominated stands of fire-origin and adjacent managed conifer stands, S. Ås (pers. comm.) found 90 species of wood-inhabiting beetles on aspen and 136 species on birch, respectively. However, total species richness of beetles did not differ much between stands of fire-origin and managed stands. Among threatened insect species, the beetles *Xyletinus tremulicola*, *Cucujus cinnaberinus*, and *Melandrya barbata* on aspen, as well as *Upis ceramboides*, *Oplocephala haemorrhoidalis* and *Leptura nigripes* on birch, seem to be confined to the deciduous stage.

The deciduous stage is also important for several bird species, e.g. white-backed woodpecker *Dendrocopos leucotos*, which has drastically declined and is now regarded as severely threatened in Fennoscandia (Aulén, 1988). Warblers increase because of a well-developed shrub layer and woodpeckers are abundant in mature stands with high numbers of wood-living insect larvae. This in turn favours hole-nesting birds such as tits *Parus* spp. and pied flycatcher *Ficedula hypoleuca*. In a study of the bird fauna in a 10-year-old post-fire succession and clear-cut areas of similar age in northern Sweden, a more diverse species composition was observed in the fire succession (Ericson, 1980). The difference was due to the burnt trees present in the fire succession, which provided nesting habitats for forest-living and hole-nesting species.

Long forest continuity

Fennoscandian conifer species can reach a great age. Norway spruce may live about 400 years in Sweden and the oldest recorded Scots pines are 711 and 810 years, in Sweden and Finland, respectively (Engelmark & Hofgaard, 1985). At least three mechanisms are responsible for a number of boreal organisms being confined to stands with a very long continuity, unbroken by fire or other major disturbances: (1) species which depend on a very long period of time to attain maximum size and to reproduce, including those with slow growth rates, e.g. lichens and some fungi; (2) species which are dependent on particular habitat structures and processes present in very late successional stands, e.g. trees of high age or large-sized tree-falls; (3) those that depend on the specific microclimatic conditions prevailing in old-growth stands. Most lichens and some fungi belong to the first group while many bryophytes, fungi, invertebrates and birds are found in the second and third groups. However, many species have links with more than one of these mechanisms.

No vascular plant species in Fennoscandia appear to be exclusively confined to very old forest stands. However, species such as *Calypso bulbosa*, *Goodyera repens* and *Listera cordata*, which usually vanish after clear-cutting, normally require a period of about 60–80 years before they can establish (Ingelög *et al.*, 1987).

Bryophytes associated with old forest stands include many species inhabiting logs of such large sizes which only occur in very late successional stands (Söderström, 1987). Most of these species also requires high air humidity. A few epiphytic bryophytes on broad-leaved trees, mainly aspen, are largely confined to old forests.

Lichens are strongly dependent on old trees and normally reach highest abundance and diversity on these. The slow growth rates and long life-spans of most lichens are the primary causes for their dependence on stable and long-lasting habitats (Topham, 1977). This seems to be the main reason for the abundance of pendent lichens in late successional

Fig. 7.5. Old-growth spruce forest in middle boreal Sweden. With abundant epiphyte vegetation, dominated by the lichen *Alectoria sarmentosa*. Photograph P.-A. Esseen.

conifer stands in boreal areas (Fig. 7.5). It has been shown that the maximum thallus size of some typical boreal lichens, i.e. *Alectoria sarmentosa* and some *Bryoria* and *Usnea* spp., increases significantly with increased tree size (Esseen, 1981). Studies of the changes in epiphytic lichen biomass during succession are scarce. However, some suggest that lichen biomass slowly increases up to a more or less constant level at tree ages of 100–150 years (Edwards *et al.*, 1960; Scotter, 1962; Stevenson, 1986). Some species even need several hundreds years of forest continuity, e.g. fire-refugia lichens such as *Usnea longissima* (Esseen *et al.*, 1981).

Many lichens are excellent indicators of natural forests characterized by a long and undisturbed forest continuity. In temperate areas, they have been used in classification systems of ancient woodland (Rose, 1976; Gilbert, 1977; Seaward, 1982). Good lichen indicators of old-growth forests in boreal Fennoscandia include *Calicium parvum*, *Chaenotheca gracillima*, *C. laevigata*, *Cybebe gracilenta*, *Cyphelium karelicum*, *Evernia divaricata*, *E. mesomorpha*, *Lobaria pulmonaria*, *L. scrobiculata*, *Ramalina thrausta*, *Sclerophora coniophaea* and *Usnea longissima* (Tibell, 1977, 1978, 1980; Esseen *et al.*, 1981; Hallingbäck, 1986; Ingelög *et al.*, 1987).

A large number of specialized insects are confined to old-growth stands. For example, the rare beetle *Pytho kolwensis* occurs exclusively in very old and fire-free spruce stands (Ehnström & Waldén, 1986; R. Pettersson, pers. comm.). Threatened noctuid moths such as *Xestia borealis* and *X. sincera* mainly occur in old, slow-growing spruce forests with an abundant growth of pendent lichens (Ehnström & Waldén, 1986). Although these moths feed on common food plants, it seems that lichens are necessary as camouflage for both larvae and adults (Imby & Palmqvist, 1978). Some bark-living beetles occurring in recently dead spruce trees, for example *Callidium coriaceum*, *Carphoborus rossicus*, *Lasconotus jelskii* and *Pytho abieticola*, are almost entirely confined to lichen-rich spruce forests (Palm, 1946).

Bird species associated with old-growth stands include species which are dependent on old trees, snags and cavities in trees, for example woodpeckers, owls, birds of prey and some tits. There is evidence that the high abundance of epiphytic lichens occurring in old conifer stands is important for the food-hoarding behaviour shown by some tits (Haftorn, 1956; Alatalo & Carlson, 1987).

Usnea longissima—*an old forest indicator.* The pendent lichen *Usnea longissima* (Fig. 7.6) is an illustrative example of a plant that has decreased because of forestry operations (Ahlner, 1931, 1948; Esseen *et al.*, 1981). This lichen is often used as a symbol of old-growth spruce forests in Fennoscandia. It is severely endangered in most of Europe due to cutting of old-growth forest and the devastating effects of air pollution. It has drastically declined during the present century in West Germany, Norway, Finland, Poland and the Soviet Union and is now extinct in some countries (Wirth, 1976; Esseen *et al.*, 1981). Of 110 Swedish sites known in the 1940s (Ahlner, 1948), *U. longissima* now remains at only seven, i.e. more than a 90% decline. This decline is due to clear-cutting of old spruce forest and various selective cutting operations. However, a number of new sites have been located in recent decades, mainly as a

Fig. 7.6. The pendent lichen *Usnea longissima*—an indicator of old-growth spruce forests in Fennoscandia. This lichen has vanished from many sites in western Europe during the present century, mainly because of its sensitivity to forest management. Photograph by P.-A. Esseen.

result of an increased focus on threatened forest plants. At present, *U. longissima* occurs on *c*. 40 sites on a total of *c*. 2000 trees in Sweden (Esseen & Ericson, 1982; Esseen, unpublished data). The lichen occurs sparsely at most of the present sites, but was evidently much more abundant in the past. In Skuleskogen National Park, one of the few rich areas, *U. longissima* has been observed on a total of 350 trees within a 25 km^2 area surveyed by a line census method.

The forests where *U. longissima* grows are naturally regenerated and slow-growing. The mean tree age of 13 sites was 194 years (range 120–260 years; Esseen & Ericson, 1982). Although the forest is of a primeval nature, some stands have signs of minor cutting operations. The lichen is largely confined to fire-free stands located on north-facing hill-slopes (Esseen *et al.*, 1981). The sites are rather well-protected from strong winds and have a microclimate characterized by a high and relatively stable air humidity (Esseen & Ericson, 1982).

In Fennoscandia *Usnea longissima* does not produce apothecia (Ahlner, 1948; Carlin & Swahn, 1977). Instead, it is almost entirely dependent on wind-dispersal of thallus fragments from one tree to another. Such fragments are mainly liberated during storms in autumn and winter and are mainly deposited within the immediate vicinity (2–3 m) of host trees (Esseen, 1985).

The present evidence suggests that very specific habitat demands and an inefficient method of long-range dispersal are responsible for *U. longissima's* sensitivity to environmental disturbance. For example, cutting a few host trees or clear-cutting nearby forest may totally destroy a site. The only way to maintain naturally functioning populations of this lichen, and of other species dependent on long forest continuity, is to preserve the habitat from all kinds of forest operations. However, it is also necessary to have buffer zones wide enough to maintain an unchanged microclimate.

Decaying wood

Tree death influences a number of patterns and processes occurring in forest ecosystems (Franklin *et al.*, 1987). Large quantities of decaying wood are one of the most characteristic features of old-growth forests (e.g. Spies *et al.*, 1988). Decaying wood provides a variety of habitats for a very large number of plants and animals (Harris, 1984; Ehnström & Waldén, 1986; Harmon *et al.*, 1986; Ingelög *et al.*, 1987). Snags, stumps, fallen logs in different stages of decay, uprooted trees and cavities in trees (Table 7.1) host a broad range of plant and animal communities. Conifer snags are important as substrates for fungi and lichens, as nutrient sources for invertebrates and as nest and feeding habitats for birds (Maser *et al.*, 1979), but appear to have less significance for bryophytes, vascular plants and mammals, at least in Fennoscandia. Logs, on the contrary, are important for most plant and animal groups. The high diversity of wood-inhabiting species in natural forest is evidently due to large microsite heterogeneity caused by variation in a number of factors, e.g. size of wood fragments, nutrient composition, bark and wood texture, water-holding capacity and microclimatic conditions. The method of wood input is also important in determining wood characteristics—a crucial factor, for example, is whether trees die standing or are alive when they fall to the ground. Large timber is usually considered more important for biodiversity than small-sized fragments.

Association with the timber of different tree species can be more or less strong (Barkman, 1958; Hamilton, 1987). Most regionally and locally common species utilize a rather broad range of tree species while those with very specific habitat demands are usually both rare and have low abundances. A characteristic feature of natural forests is the large number of species with very specialized habitat demands, i.e. confined to wood of particular decay stages, surface structures and certain tree species. Decaying wood is a patchy and unevenly distributed habitat which has a limited duration. Consequently, models of metapopulation dynamics (see chapter 6) should be of great value in the understanding of the population dynamics of many wood-inhabiting organisms, for example invertebrates.

Successional patterns on fallen timber. Plant and animal succession on logs is complicated, partly because boles may range from fresh to very decayed when reaching the forest floor. Complexity is also due to the multitude of processes involved, e.g. colonization, decomposition, fragmentation and species interactions (Harmon *et al.*, 1986). Repeated loss of bark- and wood-fragments create localized patches where succession is reinitiated. The result is that a single log may contain a mixture of early, intermediate and late successional communities. Bacteria, fungi and invertebrates are the main decomposers (Harmon *et al.*, 1986). Bryophytes, lichens and algae may also contribute to decay, mainly by increasing the water content at the surface. Unfortunately, most studies of plant and animal succession on logs are descriptive and contain only rough estimates of the time scales of the different successional stages. This is because decaying logs may persist for decades, or even centuries, and so cannot easily be aged. To circumvent these problems several systems of decay stage classification, ranging from freshly fallen logs to completely decayed ones, have been developed (e.g. Sernander, 1936; Arnborg, 1943; McCullough, 1948; Muhle & LeBlanc, 1975; Söderström, 1988*a*). However, the age of logs may be determined by dendrochronological methods. For example, Harmon (1989) dated logs up to 230 years old and Jonsson and Esseen (1990) tree-falls up to 174 years old. In spite of the scarcity of age data, it is evident that there exists an extremely broad temporal variation in plant successional patterns on logs (Raschendorfer, 1949; Philippi, 1965).

Plants. Vascular plants associated with logs include trees and forest herbs. Decayed logs (nurse-logs) are important substrates for the establishment of trees in many forests (Harmon *et al.*, 1986). In northern Sweden Arnborg (1943) estimated that a spruce log must be on the ground for about 60 years before it becomes suitable for seedling establishment. This can be compared with the very fast tree colonization, within a year or two, reported by Harmon (1989), for temperate conifer forests.

Most epiphytes, i.e. both foliose and fruticose lichens, die within a year or two after tree-fall. Exceptions include a few common foliose species such as *Parmeliopsis ambigua* and *P. hyperopta.* Bryophytes dominate the successional patterns on spruce logs, particularly in stands with high air humidity, while lichens are more abundant on pine logs (McCullough, 1948).

There is evidence that logs in intermediate stages of decay have the highest bryophyte diversity (Söderström, 1988*a*). However, many bryophytes can be present over a large part of the decay period (Söderström, 1987). Large spruce logs host a number of threatened bryophytes, mainly hepatics (Söderström, 1981, 1983, 1988*a*, 1988*b*; Gustafsson & Hallingbäck, 1988; Andersson & Hytteborn, 1991; Table 7.3). Logs positioned high above the ground have less bryophyte species than those with close contact with the forest floor, probably due to a lower moisture content (Söderström, 1988*a*). The relative rareness of large logs is a major factor which largely constrains the abundance and distribution of wood-inhabiting bryophytes.

Stumps and snags of both pine and spruce are the primary habitats for a large number of crustose lichens, for example in the genera *Lecidea, Micarea, Xylographa* and the Caliciales (Koskinen, 1955; Ahti, 1977; Tibell, 1977, 1978, 1980). Few foliose lichens are exclusively confined to decaying wood, but a number of fruticose *Cladonia* species are abundant on logs in intermediate stages of decay. Two threatened macrolichens in Fennoscandia occur on large wood-fragments, i.e. *Cladonia parasitica,* which occurs on large pine and oak logs, and *Letharia vulpina,* a conspicuous yellow species that grows on pine snags in continental areas (Ahlner, 1948; Ingelög *et al.*, 1987).

Fungi. A large number of wood-inhabiting micro- and macrofungi occur

Table 7.3
Examples of wood-inhabiting plants and fungi in boreal Fennoscandia confined to large-sized conifer logs in natural forest

Species	*Red data category*[a]	*Species*	*Red data category*
Bryophytes		**Fungi**	
Anastrophyllum hellerianum	—	*Amylocystis lapponica*	3
Blepharostoma trichophyllum	—	*Antrodia crustulina*	1
Calypogeia integristipula	—	*Fomitopsis rosea*	4
Calypogeia suecica	4	*Junghuhnia collabens*	1
Cephalozia affinis	—	*Laurilia sulcata*	3
Lepidozi reptans	—	*Phellinus ferrugineofuscus*	4
Lophocolea heterophylla	—	*Phellinus nigrolimitatus*	4
Lophozia ascendens	3	*Skeletocutis stellae*	3
Lophozia incisa	—	*Skeletocutis tschulymica*	2
Lophozia longidens	—		
Lophozia longiflora	—		
Lophozia silvicola	—	**Lichens**	
Odontoschisma denudatum	—	*Cladonia parasitic*	4
Riccardia palmata	—		
Scapania umbrosa	—		

[a]Red data categories as in Table 7.2.
Compiled from information given in Ingelög *et al.* (1987); Gustafsson & Hallingbäck (1988); Söderström (1988*a*) and Databanken för hotade arter och Naturvårdsverket (1991).

in the boreal forest. However, knowledge about their ecology and relation to forest management is extremely fragmentary. At present 436 species of macrofungi have been identified as threatened by forest operations in Sweden (Databanken för hotade arter och Naturvårdsverket (1991)). A significant proportion of the endangered macrofungi in the boreal forest is confined to large-sized spruce logs (Ingelög *et al.*, 1987; Table 7.3). This group mainly includes resupinate and hoof-shaped basidiomycetes. Some of the species listed in Table 7.3 are almost constantly present in old-growth conifer forests. Consequently, some log-inhabiting polypores are among the best indicators of pristine boreal forests characterized by a natural input of windthrows. The habitat requirements of the more large-sized wood-inhabiting fungi are rather well known (e.g. Niemelä & Kotiranta, 1982; Ingelög *et al.*, 1987), while little information is available on successional patterns. McCullough (1948) showed that fruiting bodies of fungi appear early in the succession, on recently dead fallen trees.

Typically only one or a few species of fungi were observed on each log. They remained until rather late stages of decay, but due to their unpredictable occurrence McCullough (1948) concluded that fungi are of little value in separating successional stages. An example of species replacement is that spruce trees killed by *Fomitopsis pinicola* seem to be necessary for the establishment of *Pycnoporellus fulgens*, an inhabitant of old forests (Niemelä, 1980). Some species seem to exclude one another by monopolizing a particular log, e.g. spruce logs which host *Fomitopsis pinicola* seem largely to exclude *Fomitopsis rosea* (M. Karström, pers. comm.).

Invertebrates. Decaying wood in old-growth forests hosts a very large number of invertebrates. Snags, stumps and logs in different stages of decay offer a broad variety of niches especially for insects (Hamilton, 1987). For example, in Sweden about 1000 species of beetles are dependent on dead trees, a figure which includes the majority of threatened forest invertebrates in Sweden (Andersson *et al.*, 1987). Many of these beetles do not use pure bark and wood as nutrient in the larval stages. Fungal mycelia in the bark and wood obviously play a dominant role as a nutrient source for many species (Lawrence, 1989). Most species of bark and wood-inhabiting invertebrates live in the bast (phloem plus cambium) while very few utilize the outer bark of the trees, e.g. the beetles *Microbregma emarginata* on spruce and *Notorhina punctata* on pine. The most diverse fauna on snags is found during the first two years after the trees have died. Snags occupied by bark beetles may lose most of their bark already in the first year, due to attack by woodpeckers (Bull, 1983). The sap- and heartwood of snags are more resistant and may harbour invertebrates for many decades.

Spruce logs have a more diverse invertebrate fauna than pine. This is due to a more rapid decay of the logs and to the fact that spruce hosts a higher number of bark- and wood-inhabiting fungi than pine. However, many invertebrates can occur on both conifer species, especially during late stages of log decay. Although insect succession may show a more or less typical sequence of species replacing each other as log decay proceeds, there is a very broad variation in patterns of species occurrence during succession. Important factors are regional and local variations in climate, attack by woodpeckers and the degree of fungal infection. For example, as one proceeds from the south to the north, there is a general trend of decreased rate of log decay and hence a prolonged period of invertebrate succession. The succession of invertebrates on spruce logs is briefly summarized below.

The first successional stage lasts one to two years after tree-fall (Fig. 7.7; stage I), and is dominated by different species of bark beetles that invade the cambial zone (Table 7.4). This stage mainly includes primary cambial-eaters (bark beetles, longhorn beetles and weevils) and their associated parasitoids and predators. Associated with cambial beetles are different groups of flies (Loncheidae, Dolichopodidae and Xylophagidae) and some hemipterous bugs that predate bark beetles. Some species of cerambycid beetles and weevils also belong to the primary stage together with many predators, especially beetles in the families Staphylinidae, Histeridae, Nitidulidae, Rhizophagidae and Cleridae. In contrast to the diversity of cambial-living species, very few insects exploit the sap- and heartwood during the first year after tree-fall.

The second stage lasts about five years (Fig. 7.7; stage II), and includes secondary cambial-eaters and a number of invertebrates associated with bark-inhabiting fungi. The first generation of beetles can exploit the tree three to four years after the tree death but repeated attacks may occur during more than one decade. The diverse fauna of parasitoids, especially those on bark beetles, has disappeared. However, some of the predators, especially rove beetles such as *Nudobius lentus* and *Quedius plagiatus*, can still be present. Fungi, for example, *Trichaptum abietinum*, start to attract some beetles at this stage. In the empty galleries of bark beetles a few detritophagous beetles can be found, e.g. *Corticeus* spp. Different species in the bark beetle genus *Crypturgus*, which penetrate the bark through the galleries of other bark beetles, are common at the beginning of the second stage. The faunal composition during the following years is highly dependent on the extent of insect attack in the first two years, especially to what extent this has altered the cambial structure, chemical composition and the available food content in the cambium.

The third stage begins about seven years after tree death when the cambium has been more or less consumed by insects (Fig. 7.7; stage III). Mono- or oligophagic fungal consumers, e.g. a number of beetles, have the greatest diversity during this stage, in response to fungal attacks on the outer sapwood. Windthrows and natural stumps that host the fungus *Trichaptum abietinum* have a particularly diverse beetle fauna. A specific type of succession is exhibited by beetles that live in the fruiting bodies of various Polyporaceae. Of special importance is the fact that many of these wood-inhabiting fungi have a relatively short life span. The beetle larvae mainly feed on the inner part of fruiting bodies and wood infected by mycelia, while adults mainly consume spore-layers and spores. These mycophagous beetles also have specific predators, in the orders Diptera, Coleoptera and Raphidioptera, and parasitoides in the order Hymenopt-

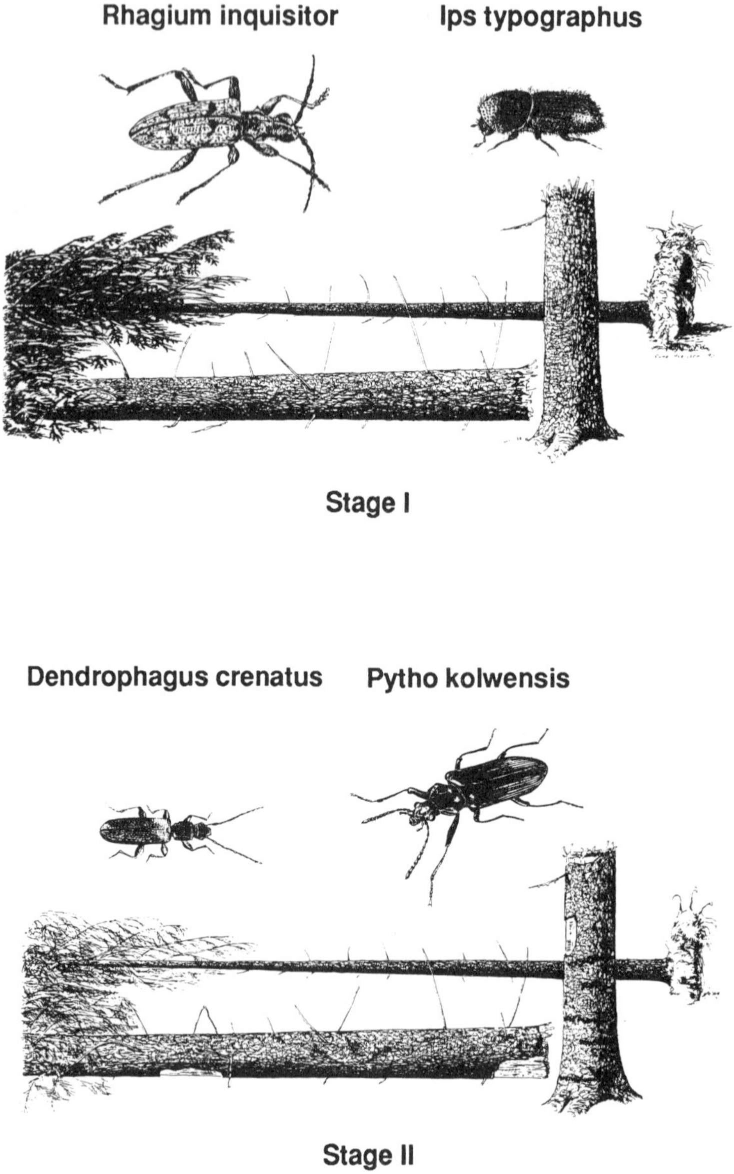

Fig. 7.7. Four typical successional stages of standing dead and fallen spruces *Picea abies* in boreal Fennoscandia. For each stage, two characteristic beetle species have been illustrated. Drawings by R. Axelsson.

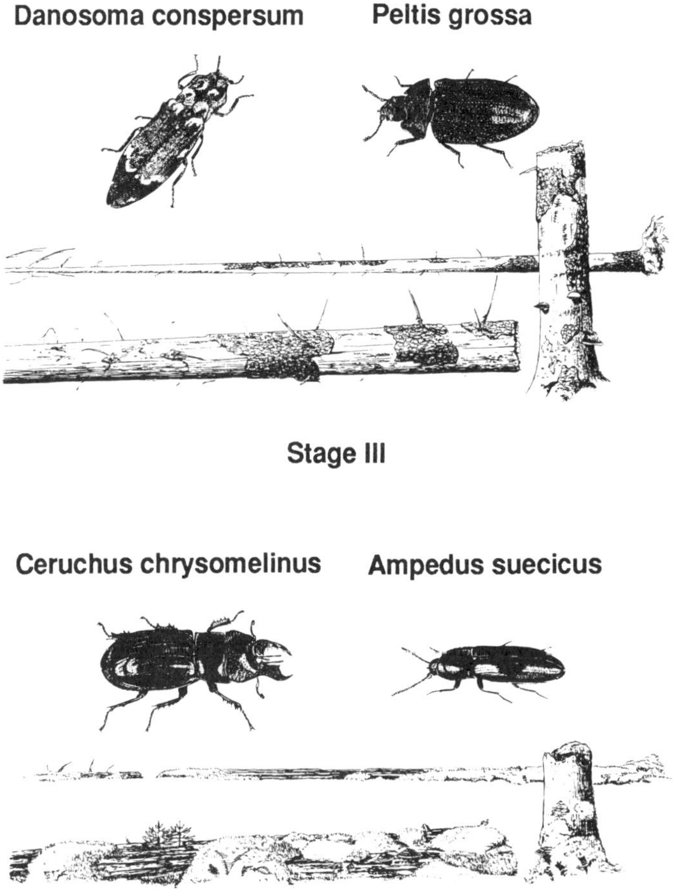

era. A general pattern seems to be that the largest number of threatened beetles occur in the second and third stages of invertebrate succession (cf. Table 7.4).

The fourth stage can be defined as a more or less bark-free spruce stem but with a well-preserved sapwood that may have started to become fragmented (Fig. 7.7; stage IV). This stage begins 10–20 years after tree death and may last more than five decades. Pine logs, on the contrary,

Table 7.4
Characteristic common and rare beetles confined to the four successional stages shown in Fig. 7.7

Common species			*Rare and threatened species*			
Scientific name	*Feeding behaviour*[a]	*Wood type*[b]	*Scientific name*	*Feeding behaviour*[a]	*Wood type*[b]	*Red data category*[c]
Stage I						
Nuobius lentus	P	S, L	*Melanophila acuminata*	C	S, B	4
Thanasimus formicarius	P	S, L	*Colydium filiforme*	W	S, L	1
Thanasimus femoralis	P	S, L	*Lasconotus jelskii*	P	S	4
Rhagium inquisitor	C	S, L	*Acmaeops septentrionis*	C	S, B	4
Tetropium castaneum	C	S, L	*Monochamus urussovi*	C	L, S	2
Tetropium fuscum	C	S, L	*Carphoborus rossicus*	C	S	4
Acmaeops pratensis	C	S (B)				
Hylurgops palliatus	C	S, L				
Polygraphus poligraphus	C	S				
Polygraphus subopacus	C	S				
Trypodendron lineatum	F	S, L				
Pityogenes chalcographus	C	S, L				
Ips typographus	C	S, L				
Stage II						
Quedius plagiatus	P	S, L	*Harminius undulatus*	C, P	L, S	4
Pytho depressus	C	S, L	*Stephanopachys substriatus*	C	S, B	2
Rhagium inquisitor	C	S, L	*Stephanopachys linearis*	C	S, B	2
			Dendrophagus crenatus	C	S	4
			Enicmus planipennis	F	S	4
			Pytho abieticola	C	L	4
			Pytho kolwensis	C	L	1
			Zilora ferruginea	C, F	S	4

Stage III						
Garbrius splendidulus	P	L, S	*Phymatura brevicollis*	F	S	4
Gyrophaena boleti	F	S, L	*Darosoma conspersum*	W, P	L, S	4
Ampedus tristis	W	L, S	*Darosoma fasciatum*	W, P	L, S	4
Ampedus nigrinus	W	L, S	*Hylis procerulus*	W, F	L, S	4
Cis glabratus	F	S, L	*Hylis cariniceps*	W, F	L, S	4
Cis punctatus	F	S, L	*Xylophilus corticalis*	W, F	S	2
Xylita laevigata	W, F	S, L	*Dorcatoma punctulata*	F	S, L	4
Anoplodera sanguinolenta	W	S, L	*Peltis grossa*	W, F	S	2
Anoplodera reyi	W	S, L	*Ostoma ferruginea*	W, F	S, L	4
Anoplodera virens	W	L	*Ipidia quadriplagiata*	F	S, L	1
			Pteryngium crenatum	F	S, L	4
			Atomaria abietina	F	S	4
			Atomaria subangulata	F	S, L	4
			Atomaria sahlbergi	F	S	4
			Atomaria lapponica	F	S	4
			Atomaria pseudaffinis	F	S	4
			Ennearthon laricinum	F	S	4
			Hadreule elongatula	F	S	4
			Xylita livida	W, F	L	1
Stage IV						
Melanotus castanipes	W	S, L	*Liodopria serricornis*	F	S, L	1
Ampedus balteatus	W	S, L	*Ampedus suecicus*	W	L, S	4
Ampedus pomorum	W	S, L	*Ceruchus chrysomelinus*	W	S	1
Oxymirus cursor	W	L, S	*Mycetochara obscura*	W	L	4
Rhyncolus chloropus	W	S	*Tragosoma depsarium*	W	L, S	1
Rhyncolus sculpturatus	W	S				

[a]P, predator; C, cambium feeder; F, fungi feeder; W, wood feeder.
[b]S, snags and stumps; L, logs; B, burnt trees.
[c]Classification for categories 1 and 2 from Andersson *et al.* (1987). Red data numbers as in Table 7.2.
Within each stage, species are arranged in systematic order. The nomenclature follows Lundberg (1986).

can be remarkably intact even after 100–200 years. Few wood-eaters are present at this stage, due to a low nutrient content in the wood. Instead, attacks of fungi, mainly brown rot, create special conditions in the sapwood that are used by several species of longhorn beetles in the tribe Lepturini. These stem-decomposers have a life cycle that lasts for two–three years (Bíly & Mehl, 1989). The largest longhorn beetle in old conifer stems in boreal Sweden is *Tragosoma depsarium*, whose larvae appear about 10 years after tree death. In suitable stems this beetle can inhabit a log during three to four decades until most of the sapwood is consumed. Species associated with resupinate Polyporaceae, are also characteristic of this stage, e.g. larvae of ostomid beetles that live within and under the fruiting bodies in the sapwood. Very old stumps and logs are important as shelter or hibernating sites for a number of ground and rove beetles (Savely, 1939; Lloyd, 1963). Not only carpenter ants *Camponotus* spp. but also many other ant species, use old logs and snags for nest building. For example, abandoned galleries of longhorn- and anobid beetles in snags can become occupied over extended periods by many species in the Hymenoptera orders Sphecidae, Apidae and Eumenidae. Within the Sphecidae alone in Sweden there are as many as 59 species that build nests in dead wood (Lomholdt, 1975, 1976).

Vertebrates. Comparatively little quantitative information is available on the importance of snags and logs for Fennoscandian vertebrates. However, hollow trees and cavities made by woodpeckers have significance for some of the bat species (Ahlén, 1977). Both snags and cavities in living trees provide roosting and nesting sites for woodpeckers, owls and other bird species (Table 7.5). Snags are also an important food source because of their rich insect fauna. It has been shown that low snag availability may adversely affect populations of birds that nest in or forage on them (Balda, 1975; Thomas *et al.*, 1979).

7.4 MAN'S IMPACT

7.4.1 The past—multiple utilization of the forest

Man has for a long time influenced and affected the Fennoscandian forests. The first step in this process was a gradual opening of the forest landscape for cattle grazing and agriculture. This process started as early as the Neolithic period (about 5000 years BP) and by the 17th century had resulted in the deforestation of large parts of southern Sweden, the

Table 7.5
Five categories of tree-nesting birds in boreal Fennoscandia

Common name	*Scientific name*	*Nesting behaviour*[a]	*Red data category*[b]
White-backed woodpecker	*Dendrocopos leucotos*	E	1
Great spotted woodpecker	*Dendrocopos major*	E	—
Lesser spotted woodpecker	*Dendrocopos minor*	E	4
Three-toed woodpecker	*Picoides tridactylus*	E	4
Grey-headed woodpecker	*Picus canus*	E	3
Green woodpecker	*Picus viridis*	E	—
Black woodpecker	*Dryocopus martius*	E	4
Willow tit	*Parus montanus*	E, C	—
Siberian tit	*Parus cinctus*	E, C	—
Crested tit	*Parus cristatus*	E, C	—
Coal tit	*Parus ater*	E, N	—
Great tit	*Parus major*	C, N	—
Blue tit	*Parus caeruleus*	C, N	—
Nuthatch	*Sitta europaea*	C, N	—
Pied flycatcher	*Ficedula hypoleuca*	C, N	—
Red-breasted flycatcher	*Ficedula parva*	C, N	4
Spotted flycatcher	*Muscicapa striata*	N	—
Tree creeper	*Certhia familiaris*	N	—
Redstart	*Phoenicurus phoenicurus*	C, N	—
Starling	*Sturnus vulgaris*	C, N	—
Wryneck	*Jynx torquilla*	C, N	—
Swift	*Apus apus*	C	—
Stock dove	*Columba oenas*	C	—
Jackdaw	*Corvus monedula*	C, N	—
Goldeneye	*Bucephala clangula*	C, N	
Smew	*Mergus albellus*	C, N	—
Goosander	*Mergus merganser*	N	—
Tengmalm's owl	*Aegolius funereus*	C	—
Pygmy owl	*Glaucidium passerinum*	C	—
Hawk owl	*Surnia ulula*	C, N	—
Great grey owl	*Strix nebulosa*	(N), S	4
Ural owl	*Strix uralensis*	N, S	4
Tawny owl	*Strix aluco*	N	—
Goshawk	*Accipiter gentilis*	O	4
Sparrow hawk	*Accipiter nisus*	O	—
Golden eagle	*Aquila chrysaetos*	O	4
White-tailed eagle	*Haliaetus albicilla*	O	2
Osprey	*Pandion haliaetus*	O	4
Honey buzzard	*Pernis apivorus*	O, S	—
Buzzard	*Buteo buteo*	O, S	—
Rough-legged buzzard	*Buteo lagopus*	O, S	—
Peregrine	*Falco peregrinus*	S	1
Kestrel	*Falco tinnunculus*	N, S	4
Hobby	*Falco subbuteo*	S	—
Merlin	*Falco columbarius*	S	—

[a]E, excavating holes in snags or living trees; C, secondary hole-nesters (using cavities of category E); N, using natural cavities in live or dead trees; O, nesting in old, large trees or trees with large branches; S, secondary usage of category O nests.
[b]Information on red data categories from Ahlén and Tjernberg (1988); red data numbers as in Table 7.2.

Bothnian coast and the Silurian area in Jämtland (Stenberger, 1969; Huttonen & Tolonen, 1972).

The second step in man's influence started during the 17th and 18th centuries, both as a consequence of the growth of the mining industry and as an effect of the increase in the human population (Heckscher, 1968). The mining industry (copper and iron), which was based on charcoal production, increased fivefold during the period 1600–1720 (Heckscher, 1968). This resulted in a local overexploitation of the forests in Bergslagen, central Sweden, and forced iron works to move from this area to other parts of central Sweden in order to secure the supply of charcoal. At the same time, the increase in human population brought about a growing need to increase fodder production for cattle. The outcome was an expansion of settlements in northern Sweden at the end of the 18th century (Bylund, 1956), with the first more extensive exploitation of the northern forests (Tirén, 1937), and an increased utilization of the forests for grazing purposes. One phenomenon was the establishment of chalets, often located 30–40 km away from the villages, for summer grazing (Montelius, 1962). Another was the increased use of fire for improving grazing conditions in the forests, favouring valuable grasses and herbs, and in southwestern Sweden also heather. Burn-beating, introduced by Finnish settlers, was also extensively utilized and in central Sweden lasted until the end of the 19th century (Heikinheimo, 1915; Arnborg, 1949; Fig. 7.8). Fire was also used by the settlers as an ecological weapon against the Lapps in order to destroy areas for reindeer grazing (Tirén, 1937; Bylund, 1956).

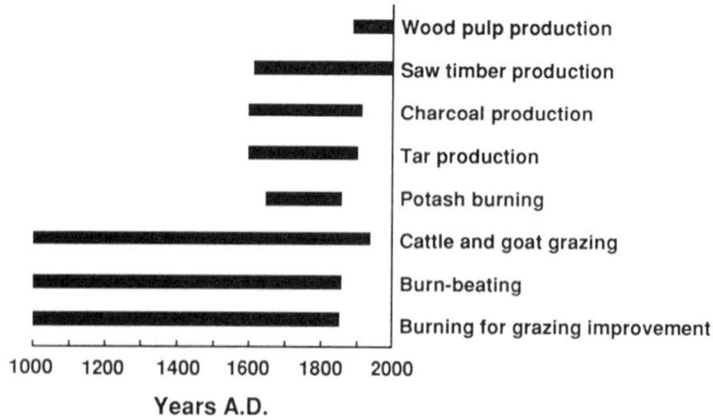

Fig. 7.8. Schematic picture of man's utilization of forest land in northern Sweden from 1000 AD up to present. Based on papers cited by Tenow (1974).

Even if fodder production for cattle was the basis of the ancient rural economy, man gradually began to utilize the boreal forest for the production of exports. Potash production had a large-scale impact from the beginning of the 17th century and reached its maximum around 1830 (Tirén, 1937; Fig. 7.8). Birch was the main species used for this purpose—it has been estimated that a total of 6 million m^3 was utilized in the county of Västerbotten, the area with largest exports (Tirén, 1937). Tar, produced from large pines, was another important export product. This peaked around 1850 but did not end until the beginning of the 20th century (Borgegård, 1973; Fig. 7.8). This extensive multiple use of the forests peaked during the middle of the 19th century, when, for example, the open heaths reached their maximum extension (Malmström, 1939). However, the picture changed at the beginning of the industrial period, in the late 19th century, even if, for example, cattle grazing was commonly practised until the 1940–50s (Fig. 7.8).

The rapid expansion of the sawmill industry in the latter part of the 19th century had a dramatic effect on the forests (Fig. 7.8). Sawmills were mainly located along river valleys as most sawlogs were floated. Pine was the preferred species. The dominant logging method was diameter-limit cutting (dimension cutting), where only the largest and healthiest trees were used. However, many areas were repeatedly logged by selecting successively smaller dimensions. At the end of the 19th century, tree removals resulted in local overexploitation of available timber and the sawmills were forced to move northwards. An increasing amount of wood, especially spruce, was harvested around the turn of the century in connection with the growth of the pulp industry (Tenow, 1974).

7.4.2 The present—the industrial forest

Sweden's forests—some statistics

Productive forest land, defined as stands with an annual production of more than 1 m^3 tree volume/ha, amounts to 23·6 million ha in Sweden (statistical information in this case and in the following according to Anon., 1989). Fifty percent of this area is private forest, mainly owned by farmers, while 23% is company forests, 19% State forests, managed by the Swedish Forest Service, and 8% other public forests. Biological and technical impediments amount to 5·9 million ha and a further 0·7 million ha constitutes high-altitude conifer forest, both with an annual production of less than 1 m^3/ha (Fig. 7.9). The total volume of the growing stock amounts to *c.* 2700 million m^3. The tree species composi-

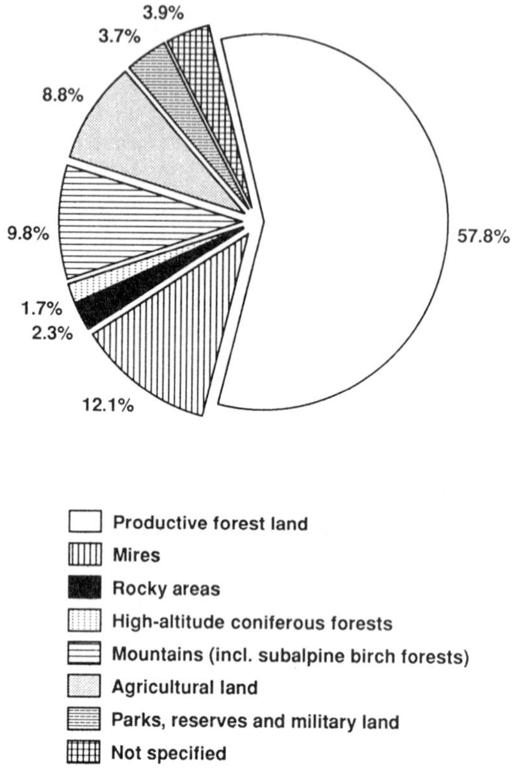

Fig. 7.9. The relative proportion of major landscape elements in Sweden. The total land area of Sweden amounts to 40·8 million ha. Data from Anon. (1989).

tion is dominated by conifers. By volume, spruce constitutes 45%, pine 37%, birch 11% and other broad-leaved trees 5%. The total tree volume logged annually has increased from *c.* 25 million m^3 in the 1850s to about 70 million m^3 in the 1980s (Fig. 7.10). At present, 48% of the total volume constitutes pulp wood, 45% sawlogs and 7% fuelwood and wood used for other purposes.

Silvicultural methods

Current forest management in Sweden is dominated by a single silvicultural method, that is clear-cutting, followed by soil scarification and planting of conifers. The widespread usage of clear-cutting emerged

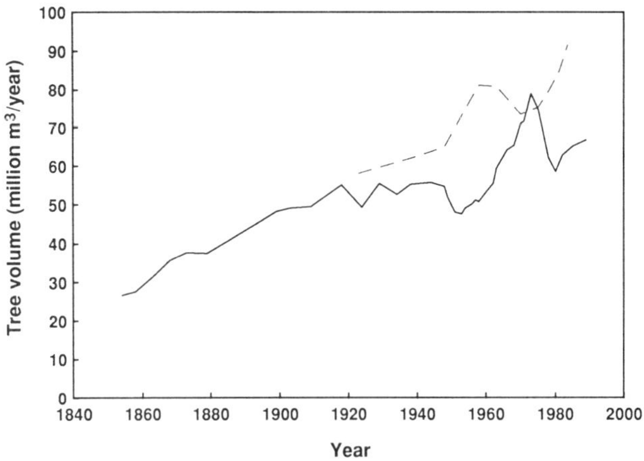

Fig. 7.10. Total annual logging volume since 1859 (——) and calculated annual increment since 1923 (- - - -) in Sweden. Simplified after Anon. (1989).

during the 1940s. At present, about 1·1% of the forest, *c.* 200 000 ha, is attributed to final cutting (including clear-cuts with or without seed trees) annually, a figure which varied somewhat during the 1980s. Thinning, mainly used to increase tree growth in the decades prior to clear-cutting, amounts to *c.* 240 000 ha annually. Additionally, about 270 000 ha is each year attributed to precommercial thinning (cleaning). Other minor cutting operations, such as cutting of a few trees, occur on *c.* 290 000 ha/year. The rotation period, i.e. time elapsed between two successive clear-cuts, varies from *c.* 80 years in Götaland, South Sweden, to about 120–130 years in Norrland, North Sweden.

Most of the clear-cut area is attributed to mechanized soil scarification. Harrowing and scarified patches amount to about 170 000–190 000 ha/year. Prescribed burning nowadays only take place at *c.* 200 ha/year. It was much more utilized during the 1950s, for example, when about 50 000 ha were burned annually. However, there has been an increased interest in using prescribed burning in recent years, especially on humid, inland sites with a thick moor layer.

Today about 75% of all clear-cuts are planted. This can be compared with the situation in the 1950s, when planting only took place on about 20% of the forest land. Natural regeneration under a sparse stand of seed trees amounts to one-third of the area cut annually, but was more common in the past. Pine is the only species used for this purpose. Artificial seeding occurs to a minor extent, at present about 2500 ha/year. Both artificial and natural regeneration is almost exclusively focused on conifers, with a roughly equal proportion of spruce and pine. The fast-growing lodgepole pine *Pinus contorta*, native to North America, has been introduced in order to truncate the long rotation periods prevailing in northern Sweden (Karlman, 1981). Another objective is to compensate for the relatively small area of middle-aged stands (Fig. 7.11) caused by the over-cutting of old-growth stands in the 1940–1970s. The area planted with lodgepole pine has been about 30 000 ha/year since the middle of the 1970s, roughly equivalent to 15% of the area regenerated annually. A total of more than 400 000 ha, about 1·7% of all Swedish forest

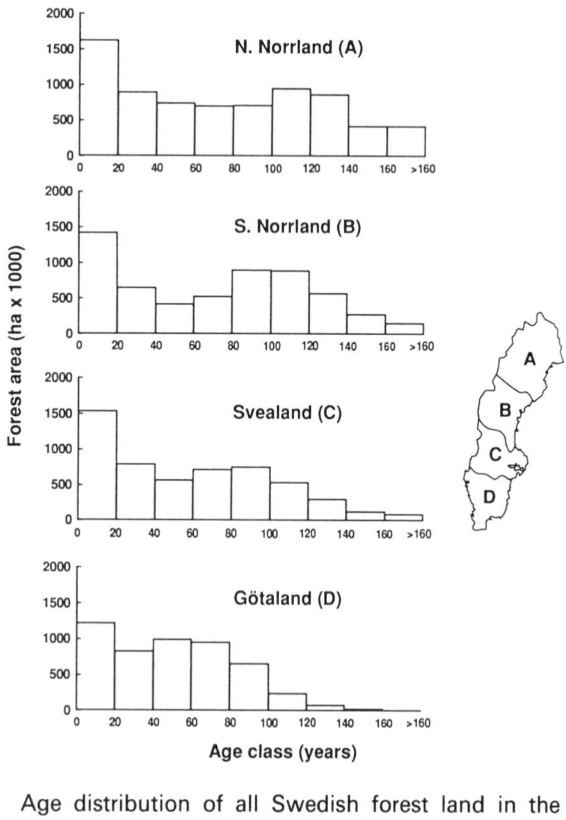

Fig. 7.11. Age distribution of all Swedish forest land in the four major forest regions during the period 1983–87. Data from Anon. (1989).

land, has been planted with this tree species (Hagner, 1985). Recently (1987), there have been restrictions in the planting of lodgepole pine.

In order to favour conifers, the presence of broad-leaved tree species, *Betula* spp. and *Populus tremula*, has been actively combated. Another reason for combating *Populus* is that this species is the alternative host for the rust fungus *Melampsora pinitorqua*, which attacks *Pinus sylvestris*. Broad-leaved trees are mainly removed by selective cutting, i.e. precommercial thinning and, sometimes, by notching stems with herbicides prior to clear-cutting and by using herbicides for suppressing deciduous shrubs on clear-cuts. Nowadays, there is a tendency to allow broad-leaved trees to grow up in the stands. The use of insecticides, fungicides and herbicides has decreased markedly during the last decade. In the 1970s around 30 000 ha/year were sprayed by airplanes; today the figure is just about 500 ha.

For improving the growth condition of the planted conifers, protective ditching of mesic and moist sites is carried out on about 20 000 ha per year. This implies a drastic change in soil moisture conditions in the future forests. Further, about 20 000 ha are drained annually in order to favour forest production on various types of peatland. Forest fertilization, utilized to enhance the growth of conifers prior to clear-cutting, has been at a high level since the early 1970s and amounts to about 120 000 ha/year. This is more than a 20-fold increase since the 1960s.

A very dense network of forest roads has been constructed for the transportation of timber. The goal is that all stands should have a minimum distance of 500 m to the nearest road. Today, 98% of all forest land in south Sweden lies within 1000 m. The corresponding figure for North Sweden is 75%.

In Sweden, forestry is heavily regulated by detailed instructions for essentially all types of forest operations, resulting in a highly mechanized and intensive silviculture with the aim of creating monospecific, even-aged and even-sized stands with a short rotation period. Efficient forest management is the main reason for the increase in the annual increment during recent decades (Fig. 7.10).

Forest trends and patterns

The age and size structure of the Fennoscandian boreal forest is largely affected by past and present forestry. The combined effect of multiple forest utilization is that most forest land in Sweden located outside the montane area has been affected by logging operations. This is most clearly seen in the general scarcity of very large conifers. However, past cutting operations have influenced spruce forests less than pine, with the result that a largely natural forest structure, with an uneven age-structure and large-sized trees, still remains in many old spruce stands. There is a general increase in stand ages from the south (Götaland) to the north (N Norrland; Fig. 7.11). This is due both to naturally lower tree ages and the greater impact of past forestry towards the south. Also, the age distribution (Fig. 7.11) clearly indicates the high proportion of young stands and the scarcity of very old stands. Past levels of over-cutting of mature forest is clearly seen in the small area of middle-aged stands (Fig. 7.11). Most stands are characterized by diameter distributions dominated by small and middle-sized trees, while very large trees are extremely rare. By volume, trees greater than 45 cm in diameter constitute 1·8 and 2·0% of pine and spruce, respectively (Anon., 1989).

7.5 EFFECTS OF FORESTRY ON FAUNA AND FLORA

A very wide range of ecological responses to forest management can be recognized. For example, some effects are more related to the degree of habitat alteration at a particular site, while others are more dependent on the spatial scale of the specific silvicultural method. Some examples of types of responses are given below.

7.5.1 Consequences of clear-cutting management

The short-term ecological effects of clear-cutting management are caused by the removal of trees, the abrupt change in the abiotic environment and the disturbance of soil and ground vegetation. In Fennoscandia, species richness of vascular plants seems to be comparatively little affected by clear-cutting, and most species survive, although with changed population sizes. However, some forest herbs, for example the orchids *Calypso bulbosa* and *Listera cordata*, normally become extinct on clear-cuts. The intensity of soil disturbance during cutting operations and subsequent scarification is a crucial factor affecting plant mortality. It has been shown, for example, that surface disturbance following cutting contributes to species losses in deciduous forest (Reader, 1987).

Rather few vascular plants seem to be threatened by clear-cutting. This is mainly because the majority of vascular plants threatened by forest management are confined to forest impediments, e.g. ravines, wet forests and steep hill-slopes (Ingelög *et al.*, 1987). However, even such habitats have come under considerable threat during recent years, for example, due to drainage activities (see Chapter 8).

It is well known that a number of vascular plants are favoured by forest management, e.g. herbs such as *Epilobium angustifolium*, *Geranium sylvaticum*, *Melampyrum pratense*, *M. sylvaticum*, *Rubus idaeus* and *Solidago virgaurea*. A similar response is also shown by grasses such as *Deschampsia flexuosa* and *Calamagrostis purpurea*, and in the south boreal zone also by *C. arundinacea* (Lindholm & Vasander, 1987). These herbs and grasses can reach high abundances some years after clear-cutting, flower frequently and may outcompete other plant species. Another effect is an increased food supply for herbivores. As a result, population sizes of microtine rodents, e.g. field vole *Microtus agrestris*, rapidly increase some years after cutting. This may result in increased levels of herbivory on preferred plant species.

By definition, clear-cutting eliminates all epiphytic lichens and bryophytes. However, most epiphytes also decline on solitary trees and snags left on clear-cuts, due to increased wind speeds and lowered air humidity. For example, large foliose lichens such as *Lobaria pulmonaria* on solitary *Salix caprea* trees disintegrate and die within a year or two because of altered climatic conditions. The epiphytic lichen *Erioderma pedicellatum*, known from a limited area in Sweden and Norway (Ahlner, 1948; Jørgensen, 1978), has vanished from both countries both as a direct and indirect effect of clear-cutting. The Norwegian localities have been clear-cut. In Sweden climatic conditions have been radically altered following cutting of the surrounding forest (Jørgensen, 1978; Ingelög *et al.*, 1987). Bryophytes growing on decaying logs rapidly dry out and die within a couple of years after cutting. For example, the specific bryophyte community that occurs on large logs in old-growth spruce forests is rapidly destroyed (L. Söderström, pers. comm.).

The effect of clear-cutting on the amount of edible mushrooms has been studied by Kardell (1980) and Veijalainen (1976). They both concluded that clear-cutting reduces mushroom diversity but that it has no great effect on the biomass (see also Wästerlund & Ingelög, 1981; Wästerlund, 1989).

Increased light availability and wind speeds on clear-cuts may result in dry conditions. This can cause drastic reductions in many drought-sensitive, ground-living invertebrates, e.g. molluscs (Ehnström & Waldén, 1986). Draining of moist clear-cuts can also be fatal for invertebrates. This applies to some threatened insects, for example, the ground beetle *Agonum mannerheimi* and the noctuid moth *Lasionycta skraelingia*, which both occur in moist spruce forests (Saalas, 1917; Palmqvist, 1974).

7.5.2 Reduced tree ages

Current forest management has altered the variable age structures of past natural forests to a more or less uniform rotation period of about 80–130 years. The major effect is a general reduction in stand ages, despite a somewhat longer interval between disturbances for pine stands characterized by a high fire frequency. At present, forest stands older than 160 years occur on less than 3% of the forest area in Sweden (Anon., 1989).

Decreased tree ages have resulted in a marked decline in abundance of many organisms requiring old-growth forest and have also caused local extinctions, especially in south and middle boreal Sweden. This is mainly a consequence of a number of natural processes and patterns, such as self-thinning, occurrence of suppressed, wounded and dying trees, almost disappearing from the managed forest (Table 7.1). An indirect effect of the general decrease in stand ages is a change in forest microclimate towards lower air humidity. Reduced tree age is a major factor responsible for the continuing decline in abundance of epiphytic lichens, as shown, for example, by the decline of *Usnea longissima* (Esseen & Ericson, 1982).

Tree diameter is an important factor for birds excavating holes in tree stems. Various species show considerable differences in this respect (Pynnönen, 1939; Haapanen, 1965). By creating younger stands, old, big trees suitable for nest construction by, for example, woodpeckers and some birds of prey (Table 7.5) have become reduced in numbers. For example, Tjernberg (1983) found that pines in northern Sweden used by the golden eagle *Aquila chrysaëtos* as nesting trees had a mean age of over 300 years. Trees younger than 225 years were hardly used at all. These ages are about two to three times greater than in managed forests. For the same reason, i.e. by creating younger stands, there are fewer dead, but still erect trees in managed forests, important for foraging by birds during winter when the ground is covered by snow, and fewer suitable trees for primary and secondary hole nesters (Table 7.5).

7.5.3 Fire elimination

The most important ecosystem change in the Fennoscandian boreal forest is probably the replacement of landscapes governed by fire perturbations with landscapes structured by cutting operations. Before man, the boreal forest landscape evidently consisted of a mosaic with very variable fire frequencies. More than 100 years of fire elimination has led to a situation where hardly any area in the Nordic countries has a natural fire-disturbance regime. The consequence is that early post-fire successional stands have become extremely rare, and are also much smaller than previously; e.g. in 1979 the average fire in Sweden covered 3 ha (Anon., 1981) . Continued fire suppression will, for example, lead to a stage with complete spruce dominance in previously mixed stands with pine, birch and aspen (Zackrisson, 1977).

The most evident effect of fire elimination is the decline of species that inhabit burnt substrates, e.g. some vascular plants and beetles (Heliövaara & Väisänen, 1984; Lundberg, 1984; Ingelög *et al.*, 1987; Table 7.2). One fire-dependent beetle, *Agonum bogemanni*, is considered to have already disappeared from Sweden (Ehnström & Waldén, 1986) and some others are close to extinction (Table 7.2). However, fire suppression has not yet forced any plant species to extinction in Sweden, despite a general decrease in population size. For example, some rare herbs, e.g. *Astragalus penduliflorus*, *Chimaphila umbellata*, *Pulsatilla patens*, and *P. vernalis*, are regarded as sensitive to decreased fire frequency.

Absence of fire disturbance will also result in a decline in the diversity of ground vegetation on lichen-rich pine heaths (Esseen, 1983). Another consequence is that only scattered post-fire successional stands, dominated by mature broad-leaved trees, remain in Sweden, mainly in the south and middle boreal zones. The suppression of the deciduous stage will probably cause one of the more drastic species alterations in the boreal forest, affecting numerous animals, plants and fungi dependent on full-grown birches, aspens and other broad-leaved trees. For example, we can expect declining population sizes for many herbs that establish or reach maximum abundance in the deciduous stage, e.g. *Coeloglossum viride*, *Platanthera bifolia* and *Pyrola media*. The same applies to a large number of invertebrates. The beetles *Upis ceramboides* and *Leptura nigripes*, which are largely confined to the deciduous stage, have apparently disappeared from south Sweden during the present century (Ehnström & Waldén, 1986).

Cessation of forest fire affects the vertebrate fauna mainly through the decrease of stands with large-sized broad-leaved trees, necessary for nest construction and as a food supply. A decreased area of such habitats will affect birds such as lesser spotted woodpecker *Dendrocopos minor*, grey-headed woodpecker *Picus canus* and long-tailed tit *Aegithalos caudatus*. The white-backed woodpecker, an endangered species in Sweden, at present occurs in small and isolated populations (Aulén, 1988). The main reason seems to be the scarcity of stands with high abundance of mature broad-leaved trees.

7.5.4 Changed tree species composition

Surprisingly little is known about the ecological consequences of a change in tree species and the impact of the introduced lodgepole pine. Planting of pine on ground that was previously spruce-dominated, often on moderately fertile soils, is a rather frequent practice that may at least affect the relative abundance of many species. According to Lundmark *et al.* (1982) lodgepole pine stands have *c.* 15% higher rate of needle litter-fall compared to Scots pine. There are indications that increased litter-fall may be detrimental to the forest floor vegetation, paticularly to bryophytes. Kardell *et al.* (1987) compared the mushroom flora in 14–37 year old plantations of lodgepole pine with that of Scots pine plantations of similar age. They found a 15% lower number of mushroom species and a lower number of fruiting bodies in lodgepole pine stands. Additionally, there were significant shifts in the relative abundance of different mushroom species.

The biomass of spiders and insects, i.e. the bulk of food for many overwintering passerine birds, in lodgepole pine plantations is significantly lower than in Scots pine stands of similar age (Sjöberg, unpublished data). One reason might be the difference in the bark structure. Rough-barked trees such as Scots pine, with abundant crevices, provide a greater supply of arthropods available to birds (Morrison *et al.*, 1985).

One reason why broad-leaved trees are important for a great number of birds species is the biomass concentration of invertebrates. Palmgren (1932) made comparative studies on the numbers of invertebrates in the branches of birch, spruce, and pine. Both the numbers and total weights of invertebrates were considerably greater in birch than in spruce and pine during summer. The bird density, however, is not much smaller in spruce than in birch stands (Haapanen, 1965). Canopy height and total leaf area/ha are noticeably greater in spruce stands than in birch stands, which possibly compensate for the difference in food per unit amount of branches (Haapanen, 1965). Nevertheless, the importance of broad-leaved trees in the boreal forest for the bird fauna is evident from a study in Swedish Lapland by Andersson *et al.* (1967), who found that bird density increased with increasing proportion of broad-leaved trees. Although hardwood comprises much less than 10% of climax spruce stands in Finland, 65% of the bird-created holes are made in hardwoods, 25% in pines and only 10% in spruce (Haapanen, 1965).

Additionally, the current practice of planting trees from remote provenances, the introduction of more foreign tree species and the development towards genetically cloned seedlings may have significant effects about which we know very little. For example, there may be a risk that the future forest will have less structural heterogeneity compared to present forests, which in turn may adversely affect biodiversity.

7.5.5 Reduced input of coarse woody debris

Decreased abundance and diversity of decaying wood have probably negatively affected more species than any other consequence of forest management in Fennoscandia. The proportion of coarse woody debris is generally low in Swedish forests and has since the 1950s varied between 1·9 and 3·1% of the growing stock (Svensson *et al.*, 1989; cf. Fig. 7.12). By contrast, as much as 40% has been recorded in an old-growth spruce forest in northern Sweden (Linder, 1984). This reflects a drastic change due to forest management. The most important changes are that large-sized snags and logs have been replaced by small-sized woody debris and that natural stumps are replaced by those with cut surfaces. For example, Kalliola (1966) has shown that in Finland the number of pine snags with

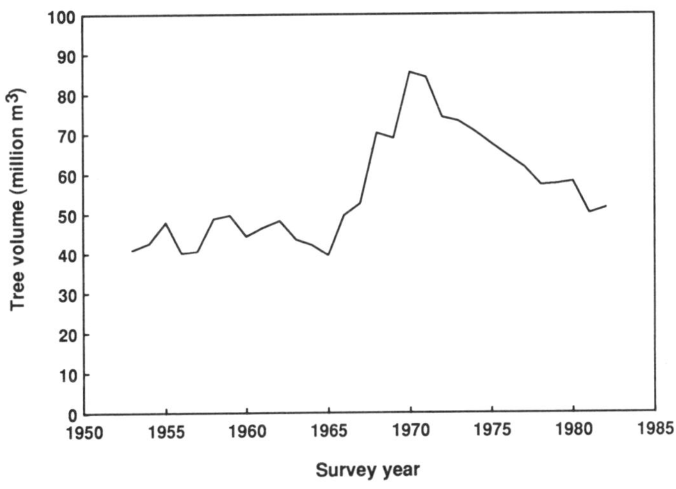

Fig. 7.12 Change in total volume of snags and fresh windthrows in Sweden from 1953 to 1982. All trees with a diameter larger than 5 cm are included. The growing stock has increased from 2100 to 2600 million m^3 during the same period. Redrawn after Svensson *et al.* (1989).

a diameter larger than 20 cm was makedly reduced from 1936–38 to 1960–63.

Although a considerable amount of small-sized wood fragments become exposed after clear-cutting (logging residues), this is mainly suitable as substrate for a number of common bryophytes and lichens with wide habitat amplitudes. The effect of reduced input of coarse woody debris is that a large number of specialized bryophytes, lichens, fungi and invertebrates that utilize wood of various tree species and stages of decay have drastically declined (Ehnström & Waldén, 1986; Ingelög *et al.*, 1987). For example, at least five species of wood-inhabiting bryophytes and lichens, i.e. *Calicium subquercinum*, *Cephalozia lacinulata*, *Cyphelium trachylioides*, *Scapania massalongi* and *Sphinctrina anglica*, have become extinct in Sweden (Floravårdskommittén för Lavar, 1987; Floravårdskommittén för Mossor, 1988).

In managed forests, most fresh windthrows and snags have been removed during the last decade. This is the main reason for the decline of some bryophytes exclusively confined to old-growth spruce stands, i.e. rare hepatics on large logs (Söderström, 1981, 1987; Gustafsson & Hallingbäck, 1988; Andersson & Hytteborn, 1991). For example, Andersson and Hytteborn (1991) have clearly shown that the frequency of wood-inhabiting bryophytes increases significantly with an increased diameter of logs. Söderström (1988*b*) compared the wood-inhabiting bryophyte and lichen flora on decaying logs and tree stumps in a managed forest with that of a nearby unmanaged, old-growth spruce forest in north Sweden. He observed a higher abundance of hepatics and a lower abundance of cup-bearing *Cladonia*-lichens in the old-growth forest. The mechanism was a combination of a lower amount of large-sized logs and a drier climate in the managed forest (Söderström, 1988*b*). A virgin spruce stand in SW Sweden had a higher species richness of bryophytes (88 species) and bryophyte cover (73%) compared to three managed stands, 56–72 species and 49–59% cover, respectively (Gustafsson & Hallingbäck, 1988). However, they found very small differences in the quantities of the more common species.

Removal of coarse woody debris causes an increasing lack of suitable food for many invetebrates in all types of managed forest. Some invertebrates can survive in logging residues such as branches and small logs. However, most specialized invertebrates, for example those associated with wood-inhabiting fungi, show marked reductions in population size. Furthermore, some insects utilize woody habitats only during a single year and are more or less dependent on a continuous input of coarse woody material.

7.5.6 Effects of forest management on avifauna

The effect of forest management on the bird fauna depends on the extent and efficiency of the management methods, but also on side effects of forest management, such as the dense network of forest roads. Some general differences are found between the bird fauna in virgin forests and managed forests in southern Finland: (1) birds of old-growth stands decrease; (2) hole-nesters decrease; and (3) some generalist species increase as a result of forest management (Haapanen, 1965). In northern Finland, however, there were no significant differences in total bird density between virgin and thinned pine forest, nor between uniform and fragmented spruce forests (Virkkala, 1987). However, the structure of the bird communities was different. The densities of generalist species, for example *Phylloscopus trochilus* and *Turdus iliacus*, were higher while those of specialists such as *Parus cinctus* and *Perisoreus infaustus* were lower in thinned forest than in virgin pine forest. Forest fragmentation had a positive effect on the densities of *P. trochilus*, *T. iliacus* and *Carduelis flammea* and a negative effect on the group of northern taiga species (*Emberiza rustica*, *Parus cinctus*, *Perisoreus infaustus*, *Picoides tridactylus*, *Pinicola enucleator* and *Tetrao urogallus*). The densiy of sedentary species was significantly higher in virgin forests than in managed forests (Virkkala, 1987).

In Finland, the foliage-gleaning guild of overwintering birds inhabiting mainly coniferous forest habitats comprise the following species: *Parus montanus*, *P. cristatus*, *P. ater* and *Perisoreus infaustus*. In northern Finland the total density of this guild decreased by about 80% from 1945 to 1975 (Järvinen *et al.*, 1977). It is supposed that the extensive cuttings of old forest in northern Finland in the late 1950s and early 1960s, which reduced the area of available habitat and the average size of the remaining forest areas, caused a fall in population density.

Overall, however, more bird species have increased in Finland than decreased due to changes in the condition of the forest and other reasons: 43% of the 86 most important land species have increased since 1945, while less than a third have decreased (Järvinen and Väisänen, 1977). One of them, *Parus cinctus*, has decreased dramatically, i.e. to less than a tenth of the numbers estimated three decades earlier. Virkkala (1990) and Virkkala and Liehu (1990) regard extensive forest management as a probable reason for the drastic decline of this species. Järvinen and Väisänen (1977) conclude that marked variation in population dynamics was found to be an essential feature of the bird fauna of the northern parts of the Palearctic region in Europe.

Helle and Järvinen (1986) found that fragmentation and changes in the age structure of the forests are mainly responsible for the recent decrease of the resident species of old forests. This seems to be true, both on a large scale and at the scale of local bird populations in the boreal forest (Väisänen *et al.*, 1986). Among the breeding birds, for example, woodpeckers tend to be scarce in small (less than 6 ha) fragments in coniferous taiga in southern Finland, and the hole-nesting passerines *Phoenicurus phoenicurus* and *Parus major* also seem to avoid small fragments. The density of hole-nesters showed a positive correlation with snag density in the fragments but not with fragment area (Haila *et al.*, 1987).

Although considerable differences were recorded in the structure of the bird community, neither Virkkala (1987) nor Haapanen (1965) found any differences between virgin and thinned forest sites as regard species richness or diversity. In southern Sweden, however, Nilsson (1979) estimated the total bird density to be three times higher in the natural forest compared with a managed forest of similar age. All groups of species had higher population densities in the natural forest, but woodpeckers and warblers much more so than other groups. Another difference compared with the more northern studies of Virkkala and Haapanen was that the hole-nesting species made up about the same proportion of the bird communities in the natural and the managed forest, while in the Finnish studies the density of this species group in managed forest was only about half of that in virgin forests. The reason for the difference may be that forest management of natural mixed forests in southern Sweden eliminates a high proportion of the deciduous tree species present. This is not so in northern taiga forests (Virkkala, 1987).

7.6 CONSERVATION STRATEGIES FOR THE FUTURE

7.6.1 How to maintain diversity?

The general goal of conservation biology is to maintain biological diversity (Soulé, 1986). This usually means preservation of all species native to the area in question as well as the maintenance of naturally functioning populations and ecosystems. When dealing with a particular environment, such as the boreal forest, these general objectives must be reformulated so that they conform to the ecological situation in that area. For example, conservation biologists often stress the importance of maximizing biodiversity. However, some pristine habitats in boreal areas characteristically have low biodiversity. Consequently, it is better to say that we want to preserve the natural pattern of plant and animal diversity, whether low or high (Noss, 1983). Another important issue is that more than 95% of Sweden's forests are already utilized for commercial forestry. This situation will probably not change much in the future because of the fundamental economic importance of the forest industry. These circumstances constrain the options of choosing among various conservation strategies compared to countries where forests have been less intensively exploited. The objectives of conservation biology in the boreal forest may be formulated as follows:

(1) to maintain the specific features of the natural forest;
(2) to ensure survival of threatened boreal organisms;
(3) to maintain naturally functioning populations, communities and ecosystems;
(4) to maintain representative forest types and successional stages;
(5) to provide a natural mosaic of boreal habitats within a landscape framework.

These points emphasize the importance of preserving diversity at all scales (cf. Noss & Harris, 1986). In order to maintain biological diversity in the boreal forest we have to preserve and maintain the patterns and processes causing diversity. This means that it is important to consider habitat quality (content; cf. chapter 2) as well as habitat quantity, as losses in both aspects can result in extirpation of species (Gilpin & Soulé, 1986). In other words, we have to preserve areas large enough to ensure operation of natural processes, such as the creation of old trees, snags, fallen logs and specific microclimates.

Natural disturbances create much of the structural and functional diversity occurring in the boreal forest. This means that it is crucial to maintain a natural disturbance regime (Pickett & Thompson, 1978), i.e. including various types of natural disturbance, frequency of disturbances, intensities and size of the disturbed area (Sousa, 1984; Pickett & White, 1985). At the landscape level, disturbances that initiate secondary succession, e.g. fire (Knight & Wallace, 1989), must be allowed to operate in a natural manner. After the local level, it is necessary to maintain natural patterns of small-scale disturbances, e.g. tree-deaths, tree-falls and gap dynamics (Shugart, 1984, 1987; Pickett & White, 1985), animal disturbance (White, 1979; Sousa, 1984) and other disturbances enhancing microhabitat diversity.

A reintroduction of forest fire in the near future is necessary in order to initiate post-fire successions, to prevent a reduction in plant and animal diversity and to ensure survival of fire-dependent organisms.

However, natural fires in boreal areas may occasionally be very large, up to 100 000 ha in some areas (Seitz, 1986; Payette *et al.*, 1989). This implies that a huge area of boreal forest is required in order to maintain an area with a natural fire-disturbance regime. Baker (1989), for example, did not find any stable, fire-regulated patch mosaic even in a 400 000 ha tract in Minnesota, USA. This can be compared with the largest forest reserve in Sweden, Muddus National Park, with an area of 50 000 ha. Consequently, it seems to be an impossible task to have whole landscapes with a natural fire mosaic in Fennoscandia. An alternative approach is to apply artificial fire management programmes founded on solid information of past fire frequencies and their regional variability (Alexander & Dubé, 1983; Engelmark, 1988; see also Schullery, 1989). This will likely result in fewer and smaller fire patches than in previous centuries. However, even widely dispersed areas of fire succession may be sufficient to maintain most fire-dependent organisms, because the majority of these species seem to have efficient mechanisms of long-distance dispersal. Programmes of artificial fire management should not totally exclude the possibility of allowing some minor natural fires to develop in the future. An increased usage of prescribed burning after the clear-cutting of stands that have a past fire history would also improve the situation for many fire-related species.

On the other hand, absence of large-scale disturbance is characteristic for some boreal habitats. For example, large portions of the high-altitude conifer forest fringing the mountain range have evidently escaped burning during past centuries. In such situations, conservation must be focused on preventing the long forest continuity from being broken by man-induced disturbances.

Two different conservation strategies must be applied in order to preserve biodiversity in the boreal forest. First, it is necessary to protect and manage remnants of natural and semi-natural forests of various successional stages. Sufficiently large forest reserves are the only option for sustaining naturally functioning populations of boreal organisms. Secondly, commercial forestry should be modified in a way to minimize detrimental effects on plants and animals.

7.6.2 Preserving and maintaning natural forests

Strategies for selecting forest reserves

Large reserves are usually considered better than small ones as they host larger populations, less sensitive to extinction. However, it has been shown that small, isolated patches of forest do not necessarily have an impoverished flora (Helliwell, 1976, Simberloff & Gotelli, 1984; Middleton & Merriam, 1985). Small forest reserves may function as source areas in managed forest landscapes from where diaspores may disperse to other sites that have the characteristics required by the species in question. However, there is generally a higher extinction risk in small reserves and hence no guarantee for long-term persistence. Counterbalancing this, a substantial number of plants and animals with very specific habitat demands have always occurred in small, isolated forest patches.

We have chosen below two examples to illustrate how to determine the size of forest reserves based on ecological knowledge of selected indicator species of old-growth forest, and to estimate the minimum area required for these species.

The first example deals with species sensitive to changes in microclimatic conditions. The lichen *Usnea longissima* inhabiting old-growth spruce forest shows a strikingly patchy occurrence even within extensive stands of suitable habitat. Each inhabited patch rarely exceeds a size of one hectare. We can assume that a buffer zone of about three times maximum tree height is necessary to exclude significant changes in forest microclimate in the patch (see chapter 2). In this case, maximum tree height is around 25 m (Esseen & Ericson, 1982), which gives a buffer zone of 75 m. If we add 50% of this distance to the buffer zone in order to have additional protection from unpredictable edge-effects, for example, from tree-falls, we arrive at an estimate of 9·0 ha for a circular reserve. Present evidence indicates that *U. longissima* and some other epiphytic lichens can survive in such small forest reserves. However, there is little information about the long-term persistence of the lichen populations.

The second example is based on the Siberian Jay *Perisoreus infaustus*, which has a home-range of *c.* 100 ha (Blomgren, 1964; Lindgren, 1975). The holders of the territory are sedentary and spend the whole year within the same area. Only the juvenile birds disperse outside the ordinary home range during autumn and winter. Although the Siberian jay does not seem to show a strict preference for pine or spruce forests, it belongs to a group of birds occurring in old-growth forests with abundant growth of pendent lichens. Thus to secure a population of ten to twenty pairs of Siberian jay we need an area of *c.* 1000–2000 ha of coniferous forest, not, or very slightly, affected by forest management.

The present system of forest reserves

Protected boreal forest occurs in a fairly large number of reserves and national parks throughout Sweden. These reserves provide habitats for a considerable number of boreal plants and animals. However, there is substantial evidence that the present system is not comprehensive enough to ensure long-term survival of all boreal organisms in Fennoscandia. The reasons for this are:

(1) The small fraction of totally protected forest. Outside the high-altitude coniferous forest these reserves only comprise 0·34% (75 000 ha) of all productive forest land (Kardell & Ekstrand, 1990). Protected conifer forests fringing the mountains, classified as productive forest, amount to *c.* 509 000 ha (Kardell & Ekstrand, 1990).
(2) The size distribution of reserves is highly skewed towards small ones. For example, Bengtsson *et al.* (1982) showed that about 75% of all Swedish nature reserves were smaller than 100 ha. Since then, a number of large forest reserves have been established, but the distribution still remains skewed towards small reserves.
(3) The largest areas of protected forest are located in the northern part of the country, and are less diverse than southern forests.
(4) Existing reserves do not provide a representative subsample of all major types of boreal forests. More productive forest types and early successional stands, for example, are highly under-represented.

In the 1970s and early 1980s an inventory of primeval forests was made in Sweden (Bråkenhielm, 1982). One of the criteria used in this inventory was that forests should have no or only minor evidence of previous cutting operations and other human impact. The inventory has resulted in a fairly good network of forest reserves. However, these reserves are biased towards forests of low productivity. The inventory considered some rare and endangered species, but there was no specific intention to select stands due to their biological diversity. At that time, it was not fully realized that the occurrence of specific habitat characteristics and natural processes are more important for biodiversity than minor human impact, which in many cases apparently has a rather small effect on plant and animal communities.

A future system of boreal forest reserves

Future forest reserves in Sweden should be created with the specific aim of preserving biological diversity. We therefore strongly recommend that a number of forest reserves should be added to the existing ones in order to create an improved network of reserves all over the country. These forest reserves should mainly be selected on the basis of the presence of the most threatened boreal organisms, i.e. to include bryophytes, fungi, lichens and invertebrates (Ehnström & Waldén, 1986; Ingelög *et al.*, 1987). The following habitats are under-represented in the existing system of forest reserves and in urgent need of protection:

(1) early to intermediate post-fire successions, particularly those dominated by broad-leaved trees
(2) fire refugia forests
(3) herb-rich conifer forests with intermediate to high productivity
(4) wet forests (see chapter 8)
(5) forested ravines
(6) forest margins along water courses, lakes and mires (see chapter 9)
(7) forested islands in archipelagoes.

In the montane area of Sweden, where some of the largest, virgin, western European forests still occur, there has been an increase in clear-cutting intensity in the last decade. Unfortunately, the discussion about how to utilize these forests, mainly of low productivity, has been largely focused on different problems of forest regeneration (see Bäckström, 1984). As a result, too little concern has been given to the unique flora, fauna and natural processes occurring in the high-altitude conifer forests. It is seldom realized that cutting of these forests eliminates structures and processes that have developed during several hundreds of years.

In spite of a decreasing area of pristine forest, it is still possible to find landscapes as large as 20 000 ha that have a natural mosaic of different biotypes, e.g. old-growth forest, wet forests, mires, lakes and water courses. It has recently been discovered that many forest plants and animals which face marked population reductions in managed forests still have large and viable populations in some northern forests, for example, the lichens *Evernia divaricata* and *E. mesomorpha* and fungi as *Haploporus odorus*. We can assume that if a major portion of the high-altitude conifer forest can be preserved, it may function as a mainland bordering a 'sea' of intensely managed forest containing an 'archipelago' of more or less isolated patches of old-growth forest in reserves. The importance of such a 'mainland' is both to maintain naturally functioning populations, communities and ecosystems and to function as a source area from where recolonization of forest reserves and managed stands may occur. However, as these questions have been little studied, we argue that forest operations should be largely constrained in the high-altitude forests until we have gained information about how large an area is necessary for maintaining biodiversity. These forests are the only options that still remain for creating some very large forest reserves in Fennoscandia.

The urgent need for rapid protection of the remaining primeval forests cannot be overstressed. During recent decades in Sweden a number of old-growth stands of great value from a conservation point of view, have been clear-cut. In some cases, such operations have even been indirectly subsidized by the government through grants for cutting and restoration of various types of understocked forests, and subsidies for the construction of forest roads.

As the fragmentation of old-growth forest is so rapid in Sweden we have to set priorities for the protection of natural forests. The first and most urgent step is to secure the remaining fragments of old-growth forest, including the protection of a number of small forest habitats with high biodiversity. The second step should be directed to reduce the regional imbalance in reserve size. More specifically, this should involve the establishment of large forest reserves in the south and middle boreal Sweden and in the coastal area in the north. This could be accomplished by buying managed forests that are allowed to develop to a natural state. In several cases, these reserves may also be formed by preserving managed stands adjacent to existing forest reserves (Ahlén, 1982).

7.6.3 A forestry that maintains biodiversity

Modification of forest management

The development and application of silvicultural methods that have little impact on biodiversity is equally important as creating forest reserves. If this is not accomplished we will face a depauperated managed forest where mainly only more or less trivial species can survive. The species composition in forest reserves will also be affected, for example, by increased extinction rates and by immigration of species favoured by forest management.

Programmes containing measures of environmental protection in forestry have existed for about a decade in Fennoscandia (Forestry Act, 1979, 1986; Lindholm, 1987; NOU, 1989). Environmental protection requirements in Swedish forestry are currently regulated by the 21st paragraph in the Forestry Act (1979, 1986), see also Ingelög (1981); Eckerberg (1986, 1988). This contains legislation and guidelines as regards nature conservation in connection with most types of forest management operations, such as regeneration, spacing, thinning, harvesting and road construction. There are instructions about the size and location of clear-cuts, about leaving snags, isolated old deciduous trees or groups of valuable trees, and the location of forest roads. For example, clear-cuts should not be larger than 20 ha in the high-altitude conifer forests. Additionally, it is stated that small patches of productive forest, sized 0·5–1·0 ha, should be excluded from cutting if they host threatened plants or animals (Forestry Act, 1979; Ingelög (1981); Ahlén *et al.*, 1979; Ehnström & Waldén, 1986). The Forestry Act also regulates which types of forest land should be excluded from forest operations, for example, forest impediments on rocky ground, steep hill-slopes, small forest islands in agricultural landscapes and small areas of wet forests. However, it is also stated that the demands of environmental protection should be compatible with a rational forestry and should not impede on current land usage. This statement may, of course, impose rather severe restrictions on the application of the legislation.

Eckerberg (1988) has clearly demonstrated that the opportunity to utilize the Forestry Act as a tool in conservation management has been far from fully realized. For example, more consideration was taken of aesthetic environmental aspects than of activities to preserve threatened plants and animals (Fig. 7.13). The most satisfactory consideration was in cases where environmental protection was in accordance with forest operations for technical and economic reasons.

In recent years there has been growing criticism of the Forestry Act, focusing on both the insufficiencies of the legislation and its poor application. For example, it includes a paragraph (no. 20) which states that recently dead or damaged conifers should be removed during the summer if their volume exceeds 4 m^3/ha. The object is to minimize the hazard of pest insect outbreaks. This paragraph is negative from a conservation viewpoint, as snags and fallen trees are essential habitats for many endangered invertebrates and plants. We argue that paragraph 20 should be changed to include more specific regulations for pest management, including measures to maintain rare and threatened species. For example, dead trees normally become unsuitable as habitat for pest insects two years after they have died, and should thus be retained. Another weakness of the legislation is that no effort is devoted to the recruitment of structural components, such as snags, fallen logs and old deciduous trees, in the rotation periods following the current one.

A new regulation system of environmental protection in forestry should be founded on the scientific knowledge of the ecology of boreal organisms and ecosystems. The specific aim should be to maintain as much as possible of those habitat and landscape characteristics that promote biodiversity and determine the persistence of populations of native boreal

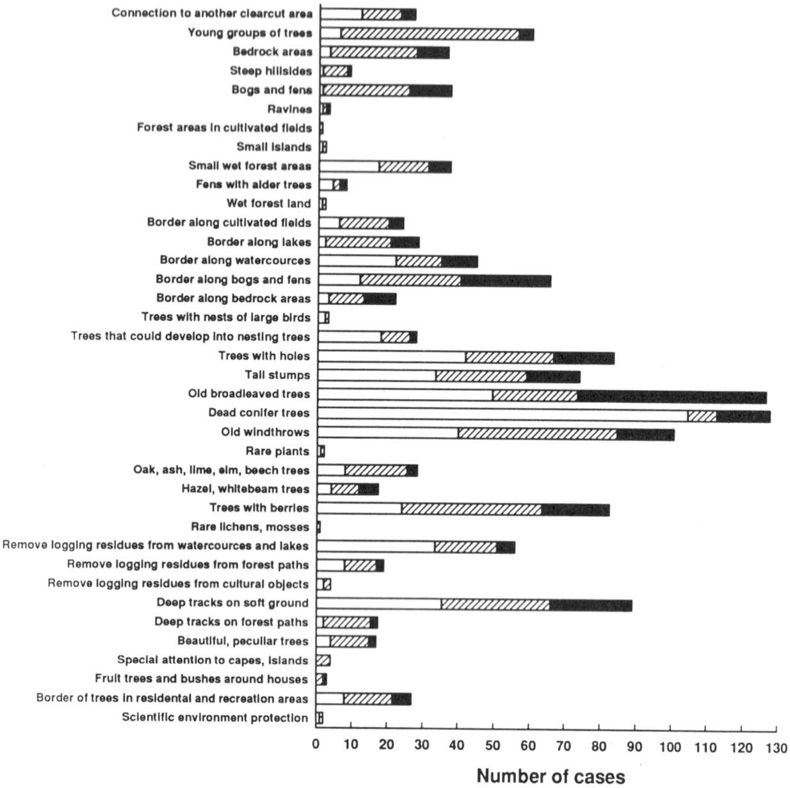

Fig. 7.13. Results of an evaluation of environmental protection in Swedish forestry. The implementation of 38 conservation measures stated in the Forestry Act (1979) were analysed in 158 clear-cuts. Modified after Eckerberg (1988). Explanations: □, no consideration; ▨, some consideration and ■, full consideration.

species. In other words, policies should be adopted that are designed to minimize the pace of extinction (Terborgh, 1974). The most important issue is the need to develop and apply forest practices that increase habitat diversity in managed forests. This view is rather similar to the ecosystem approach of Franklin (1989), in the 'New Forestry' perspective proposed for American Pacific Northwest forests, and recent developments in Canadian forestry. The most important concept of this approach is that forest operations for the production of wood pulp and timber should be viewed as only one part of the major goal of maintaining biodiversity and the long-term health of forest ecosystems. For example, it should be recognized that long-term productivity might be reduced by removing woody debris. Also, a structurally diverse old-growth forest ecosystem is important for maintaining populations of predators and parasites of pest invertebrates (Schowalter, 1989). The New Forestry concept recognizes the importance of having integrated landscapes where the effects of fragmentation are minimized and the link between different landscape elements maximized (Franklin, 1989, 1990; Hemstrom, 1990). For example, forest stands bordering streams and lakes are very important for the production of macroinvertebrates and fish, due to the input of woody debris which provide shelter and nutrients.

Environmental conservation in forestry should be broadened and apply a more landscape-oriented approach, i.e. to maintain a natural and functional mosaic of habitats (Hansson, 1977; Urban *et al.*, 1987; Turner, 1989; see also chapter 2). For example, looking at the surrounding habitat types and connectivity for a site is as important as habitat quality (Noss, 1987). The size and configuration of clear-cuts is also important to consider, as demonstrated by Franklin and Forman (1987). One attempt in this direction is the system of long-rotation forest islands proposed by Harris (1984) for Douglas fir forests in northwestern USA. In this case, a centrally located forest reserve with old-growth forest functions as a core area surrounded by managed stands, representing various successional stages that have a comparatively long rotation cycle.

In connection with forest harvesting, special attention should be paid to the question of retaining a sufficient amount of the structural components that characterize old-growth stands, i.e. biological legacies. This includes retention of snags, fallen logs and old deciduous trees, but also the retention of big, living conifer trees (green tree retention; Franklin, 1989). It is especially important to increase the amount of coarse woody debris. Leaving some large, living conifers and snags on cut areas creates an opportunity for the young stand to be inoculated by some of the old-growth species. Most of these trees should be retained over a long period, at least through the next rotation. More specific operations to sustain a continuous supply of structural components in subsequent rotation periods will be of increasing importance in the future.

The implementation of the current and future regulations of environmental protection is also a very important issue. The current educational programme of state agency personnel, private foresters and loggers in Sweden should be continued and expanded. However, it is also necessary that conservation biologists have the possibility of influencing forest management, both at the planning stage and during silvicultural field operations.

Alternatives to clear-cutting

The dominant position of clear-cutting should be reduced if we are to take the conservation of boreal organisms seriously. This calls for application of alternative forest practices, such as different systems of partial cutting, creation of mixed stands and the use of more small-scale harvest methods.

Partial cutting systems, e.g. dimension cutting and group selection, have already been used extensively, but were abandoned as they received a bad reception among Swedish foresters. However, recent research has proved that single-tree selection is an economical alternative to clear-cutting in some spruce forests (Lundqvist, 1989). In spite of insufficient knowledge of the ecological effects of this method, it seems clear that it will save a substantial portion of the qualities typical of natural forests. However, this method requires stands with a largely natural size structure and can thus only be used in rather limited areas, simply because of their scarcity.

Another alternative involves the retention of regrowth under a sheltering canopy (Hagner, 1984). Planting without previous scarification may be applied if necessary. Several such methods are now under development, suitable for various types of forest.

Research need

A new approach to forest management in the boreal forest that incorporates the maintenance of biodiversity should be based on an appropriate application of ecological knowledge (Noss, 1987). The great potential for a forestry founded on ecological principles is shown, for example, by the impact of the basic ecological research performed at the H.J. Andrews Experimental Station, in the temperate conifer forests of Oregon, northwestern USA (Franklin, 1989; Perry *et al.*, 1989; Norse, 1990). In Sweden, we would like to see a similar, broadly oriented research programme that include studies on both basic ecological processes and forest management in the old-growth forests fringing the Scandinavian mountain chain. Such research may be preferably directed to studies of indicator species (Peterken, 1974) and keystone species (Paine, 1966; see also chapter 2). Another, equally important issue is research on alternative methods of forest harvesting and the effects of other methods of environmental protection in forestry. Different methods should be developed and carefully tested. A very crucial question is to clarify how much of the original diversity that can be preserved by the application of alternative methods. Research dealing with the problem of how to maintain diversity in forest reserves in a long-term perspective is also required. However, this is probably not as urgent as the other research areas mentioned above.

In conclusion, we argue that the shortage of ecological knowledge about boreal organisms and their response to man's destruction of the environment should not constrain the necessity, possibility and urgency of performing actions to conserve biodiversity in boreal Fennoscandia. Having successful conservation and management programmes will be even more important in the future, because of the continuing development of Fennoscandian forestry.

ACKNOWLEDGEMENTS

We thank P. Angelstam, A. De Jong, M. Hagner, L. Hansson, C. Nilsson, R. Pettersson, L. Svensson and G. Thor for comments on the manuscript. This work has been supported through grants from the Swedish Environmental Protection Agency.

REFERENCES

Ågren, J. & Zackrisson, O. (1990). Age and size structure of *Pinus sylvestris* populations on mires in central and northern Sweden. *J. Ecol.*, **78**, 1049–62.

Ahlén, I. (1977). *Faunavård*. Liber, Stockholm.

Ahlén, I. (1982). Bevarande av naturskogar och urskogar, II. *Skogs- o. Lantbr.-akad. Tidskr.*, **121**, 221–6.

Ahlén, I., Boström, U., Ehnström, B. & Pettersson, B. (1979). *Faunavård i Skogsbruket*. Skogsstyrelsen, Jönköping.

Ahlén, I. & Tjernberg, M. (1988). Hotade och sällsynta ryggradsdjur i Sverige. *Sver. Nat.*, **2**, 32–42.

Ahlgren, I.F. (1974). The effect of fire on soil organisms. In *Fire and Ecosystems*, ed. T.T. Kozlowski & C.E. Ahlgren. Academic Press, New York, pp. 47–72.

Ahlner, S. (1931). *Usnea longissima* Ach. i Skandinavien. *Svensk Bot. Tidskr.*, **25**, 395–416.

Ahlner, S. (1948). Utbredningstyper bland nordiska barrträdslavar. *Acta phytogeogr. suec.*, **22**.

Ahti, T. (1977). Lichens of the boreal coniferous zone. In *Lichen Ecology*, ed. M.R.D. Seaward. Academic Press, London, pp. 145–81.

Ahti, T. & Oksanen, J. (1990). Epigeic lichen communities of taiga and tundra regions. *Vegetatio*, **86**, 39–70.

Ahti, T., Hämet-Ahti, L. & Jalas, J. (1968). Vegetation zones and their sections in northwestern Europe. *Ann. Bot. Fennici*, **5**, 169–211.

Alatalo, R.V. & Carlson, A. (1987). Hoarding-site selection of the willow tit *Parus montanus* in the presence of the Siberian tit *Parus cinctus*. *Ornis Fennica*, **64**, 1–9.

Alexander, M.E. & Dubé, D.E. (1983). Fire management in wilderness areas, parks, and other nature reserves. In *The Role of Fire in Northern Circumpolar Ecosystems*, ed. R.W. Wein & D.A. MacLean. Wiley, New York, pp. 373–97.

Andersson, G., Gerell, R., Källander, H. & Larsson, T. (1967). The density of birds in some forest habitats in Stora Sjöfallet National Park, Swedish Lapland. *Acta Univ. Lund. Sect. II.*, No. 17.

Andersson, H., Coulianos, C.C., Ehnström, B., Hammarstedt, O., Imby, L., Janzon, L.-Å., Lindelöw, Å. & Waldén, H.W. (1987). Hotade evertebrater i Sverige. *Entomol. Tidskr.*, **108**, 65–75.

Andersson, L.I. & Hytteborn, H. (1991). Bryophytes and decaying wood—a comparison between managed and natural forest. *Holarct. Ecol.* **14**, 121–30.

Angelstam, P., Lindström, E. & Widén, P. (1985). Synchronous short-term population fluctuations of some birds and mammals in Fennoscandia—occurrence and distribution. *Holarct. Ecol.*, **8**, 285–98.

Anon. (1981). *Skogsstatistisk Årsbok (1979)*. Skogsstyrelsen, Jönköping.

Anon. (1987). *Miljöstatistisk Årsbok*. Statistiska Centralbyrån, Stockholm.

Anon. (1989). *Skogsstatistisk Årsbok (1989)*. Skogsstyrelsen, Jönköping.

Arnborg, T. (1942). Lågaföryngringen i en sydlappländsk granurskog. *Sver. Skogsvårdsför. Tidskr.*, **40**, 47–78.

Arnborg, T. (1943). Granberget—en växtbiologisk undersökning av ett sydlappländskt granskogsområde med särskild hänsyn till skogstyper och föryngring. *Norrl. Handbibl.*, **14**.

Arnborg, T. (1949). Från svedjebruk till hyggesbränning. Norrlands Skogsvårdsförbunds exkursionsprogram 1949, 36 pp.

Aulén, G. (1988). Ecology and distribution history of the white-backed woodpecker *Dendrocopos leucotos* in Sweden, *Rep. Depart. Wildl. Ecol., Swed. Univ. Agric. Sci., Uppsala*, No. 14.

Bäckström, P.-O. (ed.) (1984). *Skogsföryngring i Fjällnära skogar. Forskarrapport.* Sveriges Lantbruksuniversitet, Skogsvetenskapliga Fakulteten, Umeå.

Baker, W.L. (1989). Landscape ecology and nature reserve design in the Boundary Waters canoe area, Minnesota. *Ecology*, **70**, 23–35.

Balda, R.P. (1975). The relationship of secondary cavity nesters to snag densities in western coniferous forests. *U.S. Depart. Agric. For. Serv. Wildlife Habitat Tech. Bull. No. 1, Southwestern Region. Albuquerque*, New Mexico.

Barkman, J.J. (1958). *Phytosociology and Ecology of Cryptogamic Epiphytes*. Van Gorcum, Assen.

Beatty, S.W. (1984). Influence of microtopography and canopy species on spatial patterns of forest understory plants. *Ecology*, **65**, 1406–19.

Beatty, S.W. & Stone, E.L. (1986). The variety of soil microsites created by tree falls. *Can. J. For. Res.*, **16**, 539–48.

Bengtsson, G., Herrman, J., Malmqvist, B., Nilsson, I.N. & Svensson, B.S. (1982).

Öbiogeografisk teori och bildning av naturreservatal, *Swedish Environment Protection Agency*, PM No. 1514.
Bíly, S. & Mehl, O. (1989). Longhorn beetles (Coleoptera, Cerambycidae) of Fennoscandia and Denmark. *Fauna Entomol. Scand.*, **22**.
Blomgren, A. (1964). *Lavskrika*. Bonniers, Stockholm.
Bonan, G.B. & Shugart, H.H. (1989). Environmental factors and ecological processes in boreal forests. *Ann. Rev. Ecol. Syst.*, **20**, 1–28.
Borgegård, L.-E. (1973). Tjärhanteringen i Västerbottens län under 1800-talets senare hälft. *Kungl. Skytt. Samf. Handl.*, **12**.
Boström, U. & Nilsson, S.G. (1983). Latitudinal gradients and local variations in species richness and structure of bird communities on raised peat-bogs in Sweden. *Ornis Scand.*, **14**, 213–26.
Bråkenhielm, S. (ed.) (1982). *Urskogar. Inventering av Urskogsartade Områden i Sverige, Del 1. Allmän del.* Skogsstyrelsen och Statens Naturvårdsverk, Jönköping.
Bråkenhielm, S. & Persson, H. (1980). Vegetation dynamics in developing Scots pine stands in central Sweden. In *Structure and Function of Northern Coniferous Forests. An Ecosystem Study*, ed. T. Persson. *Ecol. Bull. (Stockholm)*, **32**, pp. 139–52.
Bull, E.L. (1983). Longevity of snags and their use by woodpeckers. In *Snag Habitat Management*, ed. J.W. Davis, G.A. Goodwin & R.A. Ockenfels. *USDA For. Serv. Gen. Tech. Rep.*, RM-99. Rocky Mountain Forest and Range Experiment Station, Fort Collins, pp. 66–7.
Bylund, E. (1956). Koloniseringen av Pite Lappmark t.o.m. år 1867. *Geographica*, **30**.
Cajander, V. (1909). Über waldtypen. *Acta For. Fennicia*, **1**, 1–175.
Carlin, G. & Swahn, U. (1977). De svenska *Usnea*-arterna (skägglavar). *Svensk Bot. Tidskr.*, **71**, 89–100.
Cederlund, G. & Markgren, G. (1987). The development of the Swedish moose population, 1970–1983. *Swed. Wildl. Res., Suppl.*, **1**, 55–62.
Collins, B.S. & Pickett, S.T.A. (1987). Response of herb layer cover to experimental canopy gaps. *Amer. Midl. Nat.*, **119**, 282–90.
Collins, B.S., Dunne, K.P. & Pickett, S.T.A. (1985). Responses of forest herbs to canopy gaps. In *The Ecology of Natural Disturbance and Patch Dynamics*, ed. S.T.A. Pickett & P.S. White. Academic Press, Orlando, Florida, pp. 217–234.
Connell, J.H. (1978). Diversity in tropical rainforests and coral reefs. *Science, N.Y.*, **199**, 1302–10.
Dahl, E., (1989). Nunatakkteorin—hvilket grunnlag har den? *Blyttia*, **47**, 125–33.
Danell, K. & Hörnfeldt, B. (1987). Numerical responses by populations of red fox and mountain hare during an outbreak of sarcoptic mange. *Oecologia (Berl.)*, **73**, 533–6.
Danell, K., Ericson, L. & Jakobsson, K. (1981). A method for describing former fluctuations of voles. *J. Wildl. Manage*, **45**, 1018–21.
Databanken för hotade arter och Naturvårdsverket (1991). *Sweden's Red Lists on Vascular Plants, Bryophytes, Lichens and Macrofungi Lund* (in Swedish with English Summary).
Du Rietz, E. (1925). Die regionale Gliederung der skandinavischen Vegetation. *Sv. Växtsociol. Sällsk. Handl.*, **4**, 1–80.

Ebeling, F. (1978). Nordsvenska skogstyper. *Sver. Skogsvårdsf. Tidskr.*, **76**(4).
Eckerberg, K. (1986). Implementation of environmental protection in Swedish forestry: a policy perspective. *For. Ecol. Manage.*, **17**, 61–72.
Eckerberg, K. (1988). Clear felling and environmental protection—results from an investigation in Swedish forests. *J. Environ. Manage.*, **27**, 237–56.
Edwards, R.Y., Soos, J. & Ritcey, R.W. (1960). Quantitative observations on epidendric lichens used as food by caribou. *Ecology*, **41**, 425–31.
Ehnström, B. & Waldén, H. (1986). *Faunavård i Skogsbruket, Del 2. Den Lägre Faunan.* Skogsstyrelsen, Jönköping.
Elliott-Fisk, D. (1988). The boreal forest. In *North American Terrestrial Vegetation*, ed. M.G. Barbour & W.D. Billings. Cambridge University Press, Cambridge, pp. 33–62.
Elmqvist, T., Ericson, L., Danell, K. & Salomonson, A. (1988). Latitudinal sex ratio variation in willows, *Salix* spp., and gradients in vole herbivory. *Oikos*, **51**, 259–66.
Engelmark, O. (1984). Forest fires in the Muddus National park (northern Sweden) during the past 600 years. *Can. J. Bot.*, **62**, 893–8.
Engelmark, O. (1987). Fire history correlation to forest type and topography in northern Sweden. *Ann. Bot. Fennici*, **24**, 317–24.
Engelmark, O. (1988). Skogselden i naturvården. En studie med exempel från Muddus nationalpark. *Swedish Environmental Protection Agency*, PM no. 3441.
Engelmark, O. & Hofgaard, A. (1985). Sveriges äldsta tall. *Svensk Bot. Tidskr.*, **79**, 415–16.
Ericson, L. (1977). The influence of voles and lemmings on the vegetation in a coniferous forest during a 4-year period in northern Sweden. *Wahlenbergia*, **4**.
Ericson, L. & Oksanen, L. (1987). The impact of controlled grazing by *Clethrionomys rufocanus* on experimental guilds of boreal forest floor herbs. *Oikos*, **50**, 403–16.
Ericson, L. (1980). En kvantitativ undersökning av häckfågelfaunan i några biotoper/successionsstadier i den boreala barrskogen. Examensarbete, Inst. för Viltekologi, Sveriges Lantbruksuniversitet, Umeå.
Erlinge, S., Göransson, G., Hansson, L., Liberg, O., Nilsson, I., Nilsson, T., von Schantz, T. & Sylvén, M. (1983). Predation as a regulating factor on southern small rodent populations in Sweden. *Oikos*, **40**, 36–52.
Esseen, P.-A. (1981). Host specificity and ecology of epiphytic macrolichens in some central Swedish spruce forests. *Wahlenbergia*, **7**, 73–80.
Esseen, P.-A. (1983). Ecology of lichens in boreal coniferous forest with reference to spatial and temporal patterns. PhD, thesis, University of Umeå, Umeå.
Esseen, P.-A. (1985). Litter fall of epiphytic macrolichens in two old *Picea abies* forests in Sweden. *Can. J. Bot.*, **63**, 980–7.
Esseen, P.-A. & Ericsson, L. (1982). Granskogar med långskägglav i Sverige. *Swedish Environmental Protection Agency*, PM no. 1513.
Esseen, P.-A., Ericson, L., Lindström, H. & Zackrisson, O. (1981). Occurrence and ecology of *Usnea longissima* in central Sweden. *Lichenologist*, **13**, 177–90.
Evans, W.G. (1966). Morphology of the infrared sense organs of *Melanophila acuminata* (Buprestidae, Coleoptera). *Ann. Entomol. Soc. Amer*, **59**, 873–7.

Falinski, J.B. (1978). Uprooted trees, their distribution and influence in the primeval forest biotope. *Vegetatio*, **38**, 175–83.

Floravårdskommittén för Lavar (1987). Preliminär lista över hotade lavar i Sverige. *Svensk Bot. Tidskr.*, **81**, 237–56.

Floravårdskommittén för Mossor (1988). Preliminär lista över hotade mossor i Sverige. *Svensk Bot. Tidskr.*, **82**, 423–45.

Forestry Act (1979). *Föreskrifter mm till Skogsvårdslagen*. Skogsstyrelsen, Jönköping.

Forestry Act (1986). Skogsstyrelsens föreskrifter och allmänna råd till skogsvårdslagen (1979:429). *Skogsstyrelsens Författningssamling*, **1986:6**. Skogsstyrelsen, Jönköping.

Foster, D.R. (1988). Species and stand response to catastrophic wind in central New England, U.S.A. *J. Ecol.*, **76**, 135–51.

Franklin, J.F. (1989). Toward a new forestry. *American Forests*, November/December 1989, 1–8.

Franklin, J.F. (1990). The application of ecological principles in the new forestry: integrating ecological values with commodity production. *Forest Planning of Canada*, **6:3**, 7–15.

Franklin, J.F. & Forman, R.T.T. (1987). Creating landscape patterns by forest cutting: ecological consequences and principles. *Landscape Ecol.*, **1**, 5–18.

Franklin, J.F., Shugart, H.H. & Harmon, M.E. (1987). Tree death as an ecological process. *BioScience*, **37**, 550–6.

Gilbert, O. (1977). Lichen conservation in Britain. In *Lichen Ecology*, ed. M.R.D. Seaward. Academic Press, London, New York, pp. 415–36.

Gilpin, M.E. & Soulé, M.E. (1986). Minimum viable populations: processes of species extinction. In *Conservation Biology. The Science of Scarcity and Diversity*, ed. M.E. Soulé. Sinauer Associates, Sunderland, Massachusetts, pp. 19–34.

Granström, A. (1986). Seed banks in forest soils and their role in vegetation succession after disturbance. PhD thesis. Swedish University of Agricultural Sciences, Umeå.

Gustafsson, L. & Hallingbäck, T. (1988). Bryophyte flora and vegetation of managed and virgin coniferous forests in South-West Sweden. *Biol. Conserv.*, **44**, 283–300.

Haapanen, A. (1965). Bird fauna of the Finnish forests in relation to forest succession, I. *Ann. Zool. Fennici*, **2**, 153–96.

Haftorn, S. (1956). Contribution to the food biology of tits, especially about storing of surplus food, Part IV. A comparative analysis of *Parus atricapillus* L., *P. cristatus* L. and *P. ater* L. *Kgl. Norske Vidensk. Selsk. Skr.*, **4**, 1–54.

Hagen, Y. (1969). Norske undersökelser over avkomproduksjonen hos rovfugler og ugler sett i relasjon til smågnagerbestandens veksliger. *Fauna*, **22**, 73–126.

Hägglund, B. & Lundmark, J.-E. (1982). *Bonitering, Del 3. Markvegetationstyper-skogsmarkflora*. Skogsstyrelsen, Jönköping.

Hagner, M. (1984). Synpunkter på domänverkets skogsbruk i fjällnära områden. In *Skogsföryngring i Fjällnära Skogar. Forskarrapport*, ed. P.-O. Bäckström. Sveriges Lantbruksuniversitet, Skogsvetenskapliga Fakulteten, Umeå, pp. 1–65.

Hagner, S. (1957). Om kott- och fröproduktionen i svenska barrskogar. *Medd. Stat. Skogsforskn. Inst.*, **47**(8).
Hagner, S. (1985). Lodgepole pine management in Sweden. A strategy for higher yield. In *Lodgepole Pine. The Species and its Management*, ed. D.M. Baumgartner, R.G. Krebill, J.T. Arnott & G. F. Weetman. *Symposium, Proceedings, 8–10 May 1984, Washington, USA, and 14–16, May Vancouver, British Columbia, Canada*, pp. 211–17.
Haila, Y., Hanski, I.K. & Raivo, S. (1987). Breeding bird distribution in fragmented coniferous taiga in southern Finland. *Ornis Fennica*, **64**, 90–106.
Hallingbäck, T. (1986). Lunglavarna, *Lobaria*, på reträtt i Sverige. *Svensk Bot. Tidskr.*, **80**, 373–81.
Hämet-Ahti, L. (1963). Zonation of the mountian birch forests in northernmost Fennoscandia. *Ann. Bot. Soc. 'Vanamo'*, **34**, 1–127.
Hamilton, W.D. (1987). Evolution and diversity under bark. In *Diversity of Insect Faunas*, ed. L.A. Mound & N. Waloff. Blackwell Scientific Publications, Oxford, pp. 154–75.
Hansson, L. (1977). Landscape ecology and stability of populations. *Landscape Planning*, **4**, 85–93.
Hansson, L. & Henttonen, H. (1985). Gradients in density variations in small rodents: the importance of latitude and snow cover. *Oecologia (Berl.)*, **67**, 394–402.
Hansson, L. & Henttonen, H. (1988). Rodent dynamics as community processes. *Trends Ecol. Evolut*, **3**, 195–200.
Harmon, M.E. (1989). Effects of bark fragmentation on plant succession on conifer logs in the *Picea–Tsuga* forests of Olympic National Park, Washingtton. *Amer. Midl. Nat.*, **121**, 112–24.
Harmon, M.E., Franklin, J.F., Swanson, F.J., Sollins, P., Gerhory, S.V., Lattin, J.D., Anderson, N.H., Cline, S.P., Aumen, N.G., Sedell, J.R., Leinkaemper, G.W., Cromack, Jr, K. & Cummins, K.W. (1986). Ecology of coarse woody debris in temperate ecosystems. *Adv. Ecol. Res.*, **15**, 133–302.
Harris, L.D. (1984). *The Fragmented Forest. Island Biogeography Theory and the Preservation of Biotic Diversity*. University of Chicago Press, Chicago, Illinois.
Heckscher, E.F. (1968). *Svenskt Arbete och Liv*. Aldus/Bonniers, Stockholm.
Heikinheimo, O. (1915). Der Einfluss der Brandwirtschaft auf die Wälder Finnlands. *Acta For. Fennici*, **4**.
Heinselman, M.K. (1981). Fire and succession in the conifer forests of North America. In *Forest Succession; Concepts and Applications*, ed. D.C. West, H.H. Shugart, & D.B. Botkin. Springer Verlag, New York, pp. 374–405.
Heliövaara, K. & Väisänen, R. (1983). Environmental changes and the flat bugs (Heteroptera, Aradidae and Aneuridae). Distribution and abundance in Eastern Fennoscandia. *Ann. Ent. Fennicia*, **49**, 103–9.
Heliövaara, K. & Väisänen, R. (1984). Effects of modern forestry on northwestern European forest invertebrates: a synthesis. *Acta For. Fennicia*, **189**, 1–32.
Helle, T. & Aspi, J. (1983). Effects of winter grazing by reindeer on vegetation. *Oikos*, **40**, 337–43.
Helle, P. & Järvinen, O. (1986). Population trends in North Finnish land birds in relation to their habitat selection and changes in forest structures. *Oikos*, **46**, 107–15.

Helliwell, D.R. (1976). The effects of size and isolation on the conservation value of wooded sites in Britain. *J. Biogeogr.*, **3**, 407–16.
Hemstrom, M. (1990). New forestry—how will it look on the landscape. *Forest Planning of Canada*, **6:3**, 21–3.
Henttonen, H., Oksanen, T., Jortikka, A. & Haukisalmi, V. (1987). How much do weasels shape microtine cycles in the northern Fennoscandian taiga. *Oikos*, **50**, 353–65.
Hesselman, H. (1912). Om snöbrotten i norra Sverige vintern 1910–1911. *Sver. Skogsvårdsf. Tidskr. allm. del.*, pp. 145–72.
Högbom, A.G. (1934). Om skogseldar förr och nu. *Norrl. Handbibl.*, **13**, 1–101.
Hörnfeldt, B. (1978). Synchronous population fluctuations in voles, small game, owls, and tularemia in Northern Sweden. *Oecologia* (*Berl.*), **32**, 141–52.
Hörnfeldt, B. (1991). Cycles of voles, predators, and alternative prey in boreal Sweden. PhD thesis, University of Umeä.
Hustich, I. (1960). Plant geographical borders. In *A Geography of Norden*, ed. A. Sömme. J.W. Cappelens Forlag, Oslo, pp. 54–62.
Huttonen, P. & Tolonen, M. (1972). Pollen analytical studies of prehistoric agriculture in northern Ångermanland. *Early Norrland*, **1**, 9–34.
Hytteborn, H., Packham, J.R. & Verwijst, T. (1987). Tree population dynamics, stand structure and species composition in the montane virgin forest of Vallibäcken, Northern Sweden. *Vegetatio*, **72**, 3–19.
Imby, L. & Palmqvist, G. (1978). De svenska *Anomogyna*-arternas utseende, biologi och utbredning (Lep. Noctuidae). *Entomol. Tidskr.*, **99**, 97–107.
Ingelög, T. (ed) (1981). *Floravård i Skogsbruket, Del 1—Allman Del.* Skogsstyrelsen, Jönköping.
Ingelög, T., Thor, G. & Gustafsson, L. (eds) (1987). *Floravård i Skogsbruket, Del 2-Ar tdel*, 2nd edn. Skogsstyrelsen, Jönköping.
Järvinen, O. (1979). Geographical gradients of stability in European land bird communities. *Oecologia* (*Berl.*), **38**, 51–69.
Järvinen, O. & Sammalisto, L. (1976). Regional trends in the avifauna of Finnish peatland bogs. *Ann. Zool. Fennici*, **13**, 31–43.
Järvinen, O. & Ulfstrand, S. (1980). Species turnover of a continental bird fauna: Northern Europe, 1850–1970. *Oecologia* (*Berl.*), **46**, 186–95.
Järvinen, O. & Väisänen, R.A. (1973). Species diversity of Finnish birds, I: Zoogeographical zonation based on land birds. *Ornis Fennica*, **50**, 93–125.
Järvinen, O. & Väisänen, R.A. (1977). Long-term changes of the North European land bird fauna. *Oikos*, **29**, 225–8.
Järvinen, O., Kouki, J. & Häyrinen, U. (1987). Reversed latitudinal gradients in total density and species richness of birds breeding on Finnish mires. *Ornis Fennica*, **64**, 67–73.
Järvinen, O., Kuusela, K. & Väisänen, R.A. (1977). Effects of modern forestry on the numbers of breeding birds in Finland in 1945–1975. *Silva Fennica*, **11**, 284–94 (in Finnish with English summary).
Jonsson, B.G. & Esseen, P.-A. (1990). Treefall disturbance maintains high bryophyte diversity in a boreal spruce forest. *J. Ecol.*, **78**, 924–36.
Jørgensen, P.M. (1978). The lichen family Pannariaceae in Europe. *Opera Bot.*, **45**.
Kalela, A. (1961). Waldvegetationszonen Finnlands und ihre klimatischen Paralleltype. *Arch. Soc. 'Vanamo'* **16**, *suppl.*, 65–83.
Kalliola, R. (1966). The reduction of the area of forests in natural condition in

Finland in the light of some maps based upon national forest inventories. *Ann. Bot. Fennici*, **3**, 442–8.

Kalliola, R. (1973). *Suomen Kasvimaantiede.* Werner Söderström Osakeyhtiö, Helsinki.

Kardell, L. & Ekstrand, A. (1990). Skyddad Skogi Sverige, 1. areal och virkes förråd inon nationalparker, naturreservat och domänreservat. *Swedish University of Agricultural Sciences, Dep. of Environmental Forestry*, Report **48**.

Kardell, L. (1980). Forest berries and mushrooms—an endangered resource? *Ambio*, **9**, 241–7.

Kardell, L., Blomgren, M. & Nitare, J. (1987). Mushroom production and species composition in stands of *Pinus contorta* and *P. sylvestris. Svensk Bot. Tidskr.*, **81**, 133–42 (in Swedish with English summary).

Karlman, M. (1981). The introduction of exotic tree species with special reference to *Pinus contorta in* Northern Sweden. Review and background. *Stud. For. Suec.*, **158**.

Kershaw, K.A. (1977). Studies on lichen-dominated systems. XX. An examination of some aspects of the northern boreal lichen woodlands in Canada. *Can. J. Bot.*, **55**, 393–410.

Kinnman, (1936). Skogseldsrisken och väderleken. *Sver. Skogsvårdsf. Tidskr.*, **34**, 481–513.

Knight, D.H. & Wallace, L.L. (1989). The Yellowstone fires: issues in landscape ecology. *BioScience*, **39**, 700–6.

Korchagin, A.A. & Karpov, V.G. (1974). Fluctuations in coniferous taiga communities. In *Vegetation Dynamics*, ed. R. Knapp. Junk, The Hague, pp. 225–31.

Korpimäki, E., Huhtala, K. & Sulkava, S. (1990). Does the year-to-year variation in the diet of eagle and Ural owls support the alternative prey hypothesis? *Oikos*, **58**, 47–54.

Koskinen, A. (1955). *Über die Kryptogamen der Bäume, besonders die Flechten*, I. Privately printed, Helsinki.

Kujala, V. (1926*a*). Untersuchungen über die Waldvegetation in Süd- und Mittelfinnland, I. Zur Kenntnis des ökologisch-biologischen Charakters der Pflanzenarten unter spezieller Berücksichtigung der Bildung von Pflanzenvereinen, B. Laubmoose. *Commun. Inst. For. Fenniae*, **10**, 1–59.

Kujala, V. (1926*b*). Untersuchungen über den Einfluss von Waldbränden auf die Waldvegetation in Nordfinnland. *Commun. Inst. For Fenniae*, **10**, 1–41.

Kullman, L. (1987). Long-term dynamics of high altitude populations of *Pinus sylvestris* in the Swedish Scandes. *J. Biogeogr.*, **14**, 1–8.

Lahti, T., Kurtto, A. & Väisänen, R.A. (1988). Floristic composition and regional species richness of vascular plants in Finland. *Ann. Bot. Fenniae*, **25**, 281–91.

Larsen, J.A. (1980). *The Boreal Ecosystem.* Academic Press, New York.

Larsen, J.A. (1982). *Ecology of the Northern Lowland Bogs and Conifer Forests.* Academic Press, New York.

Lawrence, J.F. (1989). Mycophagy in the Coleoptera: feeding strategies and morphological adaptation. In *Insect–Fungus Interaction*, ed. N. Wilding, N.H. Collins, P.M. Hammond & J.F. Webber. Academic Press, London, pp. 1–23.

Lindén, H. (1988). Latitudinal gradients in predator–prey interactions, cyclicity and synchronism in voles and small game populations in Finland. *Oikos*, **52**, 341–9.
Lindén, H. (1989). Characteristics of tetraonid cycles in Finland. *Finn. Game Res.*, **46**, 34–42.
Linder, P. (1984). *Kirjesålandet. En skogsbiologisk inventering av ett fjällnära urskogsområde i Västerbottens län.* Inst. f. Skoglig ståndortslära, Sveriges Lantbruksuniversitet, Umeå.
Lindgren, F. (1975). Iakttagelser rörande lavskrikan (*Perisoreus infaustus*)—huvudsakligen dess häckningsbiologi. *Fauna och Flora*, **5**, 198–210.
Lindholm, T. (1987). Luonnonsuojelualueittemme metsäluonto—mitä se on ja millaiseksi se kehittyy. *Luonnon Tutkija*, **91**, 13–9.
Lindholm, T. & Vasander, H. (1987). Vegetation and stand development of mesic forest after prescribed burning. *Silva Fennica*, **21**, 259–78.
Linkola, P. & Myllymäki, A. (1969). Der Einfluss der Kleinsäugerfluktuationen auf das Brüten einiger kleinsäugerfressender Vögel im südlichen Häme, Mittelfinnland 1952–1966. *Ornis Fennica*, **46**, 45–78.
Lloyd, M. (1963). Numerical observations on movements of animals between beech litter and fallen branches. *J. Anim. Ecol.*, **32**, 157–63.
Lomholdt, O. (1975). The Sphecidae (Hymenoptera) of Fennoscandia and Denmark. *Fauna Entomol. Scand.*, **4-1**.
Lomholdt, O. (1976). The Sphecidae (Hymenoptera) of Fennoscandia and Denmark. Fauna Entomol. *Scand.*, **4-2**.
Loucks, O.L. (1970). Evolution of diversity, efficiency and community stability. *Amer. Zool.*, **10**, 17–25.
Lundberg, S. (1984). Den brända skogens skalbaggsfauna i Sverige. *Entomol. Tidskr.*, **105**, 129–41.
Lundberg, S. (1986). *Catalogus Coleopterorum Suecicae.* Entomologiska Föreningen, Naturhistoriska Museet, Stockholm.
Lundmark, J.-E., Berg, B. & Nilsson, Å (1982). Contortatallens inflytande på mark och markvegetation i jämförelse med sylvestristallens. *Sver. Skogsvårdsf. Tidskr.*, **1-2**, 43–8.
Lundqvist, L. (1989). Use of the selection system in Norway spruce forests—changes in the stand structure, volume increment, ingrowth and regeneration on experimental plots managed with single-tree selection. PhD thesis, Swedish University of Agricultural Science, Umeå.
MacArthur, R.H. (1965). Patterns of species diversity. *Biol. Rev.*, **40**, 510–33.
Malmström, C. (1939). Hallands skogar under de senaste 300 åren. *Medd. Stat. Skogsförs. Anst.*, **31**, 71–300.
Malmström, C. (1949). Studier över skogstyper och trädslagsfördelning inom Västerbottens län. *Medd. Stat. Skogsforskn. Inst.*, **37:11**.
Maser, C., Anderson, R.G., Cromack, Jr. K., Williams, J.T. & Martin, R.E. (1979). Dead and down woody material. In *Wildlife Habitats in Managed Forests, the Blue Mountains of Oregon and Washington, Agriculture Handbook*, No. 553, ed. J.W. Thomas. US Department of Agriculture, Forest Service, Portland, Oregon, pp. 78–95.
McCullough H.A. (1948). Plant succession on fallen logs in a virgin spruce–fir forest. *Ecology*, **29**, 508–13.

Middleton, J. & Merriam, G. (1985). The rationale for conservation: problems from a virgin forest. *Biol. Conserv.*, **33**, 133–45.
Montelius, S. (1962). Säfnäsbrukens Arbetskraft och Försörjning 1600–1865. *Geographica*, **37**.
Morrison, M.L., Timossi, I.C., With, K.A. & Marley, P.N. (1985). Use of tree species by forest birds during winter and summer. *J. Wildl. Manage.*, **49**, 1098–102.
Muhle, H. & LeBlanc, F. (1975). Bryophyte and lichen succession on decaying logs, I. Analysis along an evaporational gradient in eastern Canada. *J. Hattori Bot. Lab.*, **39**, 1–33.
Muhr, L.-E. (1986). Lavfloran i Gravbäcksravinen i Värmland. *Svensk Bot. Tidskr.*, **81**, 17–36.
Niemelä, T. (1980). On Fennoscandian polypores, 7. The genus *Pycnoporellus*. *Karstenia*, **20**, 1–15.
Nielmelä, T. & Kotiranta, H. (1982). Polypore survey of Finland, 2. The genus *Phellinus*. *Karstenia*, **22**, 27–42.
Nilsson, S.G. (1979). Effect of forest management on the breeding bird community in southern Sweden. *Biol. Conserv.*, **16**, 135–43.
Nordic Statistical Secretariat (1990). *Yearbook of Nordic Statistics 1989/90*. Nordiska Ministerrådet, Stockholm.
Nordiska Ministerrådet (1984). *Vegetationstyper i Norden*. Berlings, Arlöv.
Norse, E.A. (1990). *Ancient Forests of the Pacific Northwest*. Island Press, Covelo.
Noss, R.F. (1983). A regional landscape approach to maintain diversity. *Bio-Science*, **33**, 700–6.
Noss, R.F. (1987). Protecting natural areas in fragmented landscapes. *Natural Areas Journal*, **7**, 2–13.
Noss, R.F. & Harris, L.D. (1986). Nodes, networks, and MUMs: preserving diversity at all scales. *Environ. Manage.*, **10**, 299–309.
NOU (1989). *Flersidig Skogsbruk. Skogsbrukets Forhold til Naturmiliø og Friluftsliv. Norges Offentlige Utredninger*, 1989: 10. Statens Trykningskontor, Oslo.
Oinonen, E. (1967). Sporal regeneration of bracken *Pteridium aquilinum* (L.) kuhn in Finland in the light of the dimensions and age of its clones. *Acta For. Fennia*, **83**, 3–96.
Oinonen, E. (1969). The time-table of vegetative spreading of the lily-of-the-valley (*Convallaria majalis* L.) and the wood small-reed (*Calamagrostis epigeios* (L.) Roth) in southern Finland. *Acta For. Fenniae*, **97**, 1–18.
Oksanen, J. (1983). Vegetation of forested inland dunes in North Karelia, eastern Finland. *Ann. Bot. Fennici*, **20**, 281–95.
Oksanen, J. & Ahti, T. (1982). Lichen-rich pine forest vegetation in Finland. *Ann. Bot. Fennici*, **19**, 275–301.
Paine, R.T. (1966). Food web complexity and species diversity. *Amer. Nat.*, **100**, 65–75.
Palm, T. (1946). Coleopterfaunan i en jämtländsk lavgranskog, I. Träd- och trädsvampfaunan. *Entomol. Tidskr.*, **67**, 109–39.
Palm, T. (1951). Die Holz- und Rindenkäfer der nordschwedischen Laubbäume. *Medd. Stat. Skogsforskn. Inst.*, **40**.
Palmgren, P. (1932). Zur Biologie von *Regulus r. regulus* (L.) und *Parus atricapillus borealis* Selys. Eine vergleichend-ökologische Untersuchung. *Acta Zool. Fennia*, **14**, 1–113.

Palmqvist, G. (1974). *Lasionycta skraelingia* H.S. funnen i Torne lappmark, Kiruna-trakten (Lep., Noctuidae). *Entomol. Tidskr.*, **95**, 179–80.
Payette, S., Morneau, C., Sirois, L. & Desponts, M. (1989). Recent fire history of the northern Quebec biomes. *Ecology*, **70**, 656–73.
Perry, D.A., *et al.*, (1989). *Maintaining Long-term Productivity of Pacific Northwest Forest Ecosystems*. Timber Press, Portland.
Persson, T. (ed.) (1980). Structure and function of northern coniferous forests. An ecosystem study. *Ecol. Bull. (Stockholm)*, **32**.
Peterken, G.F. (1974). A method for assessing woodland flora for conservation using indicator species. *Biol. Conserv.*, **6**, 239–45.
Philippi, G. (1965). Moosgesellschaften des morschen Holzes und des Rohhumus im Schwarzwald, in der Rhön, im Weserbergland und im Harz. *Nova Hedwigia*, **9**, 185–232.
Pickett, S.T.A. & Thompson, J.N. (1978). Patch dynamics and the design of nature reserves. *Biol. Conserv.*, **13**, 27–37.
Pickett, S.T.A. & White, P.S. (ed.) (1985). *The Ecology of Natural Disturbance and Patch Dynamics*. Academic Press, Orlando, Florida.
Poulsen, T.L. & Platt, W.J. (1989). Gap light regime influence canopy tree diversity. *Ecology*, **70**, 553–5.
Putz, F.E., Coley, P.D., Lu, K., Montalvo, A. & Aiello, A. (1983). Uprooting and snapping of trees: structural determinants and ecological consequences. *Can. J. For. Res.*, **13**, 1011–20.
Pynnönen, A. (1939). Beiträge zur Kenntnis der Biologie finnischer Spechte, I. *Ann. Zool. Soc. 'Vanamo'*, **7:2**, 1–171.
Raschendorfer, I. (1949). Beobachtungen über die Besiedlung von moderndem Holz mit besonderer Berücksichtigung der adnaten Vereine. *Österr. Bot. Zeitschr.*, **96**, 232–80.
Reader, R.J. (1987). Loss of species from deciduous forest understorey immediately following selective tree harvesting. *Biol. Conserv.*, **42**, 231–44.
Richardson, D.H.S. & Young, C.M. (1977). Lichens and vertebrates. In *Lichen Ecology*, ed. M.R.D. Seaward. Academic Press, London, pp. 121–44.
Rose, F. (1976). Lichenological indicators of age and environmental continuity in woodlands. In *Lichenology: Progress and Problems*, ed. D.H. Brown, D.L. Hawksworth & R.H. Bailey. Academic Press, London, pp. 279–307.
Rowe, J.S. (1983). Concepts of fire effects on plant individuals and species. In *The Role of Fire in Northern Circumpolar Ecosystems*, ed. R.W. Wein & D.A. MacLean, Wiley, New York, pp. 135–54.
Riwe, J.S. & Scotter, G.W. (1973). Fire in the boreal forest. *Quat. Res.*, **3**, 444–64.
Runkle, J.R. (1985). Disturbance regimes in temperate forests. In *The Ecology of Natural Disturbance and Patch Dynamics*, ed. S.T.A. Pickett & P.S. White. Academic Press, Orlando, Florida, pp. 17–33.
Ryman, S. & Holmåsen, I. (1984). *Svampar–en Fälthandbok*. Interpublishing, Stockholm.
Saalas, U. (1917). Die Fichtenkäfer Finnlands, I. *Ann. Acad. Scient. Fenn. Ser. A*, **8**(1).
Santesson, R. (1984). *The Lichens of Sweden and Norway*. Swedish Museum of Natural History, Stockholm.
Savely, Jr, H.E. (1939). Ecological relations of certain animals in dead pine and oak logs. *Ecol. Monogr.*, **9**, 322–85.

Schoonmaker, P. & McKee, A. (1988). Species composition and diversity during secondary succession of coniferous forests in the Western Cascade mountains of Oregon. *For. Sci.*, **34**, 960–79.

Schowalter, T.D. (1989). Canopy arthropod community structure and herbivory in old-growth and regenerating forests in western Oregon. *Can. J. For. Res.*, **19**, 318–22.

Schullery, P. (1989). The fires and fire policy. *BioScience*, **39**, 686–94.

Scotter, G.W. (1962). Productivity of arboreal lichens and their possible importance to barren-ground caribou (*Rangifer arcticus*). *Arch. Soc. Zool.-bot. Fenn. Vanamo*, **16**, 155–61.

Seaward, M.R.D. (1982). Principles and priorities of lichen conservation. *J. Hattori Bot. Lab.*, **52**, 401–6.

Seitz, R. (1986). Siberian fire as 'nuclear winter' guide. *Nature, Lond.*, **323**, 116–7.

Sernander, R. (1936). Granskär och Fiby urskog. En studie över stormluckornas och marbuskarnas betydelse i den svenska granskogens regeneration. *Acta phytogeogr. suec.*, **8**.

Shafi, M.I. & Yarranton, G.A. (1973). Diversity, floristic richness, and species evenness during a secondary (post-fire) succession. *Ecology*, **54**, 897–902.

Shugart, H.H. (1984). *A Theory of Forest Dynamics*. Springer Verlag, New York.

Shugart, H.H. (1987). Dynamic ecosystem consequences of tree birth and death patterns. *BioScience*, **37**, 596–602.

Simberloff, D. & Gotelli, N. (1984). Effects of insularisation on plant species richness in the prairie–forest ecotone. *Biol. Conserv.*, **29**, 27-46.

Sirén, G. (1955). The development of spruce forest on raw humus sites in northern Finland and its ecology. *Acta For. Fennica*, **62**, 1–363.

Sjörs, H. (1963). Amphi-Atlantic zonation. Nemoral to Arctic. In *North Atlantic Biota and Their History*, ed. A. Löve & D. Löve. Pergamon Press, Oxford, pp. 109–25.

Söderström, L. (1981). Distribution of bryophytes in spruce forests on hill slopes in central Sweden. *Wahlenbergia*, **7**, 141–53.

Söderström, L. (1983). Threatened and rare bryophytes in spruce forests of central Sweden. *Svensk Bot. Tidskr.*, **77**, 4–12 (in Swedish with English Summary).

Söderström L. (1987). The regulation of abundance and distribution patterns of bryophyte species on decaying logs in spruce forests. PhD thesis, University of Umeå.

Söderström, L. (1988*a*). Sequence of bryophytes and lichens in relation to substrate variables of decaying coniferous wood in northern Sweden. *Nord. J. Bot.*, **8**, 89–97.

Söderström, L. (1988*b*). The occurrence of epixylic bryophyte and lichen species in an old natural and a managed forest stand in northeast Sweden. *Biol. Conserv.*, **45**, 169–78.

Soulé, M.E. (ed.) (1986). *Conservation Biology. The Science of Scarcity and Diversity*. Sinauer Associates, Sunderland, Massachusetts.

Sousa, W.P. (1984). The role of disturbance in natural communities. *Ann. Rev. Ecol. Syst.*, **15**, 353–91.

Spies, T.A., Franklin, J.F. & Thomas, T.B. (1988). Coarse woody debris in douglas-fir forests of western Oregon and Washington. *Ecology*, **69**, 1689–1702.

Steijlen, I. & Zackrisson, O. (1987). Long-term regeneration dynamics and successional trends in a northern Swedish coniferous forest stand. *Can. J. Bot.*, **65**, 839–48.

Stenberger, M. (1969). *Sten. Brons. Järn.* Aldus/Bonniers, Stockholm.

Stevenson, S.K. (1986). Litterfall rate as a limitation on the availability of forage lichens in second growth stands. A problem analysis prepared for MacMillan Bloedel Ltd, Woodlands Services Division, Nanaimo, BC, Canada.

Sturesson, E. (1983). Naturskogen på Brassberget—en skogsbiologisk inventering. *Rapport från Länsstyrelsen i Gävleborgs län, Naturvårdsenheten*, **1983**:**2**.

Svensson, S.A., Toet, G. & Kempe, G. (1989). Riksskogstaxeringen 1978–82. Skogstillstånd, tillväxt och avverkning. *Sveriges Lantbruksuniversitet, Institutionen för Skogstaxering. Rapport*, **47**.

Synge, H. (1981). The conservation of lower plants: report from a panel discussion. In *The Biological Aspects of Rare Plant Conservation*, ed. H. Synge. Wiley, Chichester, pp. 125–37.

Tallantire, P.A. (1972). The regional spread of spruce (*Picea abies* (L.) Karst.) within Fennoscandia: a reassessment. *Norw. J. Bot.*, **19**, 1–16.

Tamm, C.O. (ed.) (1976). Man and the Boreal Forest. *Ecol. Bull.* (*Stockholm*), **21**.

Tenow, O. (1974). Det nordiska skogslandskapets och skogsbrukets utveckling fram till 1900–talet—en kort översikt. Swedish Coniferous Forest Project. Internal Report, **2**.

Terborgh, J. (1974). Preservation of natural diversity: The problem of extinction prone species. *BioScience*, **24**, 715–22.

Thomas, J.W., Anderson, R.G., Maser, C. & Bull, E.L. (1979). Snags. In *Wildlife Habitats in Managed Forests, the Blue Mountains of Oregon and Washington*, US Department of Agriculture, Forest Service, Portland, Oregon, Agriculture Handbook, No. 553, ed. J.W. Thomas, pp. 60–77.

Thompson, J.N. (1980). Treefalls and colonization patterns of temperate forest herbs. *Amer. Midl. Nat.*, **104**, 176–84.

Tiainen, J. (1980). Regional trends in bird communities of mature pine forests between Finland and Poland. *Ornis Scand.*, **11**, 85–91.

Tibell, L. (1977). Lavordningen Caliciales i Sverige. Inledning och släktet *Calicium. Svensk Bot. Tidskr.*, **71**, 239–59.

Tibell, L. (1978). Lavordningen Caliciales i Sverige. Släktena *Chaenotheca* och *Coniocybe. Svensk Bot. Tidskr.*, **72**, 171–88.

Tibell, L. (1980). Lavordningen Caliciales i Sverige. Släktena *Cyphelium, Microcalicium, Sphaerophorus, Sphinctrina, Thelomma* och *Tholurna. Svensk Bot. Tidskr.*, **74**, 55–69.

Timdal, E. (1984). The genus *Hypocenomyce* (Lecanorales, Lecideaceae) with special emphasis on the Norwegian and Swedish species. *Nord. J. Bot.*, **4**, 83–108.

Tirén, L. (1937). Skogshistoriska studier i trakten av Degerfors i Västerbotten. *Medd. Stat. Skogsförs., Anst.*, **30**, 67–322.

Tjernberg, M. (1983). Breeding ecology of the golden eagle, *Aquila chrysaetos* (L.), in Sweden. *Rep. Dep. Wildl. Ecol., Swed. Univ. Agric. Sci.* No. 10.

Topham, P.B. (1977). Colonization, growth, succession and competition. In *Lichen Ecology*, ed. M.R.D. Seaward. Academic Press, London, pp. 31–68.

Tramer, E. (1974). An analysis of the species density of US landbirds during

winter using the 1971 Christmas Bird Count. *Amer. Birds*, **28**, 563–7.
Troedsson, T. & Lyford, W.H. (1973). Biological disturbance and small-scale spatial variations in a forest soil near Garpenberg, Sweden. *Stud. For. Suec.*, **109**.
Tuhkanen, S. (1980). Climatic parameters and indices in plant geography. *Acta phytogeogr. suec.*, **67**.
Turner, M.G. (1989). Landscape ecology: the effect of pattern on process. *Ann. Rev. Ecol. Syst.*, **20**, 171–97.
Uggla, E. (1958). Skogsbrandfält i Muddus nationalpark. *Acta phytogeogr. suec.*, **41**.
Urban, D.L., O'Neill, R.V. & Shugart, H.H. (1987). Landscape ecology. *BioScience*, **37**, 119–27.
Van Cleve, K., Chapin, F.S. III, Flanagan, P.W., Viereck, L.A. & Dyrness, C.T. (ed.) (1986). *Forest Ecosystems in the Alaskan Taiga. Ecol. Stud.*, **57**. Springer-Verlag, New York.
Väisänen, R.A, Järvinen, O. & Rauhala, P. (1986). How are extensive human-caused habitat alterations expressed on the scale of local bird populations in boreal forest? *Ornis Scand.*, **17**, 282–92.
Veijalainen, H. (1976). Effect of forestry on the yields of wild berries and edible fungi. In *Man and the Boreal Forest*, ed. C.O. Tamm. *Ecol. Bull.* (*Stockholm*), **21**, pp. 63–5.
Virkkala, R. (1987). Effects of management on birds breeding in northern Finland. *Ann. Zool. Fennici*, **24**, 281–94.
Virkkala, R. (1990). Ecology of the Siberian tit *Parus cinctus* in relation to habitat quality: effects of forest management. *Ornis Scand.*, **21**, 139–46.
Virkkala, R. & Liehu, H. (1990). Habitat selection by the Siberian tit *Parus cinctus* in virgin and managed forests in northern Finland. *Ornis Fennica*, **67**, 1–12.
von Haartman, L. (1973). Changes in the breeding bird fauna of North Europe. In *Breeding Biology of Birds*, ed. D.S. Farnes. National Academy of Sciences, Washington, DC, pp. 448–81.
von Haartman, L. (1978). Changes in the breeding bird fauna of Finland and their causes. *Fennia*, **150**, 25–32.
Wästerlund, I. (1989). How is the occurrence of mushroom fruitbodies influenced by the silvicultural treatments? *Svensk Bot. Tidskr.*, **83**, 103–12 (in Swedish with English Summary).
Wästerlund, I. & Ingelög, T. (1981). Fruit body production of larger fungi in some young Swedish forests with special reference to logging waste. *For. Ecol. Manage.*, **3**, 269–94.
Wein, R.W. & MacLean, D.A. (ed.) (1983). *The Role of Fire in Northern Circumpolar Ecosystems.* Wiley, Chichester.
Wein, R.W., Riewe, R.R. & Methven, I.R. (1983). *Resources and Dynamics of the Boreal Zone. Proceedings of a Conference held at Thunder Bay, Ontario, August 1982.* Association of Canadian Universities for Northern Studies, Ottawa.
West, D.C., Shugart, H.H. & Botkin, D.B. (1981). *Forest Succession: Concepts and Applications*, Springer-Verlag, New York.
White, P.S. (1979). Pattern, process and natural disturbance in vegetation. *Bot. Rev.*, **45**, 229–99.

Wilson, E.O. (1987). The little things that run the world. The importance and conservation of invertebrates. *Trends Ecol. Evolut.*, **1**, 344–6.

Wirth, V. (1976). Veränderungen der Flechtenflora und Flechtenvegetation in der Bundesrepublik Deutschland. *SchrReihe Vegetationsk.*, **10**, 177–202.

Zackrisson, O. (1977). Influence of forest fires on the North Swedish boreal forest. *Oikos*, **29**, 22–32.

Zackrisson, O. (1978). Vegetational successions on a shore at the lake Storvindeln, N. Sweden, during the past 200 years. *Svensk. Bot. Tidskr.*, **72**, 205–26 (in Swedish with English Summary).

Zackrisson, O. (1985). Some evolutionary aspects of life history characteristics of broad-leaved tree species found in the boreal forest. In *Broadleaves in Boreal Silviculture–an obstacle or an Asset?*, ed. B. Hägglund & G. Peterson. *Swed. Univ. Agric. Sci. Dep. Silvic. Rep.*, **14**, pp. 17–36.

Chapter 8

Forested and Open Wetland Complexes

KJELL SJÖBERG

Department of Wildlife Ecology, Swedish University of Agricultural Sciences, S-901 87 Umeå, Sweden

&

LARS ERICSON

Department of Ecological Botany, University of Umeå, S-901 87 Umeå, Sweden

8.1 INTRODUCTION

In northern Sweden 3·8 million ha of the land surface is covered by mires, while another 1·7 million ha consists of moist or wet forest (i.e. forests with a high water table) (Svensson *et al.*, 1989). The area occupied by wet forest, particularly the damp, shady type of late successional spruce, is steadily being reduced as a result of various forest practices such as clear-felling and drainage. SOU (1978) predicted that these operations would affect half of the wet forests existing in Sweden by the end of the century.

Generally speaking, the floristic and faunistic composition of these wetlands has been fairly well studied. However, little is known from a landscape ecology and wildlife point of view regarding the importance of mosaic complexes, including various types of wetlands, wet forests and mesic-dry forest.

The following features of wet forests and mires will become increasingly important as the level of exploitation of our boreal forests progresses:

(1) Wet forests constitute one of the most species-rich vegetation types in the boreal area.

(2) Wet forests have often functioned as fire refugia. This implies that species with low dispersal abilities or with specific air humidity demands will decrease if these forest types are not preserved. Further, clear-felling of the surroundings can affect species composition as a result of changes in microclimate.
(3) Wet forests characteristically contain a large proportion of shady habitat. With the ever-increasing reduction in late post-fire successional spruce stands on mesic sites, the importance of these shady areas will increase.
(4) As open wetlands constitute important foraging sites, they are of great significance for several mammal and bird species living in the surrounding forests.

8.2 DEFINITIONS

8.2.1 Wet forests

The term 'wet forest' is used here in very broad terms to refer to a heterogeneous complex of forest types characterized by a high water table of variable durations. Thus our concept is somewhat broader than the usual definition of 'wet forests' in Sweden, where wet and moist forest habitats, regardless of the height of the water table, are classified as belonging to the forest series and not the mire series, if the annual forest productivity exceeds 1 m^3 per ha (Malmström, 1928), or if the tree layer has a crown projection exceeding 30% (Nordiska Ministerrådet, 1984). In Finland, where another classification system based on botanical and hydrological criteria is applied, wet and moist forests are assigned to the mire series regardless of forest productivity and tree cover; as a result many herb-rich vegetation types are included in this category (Eurola *et al.*, 1984). The Swedish classification system is weak in that it does not adequately distinguish between sites dominated by groundwater and those dominated by surface water (i.e. mire forests and swampy forests, respectively, according to the terminology of Eurola *et al.*, 1984). In the continental phytosociological terminology (Braun-Blanquet, 1964) the corresponding categories are 'Bruchwälder' and 'Auenwälder'.

A modern classification system for wet forests specific to Sweden has yet to be developed. Thus we chose to follow the main outlines of the Finnish system, which is applicable to boreal conditions and allows category designations to be made on the basis of tree species composition.

Mire forests, in which groundwater is the dominant influence, are mainly restricted to depressions in the terrain. They include stands dominated by *Alnus glutinosa* (alder carr), *A. incana*, tall-growing *Salix* spp., *Populus tremula*, *Betula pubescens* and *Picea abies*. Swampy forests, subjected primarily to surface water influence, are found either along watercourses, e.g. alder, tall-growing *Salix* spp., *Populus tremula*, groves with *Prunus padus* and *Frangula alnus* (often close to small rapids) and *Picea abies*, or on sloping terrain where mixed stands of *Betula pubescens* and *Picea abies* prevail.

A third type is the pine *Pinus sylvestris* bog vegetation, which characteristically exists as a forest fringe at sites where ombrotrophic conditions prevail, and is the most homogeneous. It is developed on hummocks and very poor in species, viz. *Ledum palustre* (Finland, eastern Sweden) *Vaccinium uliginosum* and a number of other dwarf-shrubs and mosses commonly found in heath forests. These pine bogs may also cover the entire surface of smaller mires in the drier parts of eastern Sweden.

Stands influenced by groundwater are mainly characterized by marked variations between hummocks (often harbouring species typical of meadow and heath forests), and depressions which are periodically inundated, at least after the snowmelt, with plants habitually found in wet to moist habitats. This mixture contributes to the species richness. Characteristic features of the ground flora include tall herbs (*Filipendula ulmaria*, *Geum rivale*, *Cirsium helenioides*, *Crepis paludosa*) and grasses (*Calamagrostis purpurea* and *C. canescens*). Tussock-forming graminoids are also characteristic, with *Carex appropinquata* and *C. cespitosa* in calcareous areas, where *Cypripedium calceolus* is also often abundant.

The spruce mire forest, which fringes open mires in large parts of the boreal region with the exception of the most continental areas, contains a mixture of mire plants and species found in nearby heath and meadow forests. Where the peat layer is thick, *Equisetum sylvaticum* and *Rubus chamaemorus* are characteristic species. Thinner peat layers, which are associated with better nutrient conditions, result in an increase in herbs such as *Viola epipsila* and *Geranium sylvaticum*. Hummocks are characterized by dwarf-shrubs, especially *Vaccinium myrtillus*, *V. uliginosum* and *Carex globularis*. Where groundwater influence is strong and lasting, less frequent species such as *Ranunculus lapponicus* occur in the middle and northern boreal zones.

Extensive *Populus tremula* stands may develop, often with a sparse field layer, on sites characterized by a strongly fluctuating water table inundated during a fairly short period immediately after the snow melt—as on the better drained substrates. Here *Carex brunnescens* ssp. *vitilis* is a characteristic species.

The strongly fluctuating water table, characteristic of swampy forests, may—depending on the presence of an additional groundwater influence—result in large variations in the vegetation. These alluvial forests are above all characterized by their richness in herbs such as *Viola epipsila*, *V. palustris*, *Rubus arcticus* and *Filipendula ulmaria*. Close to small rapids or in ravine bottoms this species richness may be very high; *Viola selkirkii* and *Agrostis clavata*, on disturbed sites, exemplify species growing here. These sites are also characterized by a rich epiphytic lichen flora.

For further information on more elaborate classifications based on the composition of the field and bottom layers, see Eurola *et al.* (1984).

8.2.2 Wetlands without a tree canopy

Open wetlands without a tree canopy have been the focus of much work by Fennoscandian plant ecologists (Sjörs, 1948, 1950; Ruuhijärvi, 1960; Eurola & Ruuhijärvi, 1961; Eurola, 1962; Malmer, 1986). Thus there is a wealth of knowledge about the vegetation, including the way in which its structure depends on regional and local environmental factors, as well as its genesis. Swedish and Finnish mire terminology (e.g. Sjörs, 1950; Eurola *et al.*, 1984) differs considerably. Mires show both a zonal differentiation, with bogs being dominant in more southern and oceanic areas, and mixed mires with ombrotrophic strings and minerotrophic flarks ('aapa'-mires) dominating in more northern and continental areas (Sjörs, 1948; Eurola & Ruuhijärvi, 1961). There is also an azonal differentiation in response to variation in local nutrient conditions, as evidenced by the poor–rich gradient in the fens (e.g. poor, intermediate, rich, extremely rich (Sjörs, 1950, 1983). The size of individual wetlands and their total area increase along a gradient from south to north. For review of the Swedish and Finnish mires, see, for example, Sjörs (1983) and Ruuhijärvi (1983), respectively.

The strong position of mire ecology in Fennoscandia, which emphasizes the importance of abiotic factors, may have been partly responsible for the little interest shown by many Nordic plant ecologists in modern community ecology. For example, the theory of island biogeography, which states that there is a positive correlation between habitat size and species numbers, does not hold for mire plants (cf. Lahti & Ranta, 1985).

There is probably no other vegetation type where plant and animal ecologists differ so much in their descriptions and classifications of the vegetation—and this also applies to plant ecologists and geologists (as evidenced by quarternary maps; cf. Backéus, 1980).

8.3 ENVIRONMENTAL FACTORS

A number of environmental characteristics influence the species composition of wet forests. Besides the type and duration of water influence (previous section), soil characteristics, nutrient conditions, humidity and stand history play a significant role. On a regional scale, boreal wet forests tend to have more dwarf shrubs and fewer herbs than hemiboreal ones, which are dominated more by tall herbs and graminoids.

With regard to the differentiation of vegetation and the species composition of wet forests in the boreal area, azonal features are more evident than zonal. In areas with calcareous bedrock or loose deposits, a large number of nutrient-demanding species occur, including *Cypripedium calceolus* and *Botrychium virginianum* and, in the northern areas, *Actaea erythrocarpa* and *Calypso bulbosa*. These areas are also remarkable in that they are generally rich in herbaceous species. By contrast, in areas with more nutrient-poor conditions, herb-rich sites are mainly restricted to localities influenced by mobile water. The richer areas are located in NE Uppland–SE Gästrikland, W Medelpad–Jämtland, the area between Norsjö and Skellefteå in N Västerbotten, the Jokkmokk area in Lapland, and the area around Pajala in Norrbotten.

The fact that wet forests are often located adjacent to various types of wetlands and watercourses implies that they are characterized by a high relative air humidity. This is the main reason why a large number of lichens, bryophytes and invertebrates are confined to this forest type. A specific case in point is the dense pole thickets of spruce which earlier characterized silt-rich sites in middle boreal areas, but which nowadays only occur in one small forest reserve in northern Ångermanland.

There is reason to believe that dark coniferous forests, such as wet spruce with bilberry plants rich in nitrogen but with low concentrations of chemical defence substances (phenols, cf. Laine & Henttonen, 1987), can support a denser invertebrate population than the lighter wet pine or mesic coniferous forest surrounding or bordering the wet spruce forest. The theoretical framework for this prediction is based on a hypothesis dealing with the carbon–nutrient balance in plants characterized by a carbon-based type of chemical defence, proposed by Bryant *et al.* (1983) and developed further by Coley *et al.* (1985) (cf. also Jonasson *et al.*, 1986). The hypothesis implies that under reduced light conditions photosynthesis will be lowered and, consequently, a shortage of carbohydrates will develop. This means that resources available for growth and chemical defence will be in short supply before the uptake of nutrients. This will lead to a surplus of nitrogen in the plant tissue resulting in a food highly favoured by herbivorous invertebrates. Whether the higher degree of herbivory under reduced light conditions is due to the higher plant-nitrogen content (e.g. White, 1984), the decrease in chemical defences (Haukioja & Niemelä, 1977; Haukioja, 1980) or other reasons is an open question.

8.4 DISTURBANCE HISTORY

The dynamics of wet forests are poorly understood in Fennoscandia. Recent studies in northern Finland have, for example, indicated that some herb-rich spruce forests influenced by mobile water may be remnants of earlier human impact (Vasari, 1988). However, more detailed study is required to confirm this hypothesis. Earlier, dam-building by beaver *Castor fiber* probably played a significant role by creating new sites (cf. Naiman *et al.*, 1988; Nummi, 1989), but its importance in Fennoscandia still remains obscure. The fact that a large number of wind-dispersed species (*Eriophorum* spp., *Equisetum* spp.) rapidly establish

Fig. 8.1. Wet coniferous forests often functioned as fire refugia in the area. The long forest continuity created habitats for specialized plants and animals. Here the fire has stopped at the edge of the wet forest. Province of Västerbotten, northern Sweden.

themselves following large-scale disturbance suggests that beaver activity may have been important.

Another important point requiring clarification is the significance of forest fires (see chapter 7). In the past, fires occurred in the dry to mesic forests with a frequency of about 45–90 years in boreal Sweden (Zackrisson, 1977; Engelmark, 1984, 1987; see also Tolonen, 1983 for Finland). Studies in southwestern Finland (Tolonen, 1985) have shown that forest fires affected the wet forests bordering the mires at intervals of about 50 years (Fig. 8.1), although this probably varied greatly between regions. Frequent fires may have had a great impact on mire forests, favouring pine and juniper *Juniperus communis* as well as dioecious species in more continental areas, while in the more oceanic areas, with a lower fire frequency, spruce stands were probably favoured. In more oceanic areas, wet forests probably often functioned as fire refugia, while providing a habitat with long-term stability.

8.5 SPECIES CHANGES

8.5.1 Flora

Vascular plants

Twenty-eight of the 73 vascular plants listed as threatened or negatively affected by forestry in Sweden occur in wet forest types (Ingelög *et al.*, 1987). Of these, about 17 are only found in wet forests. Because most of these species cannot compete with tall-growing graminoids and herbs under good light conditions, they are often negatively affected by drainage operations as well as clear-felling. Furthermore, a large number of previously abundant species have shown dramatic population decreases (e.g. *Sparganium glomeratum*).

The most striking change that has occurred among boreal *mire* plants has probably been the marked decline in the number of mesotrophic species (inhabiting rich fens and moderately rich, often sloping fens), including *Carex tenuiflora, Eriophorum brachyantherum, E. gracile* and *Salix myrtilloides* (cf. Hämet-Ahti, 1984). Such species used to be fairly frequent; nowadays, however, they are absent over large areas in which drainage operations associated with forestry production were carried out during the first half of this century.

Furthermore, many rich-fen species have decreased in abundance as management practices such as hay production and grazing (trampling) have gradually been abandoned. *Ophrys insectifera*, *Nigritella nigra* and *Gymnadenia conopsea* are good examples. On the other hand, tall herbs (e.g. *Filipendula ulmaria*) and deciduous shrubs have shown a corresponding expansion. However, generalizations are difficult (Elveland, 1978), and cessation of management has certainly resulted in a short-term increase of a large number of species before litter accumulates and woody plants establish. However, *Saxifraga hirculus*, nowadays mainly confined to springs and birch-mires, represents a species whose habitat decreased considerably once scything ceased (Elveland, 1978; Ohlson, 1986, 1987).

Wet forests were frequently utilized for both hay production and grazing by cattle, goats, sheep and reindeer, but (as only reindeer occur today), the present situation, characterized by a more closed tree canopy, is probably a rather new feature. Furthermore, grazing-intolerant species such as *Cypripedium calceolus* and *Bothrychium virginianum* have probably been favoured in recent times, although, once again, there may have been shifts in their distribution among habitats.

Cryptogams

A large number of lichens are confined to wet forests. Evidently they require the presence of slow-growing trees, high air humidity and a long, undisturbed stand history. It should be noted that a large number of species with a suboceanic distribution in Fennoscandia, which are presently vanishing (or have vanished) in southwestern Sweden owing to air pollution (Löfgren & Moberg, 1984), are confined to such sites, for example, *Parmeliella plumbea* in the calcareous area in NE Uppland–SE Gästrikland. Another group is represented by *Evernia divariata*, which has disappeared from large areas of Sweden and cannot survive if the surrounding forest is clear-felled—even if the wet forest is left unaffected (cf. Fig. 8.2).

8.5.2 Mammals

Two species of mammals have increased in numbers in the boreal wetlands during recent years—the beaver and the muskrat *Ondatra zibethicus*. The beaver was reintroduced to Sweden, in the province of Jämtland, in 1922 after its local extinction in the 1870s as a result of heavy hunting (Lavsund, 1977). The muskrat was introduced to Sweden in the early 1940s, but probably also spread here from Finland, where it was introduced in 1919 (Aritmo, 1960; Marcström, 1964; Danell, 1977, 1980).

In addition a few other mammals in Sweden are more or less restricted to wet forests or wetlands of the middle and north boreal zones. The root vole *Microtus oeconomus* and field vole *M. agrestis*, which are more closely associated with boreal wetlands than other Swedish microtines, forage in mires. According to Tast (1968), behavioural interference exists between the two species. As a result *M. oeconomus* dominates the rich fens of north boreal mires, forcing *M. agrestis* to less productive habitats (cf. Boström & Hansson, 1981).

The wood lemming *Myopus schisticolor* (Kalela, 1963) is a herbivore which, like the mountain lemming *Lemmus lemmus*, feeds primarily on mosses and graminoids. It inhabits dark, old, moss-rich forests strewn with dead tree trunks, and is therfore present in some types of wet spruce forests. Its population structure is complex, and high densities are only found during peak vole years. Present data seem to indicate that high densities are only reached every second vole cycle, i.e. at intervals of 8–10 years.

8.5.3 Birds

Several ecological groups can be distinguished among bird species inhabiting forested and/or open wetlands at least during part of their life cycle. One group, the tetraonids capercaillie *Tetrao urogallus*, black grouse *Lururus tetrix*, hazel hen *Bonasa bonasa* and willow grouse *Lagopus lagopus*, are not exclusively bound to these habitats, but during the brood stage prefer wet conditions such as the ecotone between mires and forests. Capercaillie females are known to feed on the early developing floral shoots of *Eriophorum vaginatum*, which start to elongate in spring coinciding with the egg-laying period (U. Boström & P. Angelstam, pers. comm.). Even the black and willow grouse, which at other times prefer open habitat types, use wet sites as brood-rearing areas.

The curlew *Numenius arquata* and the lapwing *Vanellus vanellus* frequent agricultural areas as well as natural wetlands (Pettersson, 1988*a*,*b*). Both species are decreasing in number in Sweden, especially in the south, where agriculture is more rationalized and on a larger scale compared with the northern parts of the country.

The crane *Grus grus*, whooper swan *Cygnus cygnus* and bean goose *Anser fabalis* are part of the mire fauna. The crane does not seem to have changed in number, whereas the whooper swan has both extended its present breeding range and increased in number. The reason for its success has yet to be determined. On the other hand, the breeding population of the bean goose seems to have decreased compared with some decades ago. The cessation of hay scything on wetlands could be one reason for this decline (Mellquist, 1970; Mellquist & von Bothmer, 1984).

Another wetland species influenced by man's activities is the Temminck's stint *Calidris temminckii*. This small sandpiper used to be rather common, but nowadays there are few breeding populations left along the coastal area of the Bothnian Bay. Apparently, the cessation of cattle grazing along the coastal shorelines and rivers has been responsible for this decrease. This is probably also true for other waders, e.g. the southern subspecies of dunlin *Calidris alpina schinzii*, the great snipe *Gallinago media*, the black-tailed godwit *Limosa limosa* and the ruff *Philomachus pugnax*, whose breeding has declined or ended altogether in southern Sweden (e.g. Ahlén, 1977; Tjernberg, 1985; Jönsson, 1988).

At the present time, the strictly wetland birds, including waders such as snipe *Gallinago gallinago*, jack snipe *Lymnocryptes minimus*, wood sandpiper *Tringa glareola*, greenshank *T. nebularia*, redshank *T. totanus*, spotted redshank *T. erythropus*, ruff, broad-billed sandpiper *Limicola falcinellus*, red-necked phalarope *Phalaropus lobatus* and golden plover *Pluvialis apricaria*, do not seem to be greatly influenced by changes in habitat structure or exploitation. It should be noted, however, that past population trends in this group are not well documented and, moreover, it is not known whether the cessation of regular scything on mires serving as breeding habitats for species such as snipe, wood sandpiper, ruff and red-necked phalarope has had any sigificant impact on their populations.

Boström (1978), studying the influence of drainage on the mire bird fauna, found that the effects were related to how rapidly the changes in habitat appeared. As open water surfaces disappeared, so did ducks, grebes, some waders (spotted redshank, greenshank, and red-necked phalarope) and the red-throated diver *Gavia stellata*. In addition, the broad-billed sandpiper and jack snipe seemed to be less tolerant of drainage than the snipe and wood sandpiper (see also Simonsson, 1987).

It has been claimed that large mires house more bird species as well as higher bird densities than small ones (Boström & Nilsson, 1983). This has been shown in detail for the golden plover *Pluvialis apricaria* (Alexandersson, 1987).

As far as we know, few, if any, bird species are totally restricted to the wet forest type. Some, however, are more frequently seen here than in other habitats, for example, the rustic bunting *Emberiza rustica*. The hazel hen *Bonasia bonasia* and the green sandpiper *Tringa ochropus* are often associated with wet spruce forests, and this is probably also the case for the pygmy owl *Glaucidium passerinum*. The long-tailed tit *Aegithalos caudatus*, the lesser spotted woodpecker *Dendrocopos minor* and the white-backed woodpecker *D. leucotos* are more or less associated, mainly in southern Sweden, with wet broadleaf forests. The white-backed woodpecker is an endangered species in Sweden (Aulén, 1988; cf. chapter 7), and the lesser spotted woodpecker is regarded as decreasing in number. Less is known about trends in the long-tailed tit population.

8.5.4 Amphibians

The three widely distributed and common amphibians in the scandinavian taiga are *Rana temporaria*, *R. arvalis* and *Bufo bufo*. All breed in a variety of wetlands, from lakes and slow-moving rivers to open mires and temporary ponds (Elmberg & Ericsson, 1982; Elmberg, 1984). There seem to be a general and world-wide decline of anuran populations (e.g. Barinaga, 1990). In Sweden destruction by draning or cutting wet forests is not the only threat to boreal anurans. Andrén *et al.* (1988) demonstrated the severe effects that acid precipitation may have on reproduction, especially if a lowered pH is accompanied by increased levels of aluminium.

8.5.5 Invertebrates

According to Ehnström and Waldén (1986), 249 of the *c.* 30 000 invertebrate species in Sweden are regarded as threatened and declining, mainly as a result of agricultural and forestry activities. They caution, however, that for many species the actual situation is not well known, and the real number of endangered species could be much higher. Many of the coniferous invertebrate species use old, dead trees as substrate. It is probable that several hundreds of the bark- and wood-living species, mainly Coleoptera, are threatened by forestry activities.

Wet spruce forests are the exclusive habitats for several rare lepidopteran species, including *Anomogyna sincera* and *A. borealis*, coleopteran species such as *Pytho kolwensis* (Pythidae) and *Agonum mannerheimi* (Carabidae) and even the snail *Clausilia cruciata* (Ehnström & Waldén, 1986). Pine stands bordering wetlands are the habitats of the rare beetle *Tragosoma depsarium* (Ehnström & Waldén, 1986).

8.6 THE FUTURE

In the future we can expect a still stronger reduction of wet forests and wetlands owing to increased forestry and/or peat exploitation. If no measures are taken to counteract these trends we predict that most species-rich forest type (e.g. the herb-rich spruce forests) will decrease in area, resulting in a marked decline in herbs and herb-rich sites in the boreal region. Since the presence of flowing water is one of the main prerequisites for their development and maintenance, the first casualties will probably be various habitat-sensitive plant species and the specialist insects associated with them. Further, organisms dependent upon a long stand continuity and characterized by low dispersal abilities, specific habitat demands, and/or with specific microclimatic demands (e.g. for shade, high humidity) will become increasingly restricted to scattered, isolated sites and face an uncertain future (a good recent example is the pendent lichen *Ramalina thrausta*). Bird species likely to decline include those dependent on invertebrates and those foraging primarily in wetlands.

8.7 CASE STUDIES

Below we present three cases which exemplify the way in which species found in wet forests may be affected by various forest activities.

8.7.1 *Cypripedium calceolus*

The orchid *Cypripedium calceolus* mainly occurs in herb-rich wet forests in boreal areas, but also in herb-rich moist-mesic spruce and deciduous forests, especially in the hemiboreal area. In boreal areas of this type it has declined.

Cypripedium is a clonal plant, with slow vegetative spread. The above-ground shoots start to elongate in late spring. These are very sensitive to destruction and the plant is unable to replace lost tissue during the vegetative period. Thus it suffers heavily from competition with tall plants (herbs, grasses) favoured by increased light conditions following clear-felling, with a resultant population decline—as is apparently occurring in the southern half of Sweden (Fig. 8.2). Although this type of interspecific competition seems to decline towards the north, clear-felling at high latitudes may nevertheless result in severely reduced shoot survival and flowering due to the severe frosts in June typical of clear-felled areas (Fig. 8.3). Thus *Cypripedium* represents a species where population declines may follow clear-felling.

8.7.2 *Evernia divaricata*

Evernia divaricata is a pendent lichen growing loosely entangled around bark and twigs of a large number of tree and bush species in wet forests.

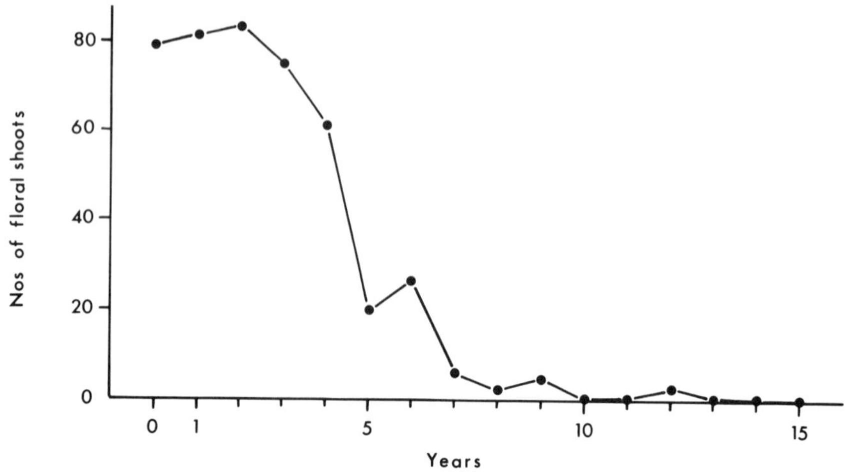

Fig. 8.2. The numbers of floral shoots of *Cypripedium calceolus* prior to (year 0) and following clear-felling (between year 0 and 1) of the forest. Data from one population in S Gästrikland which suffered competition from tall grasses (*Calamagrostis arundinacea*), herbs (*Chamaenerion angustifolium, Rubus ideaus*) and deciduous shrubs (*Betula* spp., *Viburnum opulus*) following clear-felling (L. Ericson, original data).

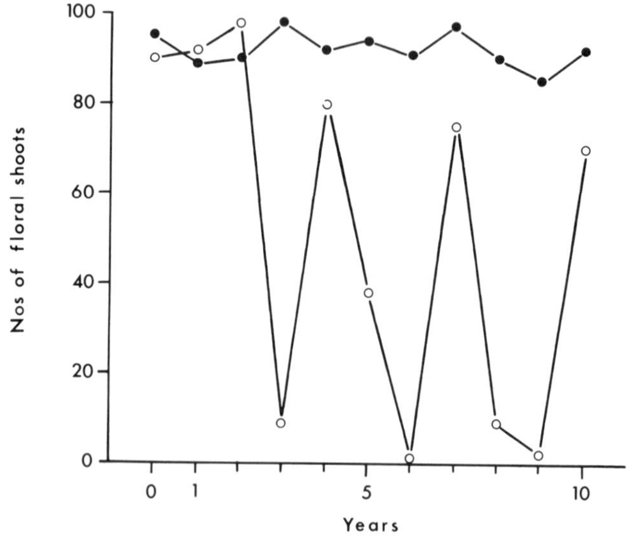

Fig. 8.3. Between-year variations in flowering for *Cypripedium calceolus* in a late-successional, herb-rich, spruce forest (●) and on the near-by clear-felled area (○). Data from two populations, 150 m apart, in the province of Västerbotten. Clear-felling took place between year 0 and 1 (L. Ericson, original data).

During recent years its populations have declined, and it has disappeared from large areas of Sweden. It is confined to sites with a higher air humidity. As shown in Fig. 8.4, this species will not survive in areas where the surrounding forests have been clear-felled, and drastic populaton declines may occur even if the wet forest is left intact. The species may also disappear on small islands in rivers if the riverine forests are clear-felled or thinned. The strong sensitivity to microclimatic change shown by *Evernia divaricata* is probably shared by a large number of other cryptogams.

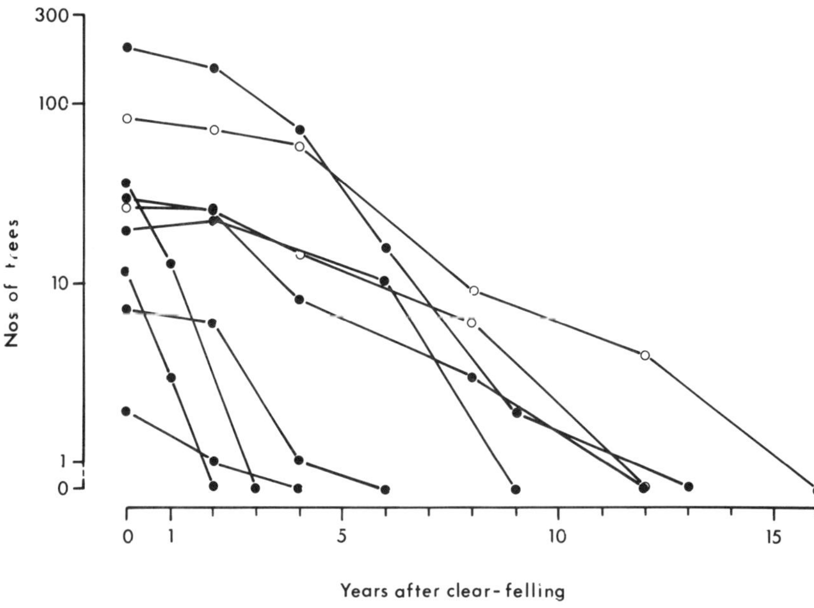

Fig. 8.4. The decline of nine populations of the pendent lichen *Evernia divaricata* (expressed as numbers of trees where the species occurred) following the clear-felling of the surrounding late successional spruce forest (year 0). ●, populations in mesic-moist mire forest; ○, populations on small islands in rivers. Data from the provinces Gästrikland and Västerbotten collected during the period 1964–1987 (L. Ericson, original data).

8.7.3 *Pytho kolwensis*

There are probably several insect species confined to fire refugia with long-term continuity (see review in Ehnström & Waldén, 1986). One is the log-inhabiting beetle *Pytho kolwensis* (family Pythidae), known in boreal Sweden from only seven patches of wet spruce forests, where the larvae live in unusually thick bark of fallen spruce trees (R. Pettersson, pers. comm.). Apparently the beetles need this type of bark to protect them from dehydration. The average age of the spruce found on the *Pytho* sites was about 200 years. This is about twice the age at which spruce is normally harvested. Thus old-growth forest, resulting from a long fire-free period (fire refugia), seems to be one of the crucial prerequisites for the survival of this species.

8.8 SIGNIFICANCE OF WET FORESTS AND WETLANDS IN A LANDSCAPE CONTEXT

8.8.1 Historical perspective

At the time when the northern interior part of Sweden was colonized by farmers, livestock constituted the basis for survival during winter (Bylund, 1956; Rudberg, 1957). Therefore, the availability of land suitable for cattle grazing and hay production for winter fodder was a primary factor used in selecting settlement sites, and also contributed to determining the density of villages in the boreal area. It is considered that almost all mires were used for hay scything, even those far from the farms, tens of kilometres away (Veirulf, 1937; Rudberg, 1957).

Both natural and forested wetlands were important—the former type as hay-producing areas and the latter as grazing for cattle, goats and sheep. These habitats were used mainly from the 17th century to the mid-20th century. Since then, they have been abandoned, with the exception of forest grazing by cattle on a very small scale along the edge of the mountainous region in northwestern Sweden. Signs of these former practices are still visible, however. For example, remains of barns and hay-drying sheds can still be found in the northern forests and on mires (Fig. 8.5). In recent years, a few nature reserves have been established and earlier management has been restored.

The open wetlands were used in different ways depending on their type and locality. Those under hay production can be divided into two main groups: (a) areas with a natural water regime, i.e. fens and moist meadows, including wetlands along rivers and lakeshores; and (b) those with a regulated water regime (Frödin, 1952; Elveland, 1983). Mires dotting the forested landscape belong to the former group. Depending on the productivity level, such mires were used every year, or perhaps more commonly, every second year, and sometimes only every third year.

Fig. 8.5. The presence of abandoned barns show the former importance of wetlands as production areas for hay as winter fodder for cattle in the old farm system. The regular scything kept the wetlands open, thereby creating habitats for waders. Cessation of scything has resulted in a rapid expansion of deciduous shrubs. Above, a mire earlier used for hay production in the province of Norrbotten, northern Sweden; below an alluvial meadow, fertilized by silt deposition during the spring flood, along a small river in the province of Västerbotten, northern Sweden.

Wetlands along rivers and lakeshores, fertilized every year with the material deposited by the spring flood, could usually be utilized every year.

To increase hay production, the water regime was often regulated, particularly, in the mid-19th century—although the practice was already in use as early as the 17th century. Two types of irrigation were employed: flowing water-meadows and catchwork water-meadows. Recent studies (Elveland & Sjöberg, 1982; Elveland, 1983) have shown that irrigated hay-meadows could be mown every year without any decrease in yield. Damming was practised in low-lying areas. These flowing water-meadows were irrigated during October–June, in order to prevent freezing and to increase the deposition of silt (inorganic and organic material). By lowering the water level just before cutting the hay, the work became easier. In order to inhibit colonization by various woody species, the water level was sometimes lowered again in winter so that the bottom substrate froze solid. Bushes were uprooted with the ice in spring, and the bottom of the lake became smooth, thereby facilitating mowing.

The other type of irrigation, catchwork water-meadows, was practised in areas with sloping terrain. During spring, water was spread out evenly over the surface of sloping mires from small canals distributed over the mire, with the result that silt was deposited over the mire and moving water rich in oxygen increased decomposition and thus productivity (Campbell, 1948). This type of irrigation, used for increasing hay production, has also been—or is still being—employed in other parts of Europe with cattle-based agriculture.

As most wetlands were utilized for hay production, this type of management, as well as its subsequent abandonment, must have had a very great influence on both the flora and fauna. Elveland (1983) summarized some of the changes in vegetation and edaphic factors occurring in wetlands after they had ceased to be managed:

(a) The biomass of various plant species changed, often considerably and relatively quickly, even when the species composition was only slightly affected.
(b) Low-growing plant species that had survived scything but were weak competitors (e.g. many herbs and small graminoids) decreased in abundance, their place being taken by tall-growing graminoids and herbs, and other species that had been repeatedly cut back by regular scything.
(c) The even surface layer became more irregular as a result of

invasion by tussock-forming plants and, in many cases, bushes and trees. This type of large-scale change is pronounced over large parts of the boreal area (Vasari, 1988).

(d) A steady increase in the accumulation of organic debris from year to year.

Probable effects on the fauna can be summarized as follows:

(a) Winter damming prevents the bottom from freezing. Thus the resumption of production/activity in the spring by plant and invertebrates would have started earlier compared with a wetland with a natural water regime. Hence birds breeding in the dammed wetlands would benefit by the early availability of food.
(b) Silt deposited as a result of damming increases the production of both plants and invertebrates, thereby increasing the availability of food for birds (cf. Fig. 8.6).
(c) Thinning out the *Carex* and *Equisetum* species and reducing the amount of litter in the regularly scythed wetlands would have had positive effects on both waders and ducks by rendering their food more easily available.

Fig. 8.6. Scything for hay in natural wetlands such as mires, riversides and lakeshores created an open landscape and reduced deposition of organic debris. Conditions for waders and herbivorous waterfowl such as dabbling ducks (e.g. wigeon) and geese (e.g. bean goose) were improved. In the hinterground a managed alluvial meadow; in the foreground a section not used for hay production.

(d) By scything, the dense vegetation belt was opened, allowing fresh shoots to emerge. This would benefit waterfowl species such as the wigeon *Anas penelope* and the bean goose.
(e) By regular scything the ecosystem was kept in a primary successional stage, characterized by low invertebrate predator pressure. This in turn would have led to a decrease in competition for food for ducks and waders.
(f) In the primary successional stages created by regular scything, invertebrates appearing in these stages in high numbers (e.g. chironomids and *Asellus*) provide a ready food source for ducks and waders (Danell & Sjöberg, 1982).
(g) The solid freezing of water-meadows during winter kills off much of the fish fauna. As a result fish predation on invertebrates is reduced, leading to a further increase in food availability for the birds.

For further details, see Elveland (1979) and Elveland and Sjöberg (1982).

8.8.2 The present situation

The mosaic of forested and open wetlands is inhabited by numerous species that are not restricted to a single habitat. They may consistently use a combination of habitats, or habitat preferences may change over time, depending on age, stage in the breeding cycle, etc. (Larsen, 1982; Speight & Blackith, 1983). It is even more difficult to determine how such species are affected by the types of habitat changes discussed above.

Based on our present knowledge it is clear that the preservation of these forest–mire complexes is of crucial importance for much of the boreal fauna (e.g. mammals and birds). Nevertheless, there is good reason to believe that the long-term survival of a large number of species with specific demands regarding air humidity, etc. (cf. the case study of *Evernia divaricata* above) can only be ensured by preserving the original mosaics.

Among mammals, moose *Alces alces* often feed on sodium-rich wetland plant species during the summer, while during part of the autumn they frequently take bilberry vegetation in dark, shady spruce forests. Other mammals, for example the field and root voles, utilize a combination of wet forests and mires, as foraging and breeding habitats.

Several species of birds are known to inhabit a combination of forest types (including wet forests) and wetlands, each of which must be important for their survival. For example, the great grey owl *Strix nebulosa*, Ural owl *S. uralensis*, and honey buzzard *Pernis apivorus* nest in the forest but forage to varying degrees in the surrounding wetlands (Mikkola, 1983).

The complex use of a combination of habitats is also exemplified by the capercaillie. Once the chicks hatch they depend on invertebrates for food, and the broods are often found in the insect-rich marginal areas. Later on, the chicks gradually change to a vegetarian diet. At this point bilberry fruit becomes an important food for the chicks, and consequently the more open forest types, e.g. mesic spruce forests, where this plant is common, become an important foraging habitat. During winter, needles of Scots pine, *Pinus sylvestris*, dominate the diet of capercallie in Fennoscandia, and thus old pine forest is preferred at this time.

In the boreal coniferous forest landscape, mires are often bordered by sunny, wet pine bogs and shady spruce mires, which also sometimes create discrete patches within homogeneous forests. Line-transect censuses in Finland showed that adult capercaillie were most common in the 'spruce–hardwood peat moors', i.e. wet forest dominated by spruce or deciduous trees (Rajala, 1962). This forest type is also preferred by capercaillie chicks. By radio-tracking broods in southern Norway, Wegge *et al.* (1982) determined that younger chicks particularly were associated with the more productive damp forests. According to Börset and Krafft (1973), capercaillie broods occur mainly in mature forests with a sparsely developed bush stratum.

It seems reasonable to assume that capercaillie chicks could increase their survival rate by searching for food in a habitat rich in arthropods, one reason being that the amount of food taken by grouse chicks is partly dependent on the weather. Erikstad and Spidsö (1982) found that rain and a decrease in the mean ambient temperature from 19·5°C to 9°C reduced the feeding time available to 2- to 5-day-old willow grouse chicks by about 60% (cf. Kastdalen & Wegge, 1985).

Another example of a complex utilization of two or more habitats is offered by the amphibians. After breeding, the three common amphibians in the Scandinavian taiga region disperse very selectively from wetlands to adjacent habitats that can provide food and shelter, i.e. mainly meadows and dense, often wet forests (Elmberg, 1990; J. Elmberg, pers. comm.). Although we still know little about the post-breeding activities of boreal anurans, Finnish and Swedish data hint at a marked site tenacity in all three species, at least during the main foraging period, June–August (Haapanen, 1970; J. Elmberg, pers. comm.). South of (but very close to) the boreal zone Loman (1978) reported a habitat segregation in adult *Rana temporaria* and *R. arvalis* in summer. The former species foraged mainly in forest habitats, whereas the latter was found mainly in meadows. Subadults, however, did not segregate spatially.

Because they are sensitive to predation as well as to over-heating and desiccation, boreal anurans have pronounced nocturnal habits in summer (Loman, 1980; Sinsch, 1984; J. Elmberg, pers. comm.). For the same reasons they mainly forage in damp and shaded locations, i.e. meadows with very tall grass, and more importantly in the taiga, in dense, often wet, forests (J. Elmberg, pers. comm.). It has been shown (Elmberg, 1990) that a disproportionally large part of the annual mortality in *R. temporaria* occurs in spring, when the frogs use open habitats for breeding.

In the boreal zone of Fennoscandia all three amphibian species mentioned usually hibernate under water (Koskela & Pasanen, 1974; Elmberg & Ericsson, 1982; Hagström, 1982; Elmberg, 1984, 1990). Streams are preferred, but ponds, lakes and deep mires suffice as long as the bottom water does not become anoxic. Hence, the three main habitat types involved in the annual cycle of a boreal anuran are running water (hibernation), stagnant water of three months' permanency (breeding and development of tadpoles), and wet forests and meadows (foraging).

8.9 RECOMMENDATIONS

(A) Management of open wetlands (scything or grazing by cattle) should be introduced (i) in calcareous areas in order to prevent local extinctions of susceptible plants, and (ii) in nature reserves, distributed over the country, to demonstrate the former utilization of wetlands. Both these measures have been initiated on a small scale during recent years.

(B) Representative mire-type complexes illustrating both regional and local differentiation should be preserved. Measures along these lines have been proposed by, for example, Sjörs and Backéus (1977).

(C) Forested wetlands generally should be preserved, since otherwise late successional forest stages will disappear from the managed forest landscape outside reserves. This is the best possibility of ensuring the survival of species dependent on forest stand continuity and characterized by low dispersal abilities.

(D) Mosaic complexes of wetlands and forested wetlands should be preserved. This is important both for species which respond negatively to changes in microclimate and hydrology and for those which need a mosaic landscape for foraging (cf. Sjögren, 1991).

ACKNOWLEDGEMENTS

We thank Johan Elmberg for contributing to the amphibian sections of this chapter, Per-Anders Esseen and Christer Nilsson for comments on an early draft of the manuscript, and the Swedish Environmental Protection Agency for financial support.

REFERENCES

Ahlén, I. (1977). *Faunavård*. Liber, Stockholm.

Alexandersson, H. (1987). Distribution and density of golden plover *Pluvialis apricaria* on raised bogs in Southwest Sweden. The importance of area and distances between bogs. *Acta Reg. Soc. Sci. Litt Gothoburgensis: Zoologica*, **14**, 9–19.

Andrén, C., Henriksson , L., Olsson, M. & Nilson G. (1988). Effects of pH and aluminium on embryonic and early larval stages of Swedish brown frogs *Rana arvalis, R. temporaria* and *R. dalmatina*. *Holarct. Ecol.*, **11**, 127–35.

Aritmo, A. (1960). The dispersal and acclimatization of the muskrat, *Ondatra zibethicus* (L.) in Finland. *Pap. Game Res.*, **21**, 1–101.

Aulén, G. (1988). Ecology and distribution history of the white-backed woodpecker *Dendrocopos leucotos* in Sweden. *Swed. Univ. Agric. Sci. Dep. Wildl. Ecol., Rep.*, No. 14.

Backéus, I. (1980). Mixed mires on the national maps Örebro NW and NE. *Geologiska Föreningens i Stockholm Förhandlingar*, **100**, 407–8 (in Swedish with English summary).

Barinaga, M. (1990). Where have all the froggies gone? *Science, N.Y.*, **247**, 1033–4.

Boström, U. (1978). Dikning av våtmarker, vad betyder det för fågelfaunan? *Viltnytt*, **9**, 18–25.

Boström, U. & Hansson, L. (1981). Small rodent communities on mires: implications for population performance in other habitats. *Oikos*, **37**, 216–24.

Boström, U. & Nilsson, S.G. (1983). Latitudinal gradients and local variations in species richness and structure of bird communities on raised peat-bogs in Sweden. *Ornis Scand.*, **14**, 213–26.

Börset, E. & Krafft, A. (1973). Black grouse *Lyrurus tetrix* and capercaillie *Tetrao urogallus* brood habitats in a Norwegian spruce forest. *Oikos*, **24**, 1–7.

Braun-Blanquet, J. (1964). *Pflanzensoziologie. Grundzüge der Vegetationskunde*, 3rd edn. Springer Verlag, Vienna.

Bryant, J.P., Chapin, F.S. III & Klein, D.R. (1983). Carbon/nutrient balance of boreal plants in relation to vertebrate herbivory. *Oikos*, **40**, 357–68.

Bylund, E. (1956). Koloniseringen av Pite Lappmark t.o.m. år 1867. *Geographica, Uppsala*, **30**.

Campbell, Å. (1948). *Från vildmark till bygd*. Bokförlaget Hermes AB, Uddevalla.

Coley, P.D., Bryant, J.P. & Chapin, F.S. III (1985). Resource availability and plant antiherbivore defence. *Science, N.Y.*, **230**, 895–9.

Danell, K. (1977). Dispersal and distribution of the muskrat *Ondatra zibethica* (L.) in Sweden. *Swed. Wildl.*, **10**, 1–26.

Danell, K. (1980). Bisam i Sverige. *Swedish Environment Protection Board*, PM No. 1292.

Danell, K. & Sjöberg, K. (1982). Successional patterns of plants, invertebrates and ducks in a man-made lake. *J. Appl. Ecol.*, **19**, 395–409.

Ehnström, B. & Waldén, H.W. (1986). *Faunavård i Skogsbruket, Del 2. Den Lägre Faunan*. Skogsstyrelsen, Jönköping.

Elmberg, J. (1984). The moor frog *Rana arvalis* Nilsson in northern Sweden. *Fauna Flora*, **79**, 69–77 (in Swedish with English summary).

Elmberg, J. (1990). Long-term survival, length of breeding season and operational sex ratio in a boreal population of common frogs, *Rana temporaria* L. *Can. J. Zool.*, **68**, 121–7.

Elmberg, J. & Ericsson, S. (1982). On the distribution and ecology of the common toad *Bufo bufo* in northern Sweden. *Fauna Flora*, **77**, 27–32 (in Swedish with English summary).

Elveland, J. (1978). Management of rich fens in North Sweden. Studies of various factors influencing the vegetational dynamics. *Swedish Environment Protection Board*, PM No. 1007 (in Swedish with English Summary).

Elveland, J. (1979). Irrigated and naturally flooded hay meadows in North Sweden—a nature conservation problem. *Swedish Environment Protection Board*, PM No. 1174 (in Swedish with English Summary).

Elveland, J. (1983). North Swedish wet hay meadows—protection of an ancient cultural landscape. *Swedish Environment Protection Board*, PM No. 1737 (in Swedish with English summary).

Elveland, J. & Sjöberg, K. (1982). Some effects of scything and other management procedures on the plant and animal life of N. Swedish wetlands formerly mown for hay.) *Swedish Environment Protection Board*, PM No. 1516 (in Swedish with English Summary).

Engelmark, O. (1984). Forest fires in the Muddus National Park (northern Sweden) during the past 600 years. *Can. J. Bot.*, **62**, 893–8.

Engelmark, O. (1987). Forest fire history and successional patterns in Muddus National Park, northern Sweden. PhD thesis, Swedish University of Agricultural Sciences, Umeå.

Erikstad, K.E. & Spidsö, T.K. (1982). The influence of weather on food intake, insect prey selection and feeding behaviour in willow grouse chicks in northern Norway. *Ornis Scand.*, **13**, 176–82.

Eurola, S. (1962). Über die regionale Einteilung der südfinnischen Moore. *Ann. Bot. Soc. Vanamo*, **33**(2), 1–243.

Eurola, S. & Ruuhijärvi, R. (1961). Über die regionale Einteilung der finnischen Moore. *Arch. Soc. Vanamo*, **16** (Suppl.), 49–63.

Eurola, S., Hicks, S. & Kakinen, E. (1984). Key to Finnish mire types. In *European Mires*, ed. P.D. Moore. Academic Press. London, pp. 11–117.

Frödin, J. (1952). Skogar och myrar i norra Sverige i deras funktioner som betesmark och slåtter. *Skrift. Inst. sammenlignende Kulturforskning, Serie B*, **46**, 1–216.

Haapanen, A. (1970). Site tenacity of the common frog *Rana temporaria* L. and the moor frog *R. arvalis* Nilss. *Ann. Zool. Fennici*, **7**, 61–66.

Hagström, T. (1982). Winter habitat selection by some north European amphibians. *Brit. J. Herpetol.*, **6**, 276–7.
Hämet-Ahti, L. (1984). Changes of the northern boreal vegetation and flora in Finland after the Second World War. *Phytocoenologia*, **12**, 359–61.
Haukioja, E. (1980). On the role of plant defences in the fluctuation of herbivore populations. *Oikos*, **35**, 202–13.
Haukioja, E. & Niemelä, P. (1977). Retarded growth of a geometrid larva after mechanical damage to leaves of its host tree. *Ann. Zool. Fennici*, **14**, 48–52.
Ingelög, T., Thor, G. & Gustafsson, L. (eds) (1987). *Floravård i Skogsbruket* 2nd edn. Skogsstyrelsen, Jönköping.
Järvinen, O. & Sammalisto, L. (1976). Regional trends in the avifauna of Finnish peatland bogs. *Ann. Zool. Fennici*, **13**, 31–43.
Järvinen, O., Kouki, J. & Häyrinen, U. (1987). Reversed latitudinal gradients in total density and species richness of birds breeding on Finnish mires. *Ornis Fennica*, **64**, 67–73.
Jonasson, S., Bryant, J.P., Chapin, F.S. III & Andersson, M. (1986). Plant phenol and nutrient in relation to variations in climate and rodent grazing. *Amer. Nat.*, **128**, 394–408.
Jönsson, P.E. (1988). Ecology of the southern dunlin *Calidris alpina schinzii*. PhD thesis, University of Lund.
Kalela, O. (ed.) (1963). Beiträge zur Biologie des Waldlemmings, *Myopus schisticolor (Lillj.) Arch. Soc. Zool. Bot. Fennicae Vanamo*, **18**, Suppl., 1–96.
Kastdalen, L. & Wegge, P. (1985). Animal food in capercaillie and black grouse chicks in south-east Norway—a preliminary report. In *Proc. Int. Grouse Symp. 3rd, York, England, 1984*, ed. P.J. Hudson & T.W. Lovel. World Phesant Association, Reading, and International Council for Game and Wildlife Conservation, Paris, pp. 26–34.
Koskela, P. & Pasanen, S. (1974). The wintering of the common frog, *Rana temporaria*, in northern Finland. *Aquilo, Ser. Zool.*, **15**, 1–17.
Lahti, T. & Ranta, E. (1985). The SLOSS principle and conservation practice: an example. *Oikos*, **44**, 369–70.
Laine, K.M. & Henttonen, H. (1987). Phenolics/nitrogen ratios in the blueberry *Vaccinium myrtillus* in relation to temperature and microtine density in Finnish Lapland. *Oikos*, **50**, 389–95.
Larsen, J.A. (1982). *Ecology of the Northern Lowland Bogs and Conifer Forests.* Academic Press, New York.
Lavsund, S. (1977). Bäverns utbredningshistoria och nuvarande förekomst i Sverige. In *Proc. Nordic Beaver Symp., 1975*, ed. S. Lavsund. *Royal College of Forestry, Institute of Forest Zoology, Research Notes*, No. 26. Stockholm.
Löfgren, O. & Moberg, R. (1984). Oceaniska lavar i Sverige och deras tillbakagång. *Swedish Environment Protection Board*, PM No. 1819 (in Swedish with English summary).
Loman, J. (1978). Macro- and microhabitat distribution in *Rana arvalis* and *Rana temporaria* during summer. *J. Herpetol.*, **12**, 29–33.
Loman, J. (1980). Annual and daily locomotor activity of the frogs *Rana temporaria* and *R. arvalis*. *Brit. J. Herpetol.*, **6**, 83–5.
Malmer, N. (1986). Vegetational gradients in relation to environmental conditions in northwestern European mires. *Can. J. Bot.*, **64**, 375–83.

Malmström, C. (1928). Våra torvmarker ur skogsdikningssynpunkt. *Medd. Stat. Skogsforsk. Inst.*, **24**, 251–372.
Marcström, V. (1964). The muskrat *Ondatra zibethicus* L. in northern Sweden. *Swed. Wildl.*, **2**, 329–407.
Mellquist, H. (1970). Sädgåsen och slåttermyrarna. *Sver. Nat.*, **61**, 65–8.
Mellquist, H. & von Bothmer, R. (1984). Effects of haymaking on bean goose, *Anser fabalis*, breeding habitats in Sweden. *Swed. Wildl. Res.*, **13**, 49–58.
Mikkola, H. (1983). *Owls of Europe*. T. & A.D. Poyser, Calton.
Naiman, R.J., Johnston, C.A. & Kelley, J.C. (1988). Alteration of North American streams by beaver. *BioScience*, **38**, 753–62.
Nordiska Ministerrådet (1984). Vegetationstyper i Norden. Berlings, Arlöv.
Nummi, P. (1989). Simulated effects of beaver on vegetation, invertebrates and ducks. *Ann. Zool. Fennici*, **26**, 43–52.
Ohlson, M. (1986). Myrbräckan—en hotad våtmarksart. *Fauna Flora*, **81**, 199–200.
Ohlson, M. (1987). *Saxifraga hirculus*–habitats, reproductive effort and ecotypic varation. PhD thesis, Swedish University of Agricultural Science, Umeå.
Pettersson, Å. (1988*a*). Tofsvipa *Vanellus vanellus* L. In *Fåglar i Jordbrukslandskapet*, ed. S. Andersson. *Vår Fågelvärld, Suppl.*, No. 12, 189–94.
Pettersson, Å. (1988*b*). Storspov *Numenius arquata* L. In *Fåglar i Jordbrukslandskapet*, ed. S. Andersson. *Vår Fågelvärld, Suppl.*, No. 12, 195–200.
Rajala, P. (1962). On the ecology of the broods of capercaillie *Tetrao urogallus*, black grouse *Lyrurus tetrix*, and willow grouse *Lagopus lagopus*. *Suomen Riista*, **15**, 28–52 (in Finnish with English summary).
Rudberg, S. (1957). Unsettled areas and frontier settlement areas in inner northern Sweden—a discussion of some causes underlying this division. *Geographica, Uppsala*, No. 33 (in Swedish with English summary).
Ruuhijärvi, R. (1960). Über die regionale Einteilung der nordfinnischen Moore. *Ann. Bot. Soc. Vanamo*, **31**(1), 1–360.
Ruuhijärvi, R. (1983). The Finnish mire types and their regional distribution. In *Ecosystems of the World, 4B. Mires: Swamp, Bog, Fen and Moor*, ed. A.J.P. Gore. Elsevier Scientific Publishers, Amsterdam, Oxford, New York, pp. 47–67.
Simonsson, P. (ed.) (1987). Environmental effects of draining wetland and forest. *Swedish Environment Protection Board*, PM No. 3270 (in Swedish with English summary).
Sinsch, U. (1984). Thermal influences on the habitat preference and the diurnal activity in three European *Rana* species. *Oecologia* (*Berl.*), **64**, 125–31.
Sjögren, P. (1991). Extinction and isolation gradients in metapopulations: the case of the pool frog *Rana lessonae*. *Biol. J. Linn. Soc.*, **42**, 135–47.
Sjörs, H. (1948). Myrvegetation i Bergslagen. *Acta Phytogeogr. Suec.*, **21**, 1–299.
Sjörs, H. (1950). Regional studies in North Swedish mire vegetation. *Bot. Not.*, **1950**, 173–222.
Sjörs, H. (1983). Mires of Sweden. In *Ecosystems of the World, 4B. Mires: Swamp, Bog, Fen and Moor*, ed. A.J.P. Gore. Elsevier Scientific Publishers, Amsterdam, Oxford, New York, pp. 69–94.
Sjörs, H. & Backéus, I. (1977). Myrinventering för skyddsändamål. Myrinventering: Örebro län. *Medd. Växtbiol. Inst. Uppsala*, 1979(9).

SOU (1978). *Skog för framtid. Betänkande av 1973 Års Skogsutredning.* Liber, Stockholm.

Speight, M.C.D. & Blackith, R.E. (1983). The animals. In *Ecosystems of the World*, **4**A, *Mires: Swamp, Bog, Fen and Moor*, ed. A.J.P. Gore. Elsevier Scientific Publishers, Amsterdam, Oxford, New York, pp 349–65.

Svensson, S.A., Toet, H. & Kempe, G. (1989). The Swedish National Forest Inventory 1978–82. State of forests, growth and annual cut. *Swed. Univ. Agric. Sci., Dep. For. Surv., Rep.* No 47, 1–236 (in Swedish with English summary).

Tast, J. (1968). Influence of the root vole, *Microtus oeconomus* (Pallas), upon the habitat selection of the field vole, *Microtus agrestis* (L.), in northern Finland. *Ann. Acad. Sci. Fennici Ser. A*, IV Biol., **136**, 23 pp.

Tjernberg, M. (1985). The southern dunlin *Calidris alpina schinzii* in Sweden—history, present occurrence, breeding biology and suggestions for conservation. *Swedish Environment Protection Board*, PM No. 1928 (in Swedish with English summary).

Tolonen, K. (1983). The post-glacial fire record. In *The Role of Fire in Northern Circumpolar Ecosystems*, ed. R.W. Wein & D.A. MacLean. *SCOPE*, **18**, 21–44.

Tolonen, M. (1985). Palaeoecological record of local fire history from a peat deposit in SW Finland. *Ann. Bot. Fennici*, **22**, 15–29.

Vasari, Y. (1988). The role of peatlands and flooded meadows in the economic history of Kuusamo. *Oulanka Rep.*, **8**, 96–102.

Veirulf, O. (1937). Skogarnas utnyttjande i Älvdalen före storskiftet. *Geographica*, Uppsala, No. 5.

Wegge, P., Storaas, T., Larsen, B.B., Bö, T. & Kolstad, M. (1982). Woodland grouse and modern forestry in Norway. In *Proc. Int. Grouse Symp., 2nd, Edinburgh, Scotland, March 1981*, ed. T.W. Lovel, pp. 117–30. World Pheasant Association, Suffolk, England.

White, T.C.R. (1984). The abundance of invertebrate herbivores in relation to availability of nitrogen in stressed food plants. *Oecologia* (*Berl.*), **63**, 90–105.

Zackrisson, O. (1977). Influence of forest fires in the North Swedish boreal forest. *Oikos*, **29**, 22–32.

Chapter 9

Conservation Management of Riparian Communities

CHRISTER NILSSON
Department of Ecological Botany, University of Umeå, S-901 87 Umeå, Sweden

9.1 INTRODUCTION

To conserve the current state of a system, or to change it in a desired direction, we need the ability to predict spatial and temporal community patterns (e.g. Grime, 1980; Peters, 1980). There are still multitudes of relationships that are insufficiently known and so far unpredictable, making it difficult to design proper management plans. Riparian communities are typical examples of inadequately known systems, partly because of their ecological complexity—these systems consist of numerous inter-related hydrological, sedimentological, morphological, and biological components. They are also situated between two major habitats, water and land. This intermediate position has tended to place riparian ecology within both limnology and terrestrial ecology.

The objective of this chapter is to propose some general principles for conservation management of natural riparian communities. Special attention is given to the remnants of natural riverbank and lakeshore communities in the forest landscape in Scandinavia. Natural riparian systems comprise the most diverse, dynamic and complex habitats in northern regions. Although the majority of such habitats have already been modified or destroyed, most of the remnants are still more or less threatened by human intervention, especially the never-failing interest in hydroelectric development.

River ecology has recently taken riparian ecology under its wing as it has progressed from a pure aquatic science, influenced by the early lake ecology, to a holistic (ecosystem) science emphasizing the importance of the terrestrial–riparian–aquatic linkages (Odum, 1971; Hynes, 1975; Cummins *et al.*, 1984; Décamps, 1984; Sedell & Froggatt, 1984; Salo *et al.*, 1986; Petersen *et al.*, 1987; Décamps *et al.*, 1988; Gurtz *et al.*, 1988; Chauvet & Décamps, 1989; Ward, 1989). This has led to a rapidly increasing interest in riparian ecology (Petts, 1984; di Castri *et al.*, 1988; Naiman & Décamps, 1990). Riparian habitats are now being seen both as systems with unique features, and as mediators of ecological communication going either from terrestrial to aquatic habitats or vice versa. This mediating role contributes to the complexity of riparian habitats, and further understanding of riparian ecology should involve an integration of theory developed for both kinds of communities.

Other recent advances in river ecology have also been mainly at the ecosystem level of organization (see, for example, Barnes & Minshall, 1983; Minshall *et al.*, 1985). Two major advances are known as the 'nutrient spiralling concept' (Wallace *et al.*, 1977; Webster & Patten, 1979), and the 'river continuum concept' (Vannote *et al.*, 1980). To summarize briefly the new ideas, rivers are now viewed as continuous, throughflow systems, with terrestrial–aquatic interplay taking place all along their lengths (Naiman *et al.*, 1988*a*; Ward, 1989). It is a tacit understanding in this approach that impact in one riverine habitat may lead to consequences in some faraway habitat of a completely different kind. River communities are also said to show predictable downstream change in various structural and functional properties, including biotic diversity, soluble organic matter, and the gross photosynthesis/respiration ratio (Vannote *et al.*, 1980). This implies that the response of a community to a specific perturbation will be different at different sites along a river, ideas summarized as the 'serial discontinuity concept' (Ward & Stanford, 1983).

9.2 THE SYSTEM

In a narrow sense, there are no natural examples of river banks and lake shores because all of them have been subjected to as least some kind of human perturbation. In other words, the remainder is near-natural (*sensu* Westhoff, 1970) rather than natural. However, for simplicity's sake, I use the term 'natural' in a broad sense, referring to both natural and near-natural habitats.

A list of residual riparian communities would include no examples along the most strongly regulated rivers, like many in Sweden. These rivers have been rebuilt to chains of reservoirs permanently subjected to completely artificial water-level fluctuations. This kind of human impact is too destructive to allow persistence of natural communities (Fig. 9.1).

Fig. 9.1. Riparian zone along a storage reservoir in the upper Ume River. This area was formerly forested, but is now barren land after having been dammed and incorporated into a reservoir. It is exposed to a reversed water-level fluctuation, with a high water period in late summer and autumn, instead of in spring. (Photo C. Nilsson.)

In contrast, slightly regulated rivers, with water still flowing in the old channel, and with near-natural water-level fluctuations operating over the original riparian zone, may maintain an almost natural riparian biotic structure, at least for some decades. Formerly, other perturbations by humans, including hay-making, cattle grazing, and timber floating, could also be carried out without causing profound change of riparian community structure. These are no longer in practice, and the affected habitats are returning to more initial stages. However, sometimes even those activities led to severe damage. For example, in connection with timber floating, some riparian habitats were laid dry because of channel operations, whereas others were covered by stone barriers to prevent logs from stranding. Today, hydroelectric schemes, clear-felling of trees, roads, fast boat traffic in narrow stretches of river, summer cottages, etc., all form examples of threats to natural riparian systems.

Most of the remnants of riparian habitats in the Northern Hemisphere are situated in the far north, and range from the scale of entire river systems to tributaries and headwaters in otherwise heavily exploited rivers. For example, the largest unregulated and unpolluted river system in Europe is the Torne and Kalix Rivers, with their natural cross-channel the Tärendö River (personal observations) (Fig. 9.2).

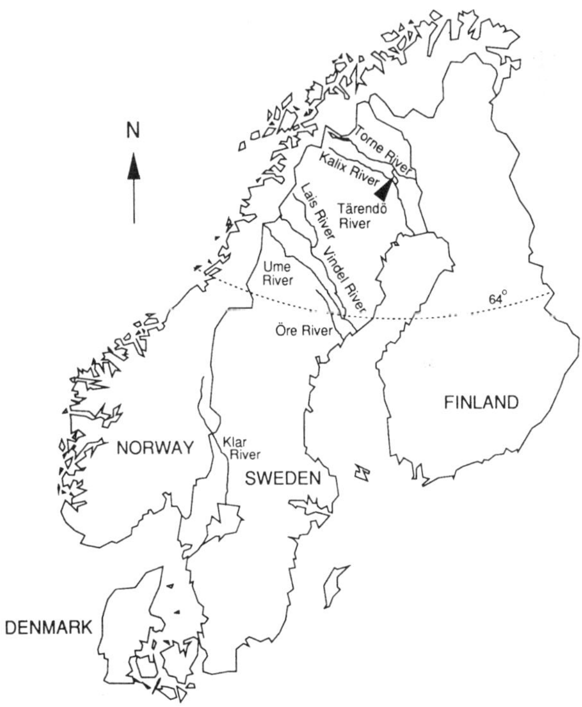

Fig. 9.2. Location of the rivers mentioned in the text.

The operative factor in riparian habitats is flooding. Natural river banks and lake shores in the Northern Hemisphere have large seasonal variation in water level. In this context the northern rivers are similar to rivers in dry regions, but the seasonality in the north is caused by thermal seasons, not so much by alternating wet and dry seasons. In northern Scandinavia, the water level is at its lowest in late winter, rapidly increases to its annual peak following snowmelt, and thereafter decreases, although with much fluctuation, during the rest of the year. The vertical range of annual water-level fluctuations exceeds 4 m in some rivers, at a few sites even more. When rivers flow through flat regions, riparian habitats can cover large tracts or entire landscapes, usually exhibiting extremely diverse and dynamic plant and animal communities (Fig. 9.3).

Fig. 9.3. A wide riparian zone along the lower Vindel River during spring flood. This area was formerly cleared of woody plants and used for hay-making, but is now being recolonized by trees and shrubs, especially *Pinus sylvestris* and *Betula pubescens* and, towards the channel, *Salix myrsinifolia.* During peak flood conditions, the entire ground shown on the photograph is covered by water. (Photo C. Nilsson.)

9.3 PATTERN AND PROCESS IN RIPARIAN HABITATS

9.3.1 Species composition

Now that we have established some abiotic characteristics of riparian systems, we need to know what their communities look like, how many species live there, and how riparian communities interact with adjacent communities. In other words, are there predictable patterns and processes in riparian ecosystems? The general vegetation structure of riparian communities is a vertical zonation from forest to shrub to herbaceous

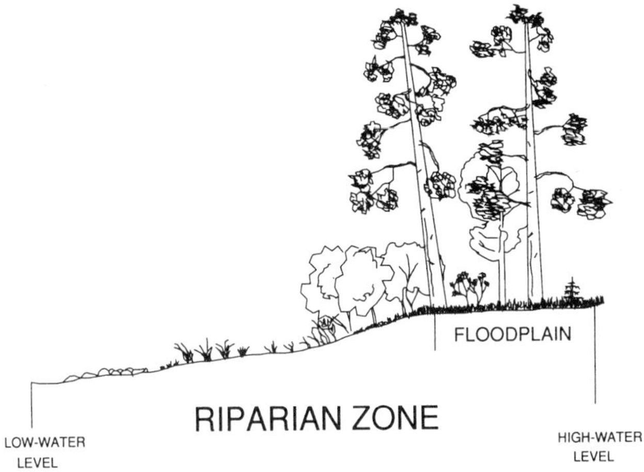

Fig. 9.4. Cross-section of a riparian zone along a river in northern Sweden showing the general vegetation structure.

communities (Wassén, 1966; Nilsson, 1983; Hupp, 1986; Shipley & Keddy, 1987) (Fig. 9.4). For example, an average river bank in northern Sweden may show the following sequence of dominants (from top to bottom): *Pinus sylvestris*, *Vaccinium vitis-idaea*, *Alnus incana*, *Calluna vulgaris*, *Molinia caerulea*, *Salix lapponum*, *Carex juncella* and *Ranunculus reptans*.

Although the loss of riparian habitat has implied a great reduction in variety, the remnants of natural river banks and lake shores still include many different types of communities. These range from the cryptogamic communities on rocky outcrops along waterfalls, to the patchy, often luxuriant and species-rich vegetation developed on boulder fields along riffles, to the dense, productive forests and meadows on fine sediments along slow glides.

The species composition and the general structure of these communities usually contrast distinctly with those adjacent, making riparian habitats into natural islands in the landscape. For example, in morainic, well-drained areas in the boreal forest landscape, the riparian forest may stand out as a dense, green strip, dominated by deciduous trees and shrubs, and with a ground vegetation of various herbs (Fig. 9.5). In contrast, the adjacent forest consists of pine trees with a low ground vegetation of lichens and dwarf shrubs. In areas composed of fine sediments the contrast between upland and riparian habitats is usually

Fig. 9.5. A pristine riparian forest, rich in herbs and deciduous trees, along a small stream in the coniferous forest landscape in northern Sweden. The photograph was taken during a flood. (Photo H. Sjörs.)

smaller in terms of general greenness, although species composition differs.

Species composition of riparian communities varies widely among different habitats, depending on variation in important environmental variables such as current velocity, wave action, and soil type. Variation occurs from the scale of local patchiness to one of entire rivers. For example, species composition changes continuously downstream. This succession is consistent with the general biogeographic pattern, but can show much irregularity. Along the Vindel River in northern Sweden (Fig. 9.2), which has a distinct series of alternating slow glides and riffles, alpine communities extend downstream via the riffles, while coastland communities extend upstream via the slow glides; the result is a discontinuous distribution of community types (Nilsson, 1986). A similar pattern can be expected also for communities of riparian beetles Coleoptera, because of their often well-developed substrate specificity (e.g. Nilsson & Lundberg, 1985).

9.3.2 Species richness

The riparian community has long been recognized for its high species richness. However, the most obvious question—How many species are there?—still awaits an answer. So far, there is no determination for any point along a river of the total number of species occurring in riparian habitats. This is remarkable, considering the great amount of such habitat that has been destroyed in recent years. The situation is best known for vascular plants, the structural components of the riparian community. Hence, along a 200 m-long stretch of river bank, comprising the area between the summer low-water and the spring high-water levels, more than 100 species of vascular plants can often be found (Nilsson *et al.*, 1988). The Swedish record for vascular plants, 131 species/200 m of river bank, is held by a section of the middle Vindel River (Nilsson & Ekblad, 1990). In comparison, the poorest record, 9 species/200 m, arises from a strongly regulated section (a storage reservoir) of the concurrent Ume River (Nilsson & Ekblad, 1990). Other species groups are likely to show even higher numbers. For example, in an investigation of riparian beetles along the Öre River (Fig. 9.2), Nilsson and Lundberg (1985) found 264 species on a river-bank locality estimated to be about 200 m in length (O. Atlegrim, pers. comm.).

On the scale of a large, seventh-order (Strahler, 1957), unregulated Swedish river like the Vindel, the number of vascular plants in the riparian zone is certainly more than 300 species. Nilsson (1983) reported 252 species from a sampling of the Lais-Vindel River (Fig. 9.2), and Nilsson *et al.* (in press *a*) found 260 species in a subsequent sampling of the entire Vindel River—about 13% of the entire Swedish flora of vascular plants (Lid, 1974). Some species groups may show an even higher representation in the riparian zone. For example, Palm and Lindroth (1936) found 981 species of beetles in the riparian zone along the Klar River (Fig. 9.2) before it was dammed. This represents 23% of the Swedish coleopteran fauna (Chinery, 1976). If the same proportion holds for other species groups, then the total number of species that occur, at least temporarily, on the bank of the Vindel River would range between 6000 and 11 000 species, most of them insects (calculation based on Nannfeldt, 1960).

An analysis of how all these species are distributed along the river would substantially improve our insight into the ecology of riparian systems, and provide better instruments for nature conservation. For example, in a current project we study similarities between the riparian floras of vascular plants in a large river and its tributaries. Such knowledge is fundamental for assessing effects of environmental impact in various parts of a watershed, and also provides means for a better understanding of the immigration history of species and of their present dispersal patterns.

A related question considers which factors govern the variation in species richness in riparian habitats. Knowledge within this field is rapidly increasing. For example, at a smaller scale, plant species richness varies with current velocity (Nilsson, 1987), standing crop and litter (Day *et al.*, 1988), litter deposition (Nilsson & Grelsson, 1990), wave exposure (Keddy, 1985), substrate heterogeneity (Nilsson *et al.*, 1989), and the duration of flooding (Bell, 1974; Bell & del Moral, 1977; Keddy, 1983). In most cases, species richness varies quadratically along these gradients, i.e. it is low at the ends of each gradient, and high in the middle.

A quadratic pattern of species richness has also been found on the scale of entire rivers. For example, Nilsson *et al.* (1989) found such a pattern for natural plant species along the river banks of the Torne and Kalix Rivers (Fig. 9.6). In contrast, species richness of ruderal plants increased towards the coast in both rivers (Fig. 9.6). While the latter pattern may reflect the degree to which the surroundings are cultivated, there are no obvious mechanisms producing the downstream pattern of natural species richness (Nilsson *et al.*, 1989). Recent results, demonstrating similar patterns also in other rivers, suggest that the post-glacial history might be an important proximate factor (Nilsson *et al.*, in press *b*).

The consistency in species richness patterns along different gradients still waits for a general explanation. Attempts to reach an understanding of the ultimate factors governing species coexistence should consider whether all gradients are in fact gradients of disturbance or productivity. Species richness has been suggested to vary quadratically along all kinds of such gradients (Connell, 1978; Huston, 1979; Sousa, 1979; Tilman, 1988).

Another question currently brought into focus is how important downstream dispersal of plant propagules is for the dynamics of riparian vegetation. Although flowing water has long been known as an important agent of plant dispersal (Guppy, 1891/93; Sernander, 1901; Ridley, 1930), the ecological implications of dispersal along water courses have never been evaluated. For example, we want to know whether input of drifted seeds occurs in a regular manner between years, and if species richness on various parts of a river bank correlates with species richness of drifted seeds.

A brief review of species richness patterns in natural riparian habitats

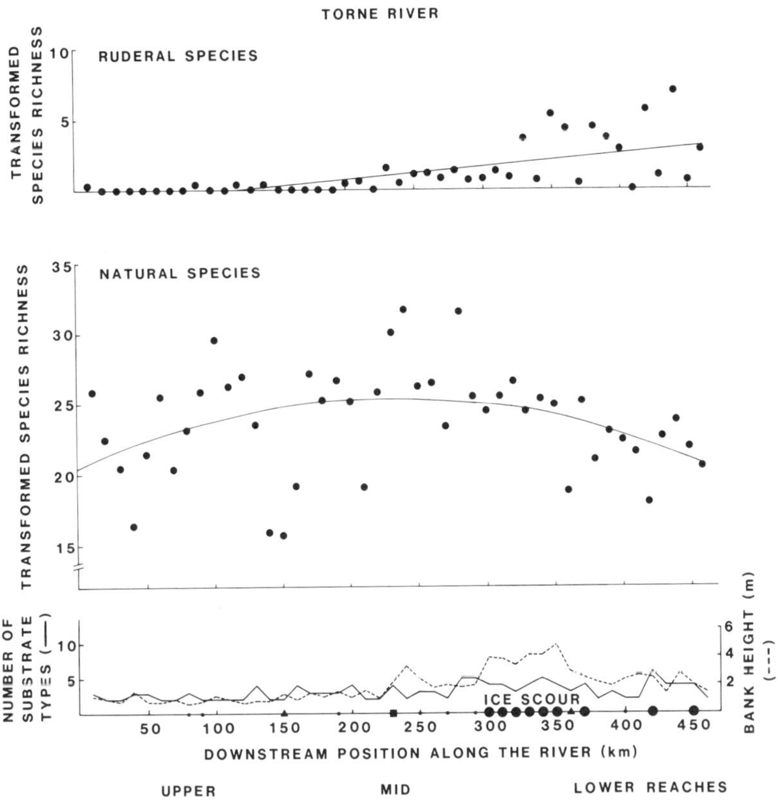

Fig. 9.6. Species richness and physical characteristics of the banks along the Torne River in northern Sweden. Ruderal species richness increases monotonically with position downstream, whereas natural species richness varies quadratically. Both regression lines are significant at $p<0{\cdot}001$. The degree of ice scour is given as ● = 1–25%, ▲ = 26–50%, ■ = 51–75%, ⬤ = 76–100%. From Nilsson *et al.* (1989), with permission of the Ecological Society of America.

would be incomplete without asking how regulation of water flow and water level affects these patterns. Although the average situation is not as bad as the above-mentioned values (131 versus 9 species/200 m of river bank) might suggest, local species richness is considerably lower along regulated rivers. In a comparison of two concurrent rivers Nilsson *et al.* (in press *a*) concluded that local species richness in the riparian zone along the regulated Ume River was only 72% of that in the natural

Vindel River. Moreover, vegetation cover in the regulated riparian zone was only 16% of that in the natural riparian zone (Nilsson *et al.*, in press *a*).

9.3.3 Interactions with adjacent habitats

A summary of interactions provided by the riparian vegetation is given in Table 9.1. Although riparian vegetation naturally interacts both with the terrestrial and the aquatic environments, the riparian–aquatic interactions are so far the most studied (Swanson *et al.*, 1982). For example, riparian vegetation supplies organic matter that sustains aquatic animal communities. Small and mid-sized water courses receive their major energy from riparian leaf litter, either by leaves falling directly into the water, or by decomposed litter washed out during floods (Fisher & Likens, 1973; Minshall *et al.*, 1983; Naiman, 1983; and others). The riparian zone also provides food and spawning ground for fishes, and for many invertebrates moving upwards from aquatic areas (Olsson & Söderström, 1978). Another, quite recently discovered, role of riparian habitats is their ability to remove nutrients from effluents. This filter function is especially important for nitrogen cycling because of the high potential for denitrification (Lowrance *et al.*, 1984; Pinay & Décamps, 1988). Riparian vegetation also stabilizes soil and reduces current velocity during floods, thereby reducing erosion and the risk of flooding of downstream areas.

A nice example of terrestrial–riparian interactions was given by Skoog (1975), who tracked animals in wintertime in northern Sweden and found an abundance of various tracks in the riparian zone along rivers, but much less in the surrounding terrestrial habitats. The species list included both herbivores—moose, hare, and grouse—and carnivores—fox, ermine and weasel. Skoog (1975) considered that the richness of animals in the riparian zone was due to the abundance of food, especially willows. The observed high primary productivity in riparian areas is generally believed to depend on seasonal flooding and the deposition of sediments, although any separation and quantification of the factors involved have not been made. Skoog (1975) and later Harris (1984) also pointed at the importance of riparian corridors as communication routes. This role is especially important during winter, because the riparian zone usually has a thinner snow cover than the surrounding terrestrial land. Knopf (1986) reported that some woodland birds have recently expanded their ranges across the North American prairie, probably by moving along riparian zones. The importance of a continuous ribbon of riparian forest has also

Table 9.1
Function of riparian vegetation with respect to aquatic and terrestrial ecosystems (modified from Swanson *et al.*, 1982)

Site	*Component*	*Function*
Aboveground/ above channel	Canopy and stems	1. Shade controls temperature and in stream primary production 2. Source of large and fine plant detritus 3. Source of invertebrates falling down 4. Provide food, habitat, and building material (beaver) 5. Provide communication routes, especially for vertebrates (riparian corridor)
In channel	Large debris derived from riparian vegetation	1. Controls routing of water and sediment 2. Shapes habitat and provide cover 3. Substrate for biological activity
Riparian zone (incl. floodplain)	Roots	1. Increase bank stability 2. Create overhanging banks which provide cover 3. Nutrient uptake from ground and streamwater
	Stems and low-lying canopy	1. Retard movement of sediment, water and organic debris in flood flows
	Ground vegetation	1. Spawning ground for fish and for insects moving up from channel 2. Source of detritus 3. Retards nutrients 4. Reduces runoff through evapotranspiration

been discussed by Décamps *et al.* (1987), who found that species richness of birds decreased following fragmentation of the riparian corridor.

Beaver is taken as a final example within this section. It operates over the entire range of aquatic, riparian, and terrestrial areas, and is probably the species next to man responsible for the most profound effects in riverine habitats. Beaver cut wood and build dams, and thereby not only modify stream morphology, hydrology and associated factors, but also change the structure and productivity of riparian plant communities (review in Naiman *et al.*, 1988*b*). They increase landscape heterogeneity by increasing the frequency of different aquatic habitats and wetland habitats. Beaver impoundments are important wetland habitats and have higher avian diversity values than adjacent upland areas (Hair *et al.*, 1979). Beaver also produce resources like dead, standing wood, thereby supplying habitat for wood-inhabiting invertebrates and hole-nesting birds. The fact that habitats produced and maintained by beaver belong to the most productive adds another dimension to their role in landscape ecology. The time scale is fascinating. Recent research (Naiman *et al.*, 1988*b*) postulates that their activity affects the landscape for centuries even after they have abandoned an area. In Sweden they were driven to extinction in the late 19th century, but are now successfully reintroduced and distributed over large areas (Ahlén, 1977). The implications for landscape and riparian ecology of the reduced, interrupted, and rapidly increasing occurrence of beaver are not yet evaluated.

9.4 CONSERVATION

9.4.1 Introduction

An important step towards the progress of conservation management of riparian ecosystems was taken with two American symposia, held in Tucson, Arizona, in 1977, and in Callaway Gardens, Georgia, in 1978. The proceedings of these symposia were entitled 'Importance, preservation and management of riparian habitat: a symposium' (Johnson & Jones, 1977), and 'Strategies for protection and management of floodplain wetlands and other riparian ecosystems' (Johnson & McCormick, 1979). These symposia provided convincing evidence of the values and important roles of riparian ecosystems. The multitude of approaches taken to management of riparian systems highlights the critical need for multidisciplinary research programmes and for improved communication between scientists, decision-makers, and the public, in developing riparian management. International cooperation is frequently demanded since many rivers are shared by two or more states (IUCN, 1980).

The question of preserving natural ecosystems has often arisen when they have been threatened by some kind of destructive exploitation. This has happened to river ecosystems in Scandinavia which have experienced a 30-year period (1950–1980) of intensive hydroelectric development. Therefore, for several decades now, many ecologists have tried to prevent decision-makers from carrying out large hydroelectric schemes by pointing at the resulting losses of riparian diversity and other values. Unfortunately, in most cases the ecologists have fought a losing battle. The reasons for these casualties are beyond the scope of this discussion. However, the objections by ecologists have probably fostered a wider recognition that monetary value is of limited application as a measurement in assessing the overall value of a river. In any case, the debates on the most recently suggested hydroelectric projects have highlighted an improved public awareness of the serious loss of both genetic, species and, so far most of all, habitat diversity, which follows hydroelectric development. At present, only a few hydroelectric projects are being carried out in Scandinavia. This has provided an opportunity to change river management objectives from the 'fire-brigade strategy', involving sudden rescue operations for acutely threatened habitats, to a more detailed planning for long-term conservation of unregulated riverine habitats.

9.4.2 General recommendations

In general, management programmes may have two major objectives: to preserve natural systems intact, and to enhance the value of ecosystems that are already damaged. We must also observe that, despite being preserved, natural ecosystems can be inadvertently damaged by air pollution and the use of fertilizers and pesticides.

The need for management could be expected to increase with the intensity of human impact. If so, the most intensively regulated rivers, i.e. the reservoir chains, would need the most comprehensive management programmes. However, as long as the artificial water-level fluctuations are operating, reclamation measures along such rivers will contribute little to conservation (Hellawell, 1988). In contrast, most of the unregulated rivers, including their riparian zones, would benefit from properly designed management programmes.

The recent advances in river ecology have increased awareness of the far-reaching complexities of riverine systems, including the interactions between water and land (Ward, 1989). The nature of these interactions awaits much further analysis. For example, we do not know in any detail at which scales different interactions operate, what magnitudes they have, etc. Therefore, so far only very general management recommendations can be given. One recommendation—and perhaps the most important—is that management should not be restricted to instream (the continuously water-filled part of the channel) habitats and to specific localities, but should take both a river system and a watershed approach (e.g. Odum, 1971). In other words, there is no point (1) in conserving instream river habitats without having also conserved the adjacent riparian habitats (and vice versa); (2) in conserving a particular stretch of river without having conserved also the upstream sections of the same river; and (3) in conserving instream and riparian habitats without having conserved the vital processes in the watershed.

Preservation

In practice, the implementation of a management programme for an entire watershed would imply preserving, i.e. leaving intact, some especially valuable areas, while allowing controlled use of others (cf. Clark, 1979; Nilsson *et al.*, 1988; Keddy & Wisheu, 1989). The strategy for the location of reserves would be largely similar to the common procedure of monitoring, evaluating and preserving. There are many criteria for the evaluation of nature (Ratcliffe, 1977; Margules & Usher, 1981; Spellerberg, 1981; Smith & Theberge, 1986; Margules, 1989). So far, there is no consensus in how such criteria can be combined to generate the 'best' evaluation (Smith & Theberge, 1987). In general, however, selecting natural areas as large as possible, with a high diversity of species and habitats, or with many rare species present, is a good guiding principle.

While a system of riparian reserves can be established largely by applying general principles, a few specific features should also be considered. For example, the scale used (that of a watershed) is usually larger than the scale of most reserves. This automatically contributes to a large biotic diversity. Unless the entire area can be preserved in one large reserve, a network of reserves is necessary to accurately represent the entire range of biotic diversity within the watershed (Margules, 1989).

Conservation

We know from observational studies that even small changes in water-level regimes are almost certain to change riparian communities (e.g. Hall *et al.*, 1946; Menges & Waller, 1983; Hupp & Osterkamp, 1985; Menges, 1986; Nilsson & Keddy, 1988) and that naturally fluctuating water levels maintain high values of vegetation diversity and plant species diversity (Keddy & Reznicek, 1986). For example, periods of high water kill dominant species and open up new habitat patches, disperse floating seeds, and leave moist sediment for seedling establishment (Gill, 1970; Harms *et al.*, 1980; Sauer, 1988). In contrast, periods of low water allow many species to regenerate from buried seeds (Salisbury, 1921; van der Valk, 1981; Keddy & Reznicek, 1986; Welling *et al.*, 1988). Therefore, one of the most important purposes in controlling the use of the areas surrounding actual reserves should be to conserve run-off patterns and water-level fluctuations. This is applicable to both the instream, riparian, and watershed levels of river system management. To attain such a goal, the ability of the watershed to dampen large natural fluctuations in rainfall and snowmelt must not be reduced. Today, development is on the wrong track, and naturally regulated run-off patterns are virtually being deregulated. For example, extensive clear-felling of trees, peat cutting, big ditches, and a net-work of roads, all help to change run-off. Hence it follows that both the flow regime and the riparian processes will be affected.

On an instream level, the conservation of natural run-off patterns could involve a restoration of channel morphology in riffles which have been deepened and narrowed in order to facilitate timber floating. This would restore water levels and reduce current velocity along the actual riffles. In general, a reclamation of run-off patterns in the watershed and in the river would increase the retention time of water in the entire river, thus reducing the risk of floods during periods of heavy rains.

Human intervention might also affect water quality, both directly by domestic and commercial wastes, and indirectly by erosion of bare soil along roads and in ditches, and by leakage from clear-felled and peat cutting areas and areas where fertilizers and pesticides have been used. Such perturbations also modify riparian communities and should be controlled for in a management programme.

In conclusion, conservation management of natural riparian communities should take into account the recent recognition that land–water interactions are fundamental for the function of natural water courses. To be able to design detailed management programmes we need to know in more detail the nature of these interactions. However, the reader should note that, although the recommendations have so far been general, action should not be postponed, while awaiting more detail. The biotic diversity of river landscapes is continuously being reduced, and there is no time to wait for sufficient data (which we never will obtain) before acting. As

Soulé (1986) puts it, 'the risks of non-action may be greater than the risks of inappropriate action'. Therefore, the survival of natural riparian communities calls for an integration of future research and management, none of which can afford to wait for the other to develop.

ACKNOWLEDGEMENTS

I thank O. Atlegrim, G. Grelsson, L. Hansson, M. Johansson, P. Keddy, and T.-B. Larsson for comments on the manuscript. This research was funded by the Swedish Environmental Protection Agency.

REFERENCES

Ahlén, I. (1977). *Faunavård*. Liber, Stockholm.

Barnes, J.R. & Minshall, G.W. (ed.) (1983). *Stream Ecology. Application and Testing of General Ecological Theory*. Plenum, New York.

Bell, D.T. (1974). Tree stratum composition and distribution in the streamside forest. *Amer. Midl. Nat.*, **92**, 35–46.

Bell, D.T. & del Moral, R. (1977). Vegetation gradients in the streamside forest of Hickory Creek, Will County, Illinois. *Bull. Torrey bot. Club*, **104**, 127–35.

Chauvet, E. & Décamps, H. (1989). Lateral interactions in a fluvial landscape: the River Garonne, France. *J.N. Amer. benthol. Soc.*, **8**, 9–17.

Chinery, M. (1976). *Nordeuropas Insekter*. Bonniers, Stockholm.

Clark, J.R. (1979). Science and the conservation of riparian systems. *USDA For. Serv. gen. tech. Rep.*, **WO-12**, 13–9.

Connell, J.H. (1978). Diversity in tropical rain forests and coral reefs. *Science, N.Y.*, **199**, 1302–10.

Cummins, K.W., Minshall, G.W., Sedell, J.R., Cushing, C.E. & Petersen, R.C. (1984). Stream ecosystem theory. *Verh. int. Verein. theor. angew. Limnol.*, **22**, 1818–27.

Day, R.T., Keddy, P.A., McNeill, J. & Carleton, T. (1988). Fertility and disturbance gradients: a summary model for riverine marsh vegetation. *Ecology*, **69**, 1044–54.

Décamps, H. (1984). Towards a landscape ecology of river valleys. In *Trends in Ecological Research for the 1980s*, ed. J.H. Cooley & F.B. Golley. Plenum, New York, pp. 163–78.

Décamps, H., Joachim, J. & Lauga, J. (1987). The importance for birds of the riparian woodlands within the alluvial corridor of the River Garonne, SW France. *Regul. Rivers*, **1**, 301–16.

Décamps, H., Fortuné, M., Gazelle, F. & Pautou, G. (1988). Historical influence of man on the riparian dynamics of a fluvial landscape. *Landscape Ecol.*, **1**, 163–73.

di Castri, F., Hansen, A.J. & Holland, M.M. (eds) (1988). A new look at ecotones:

emerging international projects on landscape boundaries. *Biology Int.*, **17**, 1–163. International Univ. of Biological Sciences, Paris.
Fisher, S.G. & Likens, G.E. (1973). Energy flow in Bear Brook, New Hampshire: an integrative approach to stream ecosystem metabolism. *Ecol. Monogr.*, **43**, 421–39.
Gill, C.J. (1970). The flooding tolerance of woody species—a review. *For Abstr.*, **43**, 145–61.
Grime, J.P. (1980). An ecological approach to management. In *Amenity Grassland: an Ecological Perspective*, ed. I.H. Rorison & R. Hunt. Wiley, Chichester, pp. 13–35.
Guppy, H.B. (1891/93). The River Thames as an agent in plant dispersal. *J. Linn. Soc.*, **29**, 333–45.
Gurtz, M.E., Marzolf, G.R., Killingbeck, K.T., Smith, D.L. & McArthur, J.V. (1988). Hydrologic and riparian influences on the import and storage of coarse particulate organic matter in a prairie stream. *Can. J. Fish. aquat. Sci.*, **45**, 655–65.
Hair, J.D., Hepp, G.T., Luckett, L.M., Reese, K.P. & Woodward, D.K. (1979). Beaver pond ecosystems and their relationships to multi-use natural resource management. *USDA For. Serv. gen. tech. Rep.*, **WO-12**, 80–92.
Hall, T.F., Penfound, W.T. & Hess, A.D. (1946). Water level relationships of plants in the Tennessee Valley with particular reference to malaria control. *J. Tenn. Acad. Sci.*, **21**, 18–59.
Harms, W.R., Schreuder, H.T., Hook, D.D., Brown, C.L. & Shropshire, F.W. (1980). The effects of flooding on the swamp forest in Lake Ocklawaha, Florida. *Ecology*, **61**, 1412–21.
Harris, L.D. (1984). *The Fragmented Forest. Island Biogeographic Theory and the Preservation of Biotic Diversity.* University of Chicago Press, Chicago, Illinois.
Hellawell, J.M. (1988). River regulation and nature conservation. *Regul. Rivers*, **2**, 425–43.
Hupp, C.R. (1986). Upstream variation in bottomland vegetation patterns, northwestern Virginia. *Bull. Torrey bot. Club*, **113**, 421–30.
Hupp, C.R. & Osterkamp, W.R. (1985). Bottomland vegetation distribution along Passage Creek, Virginia, in relation to fluvial landforms. *Ecology*, **66**, 670–81.
Huston, M. (1979). A general hypothesis of species diversity. *Amer. Nat.*, **113**, 81–101.
Hynes, H.B.N. (1975). The stream and its valley. *Verh. int. Verein. theor. angew. Limnol.*, **19**, 1–15.
IUCN (1980). *World Conservation Strategy.* International Union for Conservation of Nature and Natural Resources (IUCN), Gland.
Johnson, R.R. & Jones, D.A. (ed.) (1977). Importance, preservation and management of riparian habitat: a symposium. *USDA For. Serv. gen. tech. Rep.*, **RM-43**.
Johnson, R.R. & McCormick, J.F. (ed.). (1979). Strategies for protection and management of floodplain wetlands and other riparian ecosystems. *USDA For. Serv. gen. tech. Rep.*, **WO-12**.
Keddy, P.A. (1983). Shoreline vegetation in Axe Lake, Ontario: effects of exposure on zonation patterns. *Ecology*, **64**, 331–44.

Keddy, P.A. (1985). Wave disturbance on lakeshores and the within-lake distribution of Ontario's Atlantic coastal plain flora. *Can. J. Bot.*, **63**, 656–60.

Keddy, P.A. & Reznicek, A.A. (1986). Great Lakes vegetation dynamics: the role of fluctuating water levels and buried seeds. *J. Great Lakes Res.*, **12**, 25–36.

Keddy, P.A. & Wisheu, I.C. (1989). Ecology, biogeography, and conservation of coastal plain plants: some general principles from the study of Nova Scotian wetlands. *Rhodora*, **91**, 72–94.

Knopf, F.L. (1986). Changing landscapes and the cosmopolitism of the eastern Colorado avifauna. *Wildl. Soc. Bull.*, **14**, 132–42.

Lid, J. (1974). *Norsk og Svensk Flora*, 2nd edn. Norske Samlaget, Oslo.

Lowrance, R., Todd, R., Fail, J., Hendrickson, O., Leonard, R. & Asmussen, L. (1984). Riparian forests as nutrient filters in agricultural watersheds. *BioScience*, **34**, 374–77.

Margules C. (ed.). (1989). Australian developments in conservation evaluation. *Biol. Conserv.* (*Special Issue*), **50**, 1–303.

Margules, C. & Usher, M.B. (1981). Criteria used in assessing wildlife conservation potential: a review. *Biol. Conserv.*, **21**, 79–109.

Menges, E.S. (1986). Environmental correlates of herb species composition in five southern Wisconsin floodplain forests. *Amer. Midl. Nat.*, **115**, 106–17.

Menges, E.S. & Waller, D.M. (1983). Plant strategies in relation to elevation and light in floodplain herbs. *Amer. Nat.*, **122**, 454–73.

Minshall, G.W., Petersen, R.C., Cummins, K.W., Bott, T.L., Sedell, J. R., Cushing, C.E. & Vannote, R.L. (1983). Interbiome comparison of stream ecosystem dynamics. *Ecol. Monogr.*, **51**, 1–25.

Minshall, G.W., Cummins, K.W., Petersen, R.C., Cushing, C.E., Bruns, D.A., Sedell, J.R. & Vannote, R.L. (1985). Developments in stream ecosystem theory. *Can. J. Fish. aquat. Sci.*, **42**, 1045–55.

Naiman, R.J. (1983). The annual pattern and spatial distribution of aquatic oxygen metabolism is boreal forest watersheds. *Ecol. Monogr.*, **53**, 73–94.

Naiman, R.J. & Décamps, H. (ed.) (1990). *The Ecology and Management of Aquatic–Terrestrial Ecotones.* UNESCO, Paris.

Naiman, R.J., Décamps, H., Pastor, J. & Johnston, C.A. (1988*a*). The potential importance of boundaries to fluvial ecosystems. *J.N. Amer. benthol. Soc.*, **7**, 289–306.

Naiman, R.J., Johnston, C.A. & Kelley, J.C. (1988*b*). Alteration of North American streams by beaver. *BioScience*, **38**, 753–62.

Nannfeldt, J.A. (1960). The mycofloristical exploration of Scandinavia, especially Sweden. *Friesia*, **6**, 167–213.

Nilsson, C. (1983). Frequency distributions of vascular plants in the geolittoral vegetation along two rivers in northern Sweden. *J. Biogeogr.*, **10**, 351–69.

Nilsson, C. (1986). Change in riparian plant community composition along two rivers in northern Sweden. *Can. J. Bot.*, **64**, 589–92.

Nilsson, C. (1987). Distribution of stream-edge vegetation along a gradient of current velocity. *J. Ecol.*, **75**, 513–22.

Nilsson C. & Ekblad, A. (1990). Utbyggd älv får kala stränder. *Svet. Nat.*, **81** (4), 12–16.

Nilsson, C. & Grelsson, G. (1990). The effects of litter displacement on riverbank vegetation. *Can. J. Bot.*, **68**, 735–41.
Nilsson, C. & Keddy, P.A. (1988). Predictability of change in shoreline vegetation in a hydroelectric reservoir, northern Sweden. *Can. J. Fish. aquat. Sci.*, **45**, 1896–904.
Nilsson, C., Grelsson, G., Johansson, M. & Sperens, U. (1988). Can rarity and diversity be predicted in vegetation along river banks? *Biol. Conserv.*, **44**, 201–12.
Nilsson, C., Grelsson, G., Johansson, M. & Sperens, U. (1989). Patterns of plant species richness along riverbanks. *Ecology*, **70**, 77–84.
Nilsson, C., Ekblad, A., Gardfjell, M. & Carlberg, B. (in press *a*). Long-term effects of river regulation on river margin vegetation. *J. appl. Ecol.*
Nilsson, C., Grelsson, G., Dynesius, M., Johansson, M.E. & Sperens, U. (in press *b*). Small rivers behave like large rivers: effects of postglacial history on plant species richness along riverbanks *J. Biogeogr.*
Nilsson, O. & Lundberg, S. (1985). Strandlevande skalbaggar vid Öre älv. *Entomol. Tidskr.*, **106**, 27–37.
Odum, E.P. (1971). *Fundamentals of Ecology*, 3rd edn. Saunders, Philadelphia, Pennsylvania.
Olsson, T. & Söderström, O. (1978). Springtime migration and growth of *Parameletus chelifer* (Ephemeroptera) in a temporary stream in northern Sweden. *Oikos*, **31**, 284–9.
Palm, T. & Lindroth, C.H. (1936). Coleopterfaunan vid Klarälven, I. Allmän del. *Ark. Zool.*, **28A**(19), 1–42.
Peters, R.H. (1980). Useful concepts for predictive ecology. In *Conceptual Issues in Ecology*, ed. E. Saarinen. Reidel Dordrecht, pp. 215–27.
Petersen, R.C., Madsen, B.L., Wilzbach, M.A., Magadza, C.H.D., Paarlberg, A., Kullberg, A. & Cummins, K.W. (1987). Stream management: emerging global similarities. *Ambio*, **16**, 166–79.
Petts, G.E. (1984). *Impounded Rivers. Perspectives for Ecological Management.* Wiley, Chichester.
Pinay, G. & Décamps, H. (1988). The role of riparian woods in regulating nitrogen fluxes between the alluvial aquifer and surface water: a conceptual model. *Regul. Rivers*, **2**, 507–16.
Ratcliffe, D. (ed.) (1977). *A Nature Conservation Review, Vol. 1.* Cambridge University Press, Cambridge.
Ridley, H.N. (1930). *The Dispersal of Plants Throughout the World.* Reeve & Company, Ashford.
Salisbury, E.J. (1921). Vegetation of drying mud and retarded germination. *Naturalist, Hull*, **1921**, 329–32, 365–6.
Salo, J., Kalliola, R., Häkkinen, I., Mäkinen, Y., Niemelä, P., Puhakka, M. & Coley, P.D. (1986). River dynamics and the diversity of Amazon lowland forest. *Nature, Lond.*, **322**, 254–8.
Sauer, J.D. (1988). *Plant Migration. The Dynamics of Geographic Patterning in Seed Plant Species.* University of California Press, Berkeley, California.
Sedell, J.R. & Froggatt, J.L. (1984). Importance of streamside forests to large rivers: the isolation of the Willamette River, Oregon, USA, from its flood-

plain by snagging and streamside forest removal. *Verh. int. Verein. theor. angew. Limnol.*, **22**, 1824–34.
Sernander, R. (1901). *Den Skandinaviska Vegetationens Spridningsbiologi.* Lundequistska Bokhandeln, Uppsala.
Shipley, B. & Keddy, P.A. (1987). The individualistic and community-unit concepts as falsifiable hypotheses. *Vegetatio*, **69**, 47–55.
Skoog, P. (1975). The watercourses in northern Norrland, zoological conservation values. Department of Zoology, University of Stockholm, Stockholm. Unpublished report, in Swedish.
Smith, P.G.R. & Theberge, J.B. (1986). A review of criteria for evaluating natural areas. *Environ. Manage.*, **10**, 715–34.
Smith, P.G.R. & Theberge, J.B. (1987). Evaluating natural areas using multiple criteria: theory and practice. *Environ. Manage.*, **11**, 447–60.
Soulé, M.E. (1986). Conservation biology and the 'real world'. In *Conservation Biology. The Science of Scarcity and Diversity*, ed. M.E. Soulé. Sinauer Associates, Sunderland, Massachusetts, pp. 1–12.
Sousa, W.P. (1979). Disturbance in marine and intertidal boulder fields: the nonequilibrium maintenance of species diversity. *Ecology*, **60**, 1225–39.
Spellerberg, I.F. (1981). *Ecological Evaluation for Conservation.* Edward Arnold, London.
Strahler, A.N. (1957). Quantitative analysis of watershed geomorphology. *Trans. Amer. geophys. Un.*, **38**, 913–20.
Swanson, F.J., Gregory, S.V., Sedell, J.R. & Campbell, A.G. (1982). Land–water interactions: the riparian zone. In *Analysis of Coniferous Forest Ecosystems in the Western United States*, ed. R.L. Edmonds. *US/IBP Synthesis Ser.*, No. 14. Dowden, Hutchinson & Ross, Stroudsburg, Pennsylvania, pp. 267–91.
Tilman, D. (1988). *Plant Strategies and the Dynamics and Structure of Plant Communities.* Princeton University Press, Princeton, New Jersey.
van der Valk, A.G. (1981). Succession in wetlands: a Gleasonian approach. *Ecology*, **62**, 688–96.
Vannote, R.L., Minshall, G.W., Cummins, K.W., Sedell, J.R. & Cushing, C.E. (1980). The river continuum concept. *Can. J. Fish aquat. Sci.*, **7**, 130–7.
Wallace, J.B., Webster, J.R. & Woodall, W.R. (1977). The role of filter feeders in flowing waters. *Arch. Hydrobiol.*, **79**, 506–32.
Ward, J.V. (1989). The four-dimensional nature of lotic ecosystems. *J.N. Amer. benthol. Soc.*, **8**, 2–8.
Ward, J.V. & Stanford, J.A. (1983). The serial discontinuity concept of lotic ecosystems. In *Dynamics of Lotic Ecosystems*, ed. T.D. Fontaine & S.M. Bartell. Ann Arbor Science, Ann Arbor, Michigan, pp. 29–42.
Wassén, G. (1966). Gardiken. Vegetation und Flora eines lappländischen Seeufers. *K. Svenska Vetensk. Akad. Avh. Naturskydd.*, **22**.
Webster, J.R. & Patten, B.C. (1979). Effects of watershed perturbation on stream potassium and calcium dynamics. *Ecol. Monogr.*, **49**, 51–72.
Welling, C.H., Pederson, R.L. & van der Valk, A.G. (1988). Recruitment from the seed bank and the development of zonation of ermergent vegetation during a drawdown in a prairie wetland. *J. Ecol.*, **76**, 483–96.
Westhoff, V. (1970). New criteria for nature reserves. *New Scient.*, **46**, 108–13.

Chapter 10

Forests in the Temperate–boreal Transition—Natural and Man-made Features

SVEN G. NILSSON

Lund University, Department of Ecology, Animal Ecology, Ecology Building, S-223 63 Lund, Sweden

10.1 INTRODUCTION

To fit nature into categories is always difficult, especially with complex vegetation. In the northern hemisphere forest vegetation in the temperate zone usually consists of deciduous forests and in the boreal zone often of coniferous forests. In the transition zone there are forests with deciduous trees, coniferous trees and also very often deciduous and coniferous trees mixed (later in this chapter referred to as mixed forests). In the Nordic countries this transition is termed the 'hemiboreal' zone (Ahti *et al.*, 1968; see map in chapter 7). In Sweden this zone is about 600 km wide and in the western Soviet Union even wider. Thus, the hemiboreal zone covers large areas. In spite of this its main ecological features are less well understood than those of more southern and northern zones. This is due to intensive exploitation in historic times and the complex dynamics between deciduous and coniferous trees. The separation of natural and man-made factors affecting vegetation dynamics has been especially difficult. In this short presentation of the hemiboreal zone in Sweden the description of some important features and processes must be very brief due to space limitation. I present major landscape changes and their main causes over the last 1000 years. Keystone species, i.e. species whose presence affects the distribution and/or abundance of several other species or whose absence would cause cascading effects throughout a community, are sometimes treated in more detail. Examples of such species are major forest trees and shrubs, common vertebrate predators and herbivores, pollinators and seed dispensers.

In addition, little known features and some neglected processes and groups of organisms are treated in greater detail. Thus the contents of this review of forest in the hemiboreal zone outlines my personal view of what is important in conservation. When selecting references I have where possible mainly used studies performed in the hemiboreal zone, even if a more general process is involved. However, many important processes have not been studied systematically in this zone. Therefore, many statements in this chapter are hypotheses based on general observations but in need of testing.

10.2 LANDSCAPE CHANGES AND VEGETATION HISTORY

About 1000 years ago the impact of man on the south Swedish landscape was already high, especially along the sea, big lakes and water courses. Nevertheless, there were probably many large tracts with virgin high forest. These were considerably reduced over the following centuries. Man-made fires and grazing by domestic animals altered most of those remaining into grazing forests with scattered large trees and much young successional woody growth, usually dominated by birches *Betula* spp. However, stands with high forest dominated by oaks *Quercus* spp. and beech *Fagus sylvatica* were retained scattered over the landscape because acorns and mast were very important as food for domestic pigs. On large estates high forests were also protected as hunting grounds for the wealthy.

The grazing pressure of domestic animals increased to reach a peak around the middle of the 19th century. Sheep and goats were especially damaging and they apparently prevented the regeneration of many species of shrubs and trees over large tracts. This can be deduced from the present distribution of old individuals of these sensitive species which survive near the villages (personal observations). Here, the land was used for mowing winter fodder and for growing crops. On the meadows, which were mowed annually in July or August, domestic animals were only allowed to graze in autumn and sometimes for a short period in spring. Species such as lime *Tilia cordata*, ash *Fraxinus exelsior*, maple *Acer platanoides*, hornbeam *Carpinus betulus*, elm *Ulmus glabra* and hazel *Corylus avellana* were almost restricted to these meadows near villages. Another important factor shaping this distribution was man-made fires. With intervals of 20–30 years most forest land in the hemiboreal zone in Sweden seems to have been used for growing crops for one or a few years (e.g. Weimarck, 1979). To prepare the land, the forest, usually consisting of a young birch-dominated stand, was cut down and burned when the branches were dry. The fields were usually small (less than 1 ha) but numerous. After soil fertility was reduced by collecting the crops, the fields were used for grazing and then the forest was allowed to come back. This type of shifting agriculture reached its highest intensity in Sweden around 1800 but persisted in some districts until the turn of the present century (Larsson, 1989).

The fact that most of the existing old forests in the hemiboreal zone in south Sweden are dominated by spruce *Picea abies* and pine *Pinus sylvestris* is most likely due to these species' resistance to fire (pine) and heavy grazing (spruce). Birches, which a hundred years ago were equally important in old forests, have recently been removed by forest managers for pulp and firewood. More natural mixed and mature forests have mainly survived to the present time only on the very small fraction of land which could not be used for practical reasons—isolated islands, steep slopes and very stony ground. In some such places we can still find stands whose tree composition is probably as it would be over most of the land in the absence of management. The value of these stands for science and nature conservation is immense.

Over the last 150 years enormous and rapid changes have taken place in the south Swedish landscape. For plants and animals the most important are

(1) drainage of almost all fertile swamps and fens;
(2) reduction in numbers of very old, often hollow, trees;
(3) almost total elimination of the formerly extensive, very species-rich, mown meadows with scattered trees and shrubs;
(4) altered species composition of many high forests from deciduous and mixed deciduous/coniferous forests to coniferous forests—most rapid in the hemiboreal zone (Malmer, 1973);
(5) almost total elimination of forest fires (man-made and natural) and forest grazing by domestic animals;
(6) first increased, and then decreased, hunting pressure affecting especially big birds and mammals (see below).

Contrary to common opinion, each of these changes has up to the present time affected more plant and animal species than pesticides and industrial emissions. This is not to deny the importance of such recent threats, especially for the future, but to emphasize these six other revolutionary changes. We still are a long way from understanding in any detail the enormous consequences of these changes. This is so, mainly because of our present ignorance about the ecology of major groups of organisms, for example insects and fungi. Below I will summarize some of the effects of the changed landscape on forest trees, mammals, birds and some other organisms.

10.3 VEGETATION DYNAMICS

As indicated above, in the hemiboreal zone of Sweden man has been the most important factor affecting the vegetation (directly and indirectly by his domestic animals) over the last 1000 years. The relative importance of natural factors is therefore difficult to evaluate. Fires started by lightning were formerly perhaps of less importance than in more northern forests. It is likely that pine was favoured over spruce on dry and sandy ground, where fires were most frequent, since pine can often survive forest fires (e.g. Zackrisson, 1977). Some early-successional trees with small, wind-dispersed seeds, for example birch, aspen *Populus tremula* and sallow *Salix caprea*, are probably also favoured by forest fires because they can rapidly invade open ground. The natural fire-interval in the hemiboreal zone is unknown but is probably longer than the 20–30 year interval for shifting agriculture (see above). It is also likely that the areal extent of each natural forest fire was orders of magnitude higher than man-made fires in south Sweden. Still, many fire-dependent plants and insects (see chapter 7) probably survived in forests used for shifting agriculture.

Beech has a relatively thin bark and the species is therefore very sensitive to high temperatures. It can even die due to overheating if it is exposed after felling of surrounding trees (Nicolai, 1986). Beech is also very sensitive to forest fire and large trees can die even if the fire only burns off the dry leaves on the ground (personal observations). Fire is probably a particularly important reason both for the enormous decrease of the area of beech forests during recent centuries (e.g. Wibeck, 1909; Malmström, 1939) and the present distinct borders between beech-dominated and other forests. Following the recent suppression of fire frequency and elimination of forest grazing both beech and oak are dispersed by the jay *Garrulus glandarius* to old coniferous forests (Nilsson, 1985). This bird must be considered a keystone species in European forests because it seems to be the most important dispersal agent for both oaks *Quercus robur* and *Q. petraea* (Bossema, 1979), and beech (Nilsson, 1985).

On moist and wet grounds, windfalls followed by increased growth of suppressed saplings seems to be the dominating successional process in hemiboreal forests (Skoglund & Verwijst, 1989). This has been studied in detail in Poland by Falinski (1986), but his results are probably also valid further north in the hemiboreal zone. When there is little disturbance, mainly lime increases in importance due to its dense canopy and high age. It can also reproduce below a closed canopy, which is also true for ash, elm, maple, beech and spruce. In the absence of disturbance these late successional species would dominate, ash and elm on the most fertile soils, beech and spruce on poor soils and lime and maple on intermediate soils. Of course, there could be many local deviations due to slope, stoniness, aspect, etc.

The proper place for the oaks in this gradient is not clear. Due to their large size and high potential ages they can persist even in late successional stages, but fail to reproduce in many mature stands where they are at present dominant (e.g. Nilsson, 1985; Falinski, 1986). They seem to need higher light levels than in closed stands for establishment (e.g. Jarvis, 1964; Shaw, 1974). Thus, the oaks have features of both early- and late-successional tree species. They seem to need some type of disturbance (fires? windfalls?) to reproduce but may then persist for several hundred years.

Oaks grown up in open man-made grazing forests have long horizontal branches low down on the trunk. Such individuals cannot survive for long when spruce grow up among the branches. In contrast, oaks growing up in denser forests have few branches on the lower trunk and can probably withstand competition from spruce and beech better than open-stand oaks. This hypothesis has not been yet tested.

Mature riparian forests with a natural water regime are now very rare in southern Sweden. In such forests spruce is killed by high water levels and pines only occur as scattered individuals. Alder *Alnus glutinosa* often dominates, but birch *Betula pubescens* and aspen are also frequent. Oak is also regular on slightly drier ground. In these riparian forests insect and bird life is very rich, especially those species dependent on dead and dying trees (see below). Many threatened animal species would survive locally if riparian forests could be left for free development; if this happens they seem to reach a natural appearance comparatively quickly, perhaps in 100 years. In forests on drier soils it will take at least 200 years to reach a similar natural state (Falinski, 1988).

One fascinating possibility in natural tree dynamics is alternating tree species replacement over time at a given point in some mixed mature forests (Fox, 1977). In the hemiboreal zone spruce–beech is a probable pair (Nagel, 1950). Spruce can establish successfully in old beech forests and the converse is also true (Nilsson, 1985). Another indication of the very old age of mixed beech–spruce forests is that the few old undisturbed forests of this type contain several Urwaldrelicte among the beetles, for example *Ceruchus chrysomelinus*, *Lathridius brevicollis* and *Corticeus unicolor* (R. Baranowski & S.G. Nilsson, unpublished data).

10.4 FORESTRY AND FOREST PROCESSES

Forestry has been the dominating form of land use in the hemiboreal zone of Sweden for about 100 years. During this short period the recommended methods have changed several times. Presently, similar forestry methods are used in the hemiboreal and northern boreal zone (see chapter 7). In dry, especially pine-dominated, forest types, the natural processes in these vegetation zones were probably similar (see above and chapter 7). However, on mesic, fertile ground small-scale windfalls were probably the dominant disturbance process in the south, in contrast to large-scale forest fires in the north. Using clear-cuttings in both zones will have different effects in the hemi- and north-boreal zones, respectively. More species in the south are expected to have a low dispersal ability and therefore a higher risk of going extinct locally after a large-scale disturbance (see chapter 3). Similarly, to plant one species of conifer (mainly pine in the north but spruce in the south) on the clear-cut areas will alter forest structure and animal communities more in the south with its natural mixed forests (Nilsson, 1979*a*). Furthermore, only few and small forest areas are protected in the south of Scandinavia (see chapter 2), providing few sources for dispersal of old-growth species. Summarizing, forest processes have been altered more drastically by forestry in the hemiboreal zone than further north. The small-scale successional kaleidoscope with many tree species involved has been changed to much larger clear-cuts followed by stands with a dense canopy of spruce shading out almost all other plants.

Table 10.1
Maximum sizes (diameter at 1·3 m height and total height) and ages of some major forest trees in unmanaged forests

Species	*Diameter (cm)*	*Height (m)*	*Age (years)*
Pinus sylvestris	160	42	350
Picea abies	140	55	300
Quercus robur	230	43	400+
Tilia cordata	200	42	—
Fagus silvatica	160	—	300+
Ulmus glabra	150	37	—
Fraxinus excelsior	130	40	—
Acer platanoides	110	37	—
Carpinus betulus	100	30	—
Alnus glutinosa	100	40	—
Populus tremula	100	36	180
Betula pendula	100	30	—
Betula pubescens	70	30	—

Data from Falinski (1986) and for *Fagus* Lemeé (1989) and S.G. Nilsson (unpublished data from Sweden).

The second, equally important, change in a forest process is the fate of large stems. On managed forest land they are of course harvested at less than half the normal age of the canopy trees (see Table 10.1). In natural forests (those with natural regeneration and no management in recent decades) large trunks with rough bark, dying and dead stems provide habitats for the majority of forest organisms (Elton, 1966). The very important pathway in natural forests for the recycling of nutrients in wood and bark when these break down, involving mainly fungi, insects and bacteria, is drastically altered in managed forests. The natural web of organisms with its bewildering complex interactions is still more simplified. The long-term consequences of present forestry practices are unknown.

The ground vegetation in virgin deciduous and mixed hemiboreal forests is species-rich with many perennial herbs. In secondary—but high—forests, species richness is reduced, affecting mainly plants of mesic soils (Falinski *et al.*, 1988). These herbs are generally very long-lived (Tamm, 1956, 1972) and apparently favoured by canopy gaps due to single windfalls (Falinski, 1978). A few vascular plants, for example *Festuca altissima*, seem to require forest continuity for survival (Gustafsson, 1988). Otherwise, it is mainly specialized fungi, mosses and lichens among plants that require old-growth forests for survival. Among lichens, *Lobaria* spp. grow on old trees and have decreased much this century (Hallingbäck, 1986, 1989; Hallingbäck & Olsson, 1987; Hallingbäck & Thor, 1988). The reason for the decrease is apparently not only the clear-cutting of old-growth but also air pollution (Hallingbäck & Olsson, 1987).

The role of recently extinct large herbivorous mammals, for example wild horse *Equus caballus*, bison *Bison bonasus* and aurochs *Bos taurus*, in natural forest processes is difficult to evaluate. Possibly, they could slow down vegetation succession along shores and alter forest fires, creating habitats for herbs which are now dependent on domesticated animals for survival on grazing meadows. Large-scale experiments with mixed herds of bison and domesticated but 'primitive' horses and cattle are necessary to throw light on this hypothesis.

10.5 ACIDIFICATION AND POLLUTION—NEW THREATS TO HEMIBOREAL FORESTS

During the last 50 years the pH of forest soils in southwestern Sweden has decreased about 0·5 unit (Hallbäcken & Tamm, 1986; Falkengren-Grerup, 1987). The causes of this acidification have been much debated but seem to be a combination of air pollution, succession and planting of coniferous trees (Hallbäcken & Tamm, 1986; Wittig, 1986). The acidification of soils threatens their future productivity since in some places essential plant nutrients are now leaching out from the soils and aluminium is becoming soluble.

The effects of acidification are generally negative for forest organisms, especially for many oceanic lichens, which in Sweden mainly occur in the southwest with its high precipitation and pollution levels (Löfgren & Moberg, 1984). There are also alarming indications of negative effects on ectomycorrhizal fungi from acid rain and nitrogen deposition (Nitare, 1988). However, a review of the effects of air pollution is outside the scope of the present survey of the hemiboreal zone.

10.6 ANIMAL COMMUNITIES

10.6.1 Mammals

In Europe the densities of large mammals, both herbivores and carnivores, have been profoundly affected by hunting. In Sweden, bear *Ursus arctos*, wolf *Canis lupus* and lynx *Lynx lynx* were severely persecuted in the 19th century and went extinct in southern Scandinavia. Some smaller carnivores, like the pine marten *Martes martes*, were probably also reduced in numbers but not so severely as, for example, in Great Britain (Langley & Yalden, 1977). The pine marten has recovered and is now a dominant predator on nests of the black woodpecker *Dryocopus martius* (Nilsson *et al.*, 1991) and the hole-nesting Tengmalms' owl *Aegolius funereus* (Sonerud, 1985). Thus, the present mammal community is unbalanced due to lack of large predators.

Hunting was also responsible for the extinction of beaver *Castor fiber* and the severe reduction in range of the moose *Alces alces* and roe deer *Capreolus capreolus* in Sweden in the 19th century (Ahlén, 1977). Now the beaver has been reintroduced in many places, but its ecological role (see Naiman *et al.*, 1988) is still restricted to very local areas. Conversely, the density of moose and roe deer is now probably higher than any time previously due to comparatively low hunting pressure, an abundance of early successional stages (see above) and the previous elimination of big carnivores. The consequence of this high density of wild browsers is spruce being favoured at the expense of pine and, especially, many deciduous tree species due to selective browsing during forest succession. Highly preferred species like rowan *Sorbus aucuparia*, sallow *Salix caprea* and aspen can now mainly establish successfully near human habitations (personal observations). On poor soils, which previously supported pine-dominated forests, spruce is now often planted after clear-cutting because planted pine is often severely damaged, particularly by moose. Thus, the high density of wild browsers is presently acting in concert with forest management, both directly and indirectly, to increase the dominance of spruce in the Swedish hemiboreal zone.

Folivorous small mammals fluctuate less in density between years in the temperate than in the boreal zone (Hansson & Henttonen, 1985). However, the granivorous bank vole *Clethrionomys glareolus* increases greatly in density following years with abundant beech mast supply (Jensen, 1982) in beech forests. Therefore, the food supply for small mammal predators like the tawny owl *Strix aluco* can change considerably from year to year in temperate deciduous forests also (Southern, 1970).

It has been claimed that the activities of seed-storing small mammals are necessary for successful establishment of beech seedlings (Jensen, 1985), but this has been questioned (Nilsson, 1985). In any case, it is clear that the common small mammals are keystone species in both temperate and boreal forests because of their importance as predators of seeds and their role as prey for predators.

The factors influencing bat population densities are poorly understood. However, several species use big, hollow deciduous trees—a declining resource. Stands with such trees support a species-rich bat community (Gerell, 1988). The impact of bats on their food supply, nocturnal insects, is unknown, nor, as a consequence, whether bat communities are food-limited.

10.6.2 Birds

Man's enormous transformation of the forests in the hemiboreal zone of Sweden has had serious consequences for the bird populations, the most important changes having been the reduction of mixed coniferous/deciduous stands and of very old stands. The effects of this are briefly described below.

Very few European bird species are dependent on a single deciduous tree species. The black grouse *Tetrao tetrix* mainly eat buds of birch in winter. With the suppression of forest fires the black grouse formerly decreased, but it can also utilize large clear-cut areas and so has been able to recover (Angelstam, 1983). The thick-billed race of the nutcracker *Nucifraga c. caryocatates* seems to be totally dependent on hazel nuts, which it collects in autumn and stores in the soil of spruce-dominated forests, sometimes over 10 km away from the source (Turcek & Kelso, 1968). The nuts are used as winter food and even as nestling food the following spring. Breeding success is highest after years with high hazel nut production (Swanberg, 1981). In western Europe the nutcracker is almost restricted to the mixed forests of the temperate–boreal transition since it is dependent on the combination of hazel and spruce stands.

The brambling *Fringilla montifringilla* is highly dependent on beech seeds as winter food. Its winter distribution is determined by the occurrence of abundant beech mast and to a lesser extent by the snow cover (Nilsson, 1984; Jenni, 1987). Even within a beech forest its density was correlated with seed supply (Nilsson, 1979*c*).

There are several other bird species which mainly take their food from one plant species, but a large majority of forest birds can use several tree species even in pure stands. However, there is a clear distinction between birds dependent on deciduous trees and those dependent on coniferous trees. The severe reduction of natural mixed stands, which have been replaced with coniferous stands, has obviously hit bird species dependent on deciduous growth most strongly. Surprisingly, several species considered to be typical for coniferous forests, for example the willow tit *Parus montanus*, tree creeper *Certhia familiaris*, song thrush *Turdus philomelos* and even the goldcrest *Regulus regulus*, had a higher density in a mixed natural forest compared to a managed coniferous forest of the same age (Nilsson, 1979*a*). The cause of this is unknown but results in the total bird population density in a managed forest being only one-third that of a natural mixed forest. It must be stressed that all families of birds had higher combined population densities in the natural forest, but especially woodpeckers (see below) and warblers (Nilsson, 1979*a*). This contrasts strongly with a study of Finnish boreal forests where the higher number of birds in natural compared with managed forests was entirely due to a higher density of hole-nesting species (Haapanen, 1965).

The higher combined bird population density in mixed forests is not due to a pure deciduous forest effect because deciduous forests in the same study area had a lower bird density than mixed forests (Nilsson, 1979*b*). Many species occurring both in pure coniferous and pure deciduous forests seem to reach their highest density in mixed forests (Table 10.2). Even in the absence of forest management mixed forests have higher bird density and species richness than nearby coniferous forests in the temperate–boreal transition zone (Tomialoj'c *et al.*, 1984). I propose that the higher bird density in mixed forests compared with deciduous or coniferous forests have two main causes: (1) a higher insect density in mixed than in coniferous forest due to a warmer microclimate at trunks, branches, and the ground in spring before the leaves unfold; and (2) safer hiding places from predators in mixed than in deciduous forests. These suggestions still have to be critically examined but the conclusion remains that old, mixed deciduous/coniferous forest in the hemiboreal zone is a haven for birds. The optimal mix is not precisely known but bird density seems to be maximal when deciduous trees are dominant (Table 10.2), perhaps with 10–30% coniferous trees.

Following natural succession, snags and big hollow, but still living, trees will increase in abundance. This provides greater opportunities for two important groups of forest birds, viz. primary hole-nesters excavating their own nests and secondary hole-nesters using both holes made by the former and those caused by wood fungi. With few exceptions forest management will decimate and even eliminate species in both groups.

Primary hole-nesters, in the forests considered here mainly woodpeckers, are hit hardest because most are dependent on mature trees not only for hole excavation but also as feeding sites. Old, mixed forests in the hemiboreal zone are optimal woodpecker habitat, where seven species can breed in the same forest (Tomialoj'c *et al.*, 1984). Following forest management, as now practised in southern Sweden, only two species

Table 10.2
Densities (territories/km^2) of selected bird species in relation to proportion of deciduous trees in the stands (data from Nilsson 1979*b*)

Proportion of deciduous trees[a]	4	13	78	82	85	95	100
Dominant tree	Spruce	Spruce	Birch	Oak	Alder	Beech	Oak
Bird species							
Parus cristatus	5	5	0	0	0	0	0
Pyrrhula pyrrhula	11	10	0	6	0	0	3
Parus ater	21	20	19	6	0	4	0
Carduelis spinus	21	10	19	13	0	0	0
Parus montanus	0	6	19	19	0	0	0
Regulus regulus	27	46	58	44	0	0	0
Garrulus glandarius	11	10	0	16	9	4	9
Prunella modularis	21	20	19	19	18	0	6
Troglodytes troglodytes	22	15	19	19	18	14	12
Columba palumbus	5	10	19	19	18	8	12
Turdus philomelos	26	35	19	44	27	18	38
Certhia familiaris	11	26	19	60	27	19	31
Anthus trivialis	11	15	0	35	18	15	31
Ficedula hypoleuca	16	15	38	38	18	19	25
Erithacus rubecula	53	61	77	157	89	53	43
Dendrocopos major	0	10	10	28	18	11	12
Turdus merula	20	29	48	63	44	36	52
Parus major	11	26	58	85	53	46	59
Fringilla coelebs	43	61	134	170	107	87	105
Phylloscopus trochilus	16	41	173	220	213	53	105
Sylvia atricapilla	15	19	38	38	36	19	22
Phylloscopus sibilatrix	0	15	38	113	53	38	49
Parus palustris	0	15	19	19	27	23	25
Sitta europaea	0	15	19	38	18	8	28
Muscicapa striata	0	0	19	13	18	15	19
Sylvia borin	0	0	58	47	71	23	56
Parus caeruleus	0	0	38	57	27	27	40
Coccothraustes coccothraustes	0	0	0	6	0	11	15

[a]Calculated from basal area of all trees.

persist (Nilsson, 1979*a*)—the great spotted woodpecker, because it subsists on spruce and pine seeds in winter, and the black woodpecker, because it can feed on wood-living beetles in stumps left on clear-cut areas. Nevertheless, this latter species is not safe if the wood in the stumps is also collected, as practised in some parts of Sweden. Furthermore, this large woodpecker depends on nesting trees considerably older than the present rotation age of the forests (Nilsson, Johnsson & Tjernberg, in prep.). The middle spotted woodpecker *Dendrocopos medius* was previously fairly widespread in southern Sweden but became extinct mainly due to cutting of big oaks (Pettersson, 1985). The white-backed woodpecker *Dendrocopos leucotus* was a typical and formerly fairly common bird of old, mixed forests but by about 1975 had decreased enormously in Sweden to about 100 pairs and in 1984 only 50 pairs were recorded (Aulén, 1988). Each pair forages over an area of at least 100 ha and subsists mainly on wood-living insects in deciduous trees (Aulén, 1988). Other woodpecker species now also seem to be declining but they have not been closely studied. What we are now witnessing is not only a severe reduction of a specialized group, it is also an indication of something more serious. A highly diverse group, mainly consisting of insects, lichens, and fungi is dependent on snags and big trees with dying branches. The decline of woodpeckers indicates the problems for those other neglected species. Furthermore, woodpeckers provide secondary hole-nesters with suitable breeding sites. The black woodpecker especially must be considered a keystone species because it is the only one which provides suitable breeding holes for several other species, for example the marten, Tengmalm's owl, goldeneye *Bucephala clangula* and stock dove *Columba oenas*.

Secondary hole-nesters often seem to be limited by hole availability in managed forests. For example the great tit *Parus major* is limited by hole availability in managed forests (Löhrl, 1977; Dhondt & Eyckerman, 1980; van Balen *et al.*, 1982) but not in natural forests with a higher abundance of holes (Tomialoj'c *et al.*, 1984; Nilsson *et al.*, 1985). However, in natural, unmanaged forests only the pied flycatcher *Ficedula hypoleuca* has been shown to be limited by hole availability (Nilsson *et al.*, 1985; for mountain birch forests see also Enemar & Sjöstrand, 1972). The autumn density of the hole-nesting nuthatch *Sitta europaea* varies with the food supply because pairs defend smaller territories when the food supply is high (Nilsson, 1987; Enoksson, 1988), but its breeding density depends mainly on winter weather and food (Nilsson, 1987) and not on hole density (Nilsson *et al.*, 1985). Perhaps this is also true for other hole-nesting species in virgin forests.

Comparing forests of similar age, all major groups of birds increase in density and species richness with snag density in the hemiboreal zone (Nilsson, 1979*b*). The explanation for this is probably complex and still partly unresolved. One factor is that unmanaged forests were mixed deciduous–coniferous in the study area (see above). The importance of snags for hole-nesters has been discussed above, but it is still difficult to understand why, for example, the wood warbler *Phylloscopus sibilatrix* occurs with such high densities in mixed forests.

10.6.3 Insects

The insect community in natural, mixed forests is highly diverse but has not been studied in detail. However, it has been estimated that species richness of insects and other invertebrates decreases from 8000 to 2000 following modern forest management (Ehnström, in Ahlèn *et al.*, 1979).

Among over 4300 beetle species recorded from Sweden, about 1000 are dependent on wood and bark including associated fungi (Ehnström & Waldén, 1986; R. Baranowski pers. comm.). Most of these beetles can use several tree species at a given stage of decay but many are dependent on particular tree species (Palm, 1959). Among threatened Swedish beetle species, 39 are associated with oak, 26 with beech, nine with lime and so on, while 27 are dependent on deciduous trees in general (Ehnström & Waldén, 1986).

Wood-living beetles and tipulids are important agents in the decomposition of wood and the circulation of basic nutrients. Beetles also seem to be important dispersal agents for fungi (e.g. Anderbrant & Schlyter, 1987; Wilding *et al.*, 1989). Since wood fungi are important as food for many beetles (Palm, 1959) there is potential for coevolution between these groups of organisms (Wilding *et al.*, 1989).

Many wood beetles have formerly been recorded in the hemiboreal transition zone in southern Sweden but now occur only in the northern boreal zone or have a much reduced distribution (Palm, 1959; B. Ehnström, pers. comm.). Apparently, despite the fact that wood biomass was considerably lower in the 19th century that it is now, the density of snags and very old, living trees was apparently higher (see also above). It is becoming apparent that wood-living insects are the most vulnerable category of forest organisms, with many species already extinct in Sweden. Most of these occurred in deciduous trees in the south (Ehnström & Waldén, 1986). We know very little about the dispersal ability and area requirement of these wood beetles.

Several very rare beetles occur on a patch of about 5 ha with about 20 very old oak trees at Halltrop on Öland (Table 10.3). In about 1910 over 1000 similar oaks in the surrounding area were cut down (Leiler, 1973). Nevertheless, the magnificent longhorn beetle *Cerambyx cerdo*, which has its only confirmed Scandinavian occurrence here, persists in the few

Table 10.3
Rare species of wood beetles recorded in a 5-ha patch of very old oak forest at Halltorp, Öland

Swedish red data category[a]	*Species*	*Other localities in Sweden*
Extinct (0)	*Teredus cylindicus*	None
Endangered (1)	*Lacon lepidopterus*	Only old finds
	Lacon quercus	None
	Brachygonus megerli	None
	Cerambyx cerdo	Only old finds
	Rhopalopus femoratus	Strömserum
	Scolytus carpini	A few
Vulnerable (2)	*Tachysida gracilis*	A few
	Ampedus nigerimus	Only near Halltorp
	Cardiophorus gramineus	Strömserum
	Attagenus punctatus	Honsö, Nybro
	Globicornis nigripes	A few
	Oligomerus brunneus	Several

[a]From Ehnström and Waldén (1986).

surviving big oak trees (Ehnström & Waldén, 1986). However, some other very rare beetles previously recorded from the area may now be extint (R. Baranowski, pers. comm.). We can conclude that even small patches of very old deciduous forest can be valuable for vulnerable insects. This contrasts with studies of birds, where isolated forest areas of less than 10 ha lack vulnerable species (Ahlén & Nilsson, 1982; Nilsson, 1986).

Many click beetles (Elateridae) are dependent on old deciduous trees for survival. The most severely threatened mainly occur in living, but hollow, stems over 200 years old (Martin, 1989). Several are very suitable as indicator species of long forest continuity since they show a very low vagility. Some of the species are hardly ever found outside the big stems, where the same species can breed for over a hundred years (Martin, 1989). Such click beetles can be considered as relicts of the virgin forest, which disappear with the introduction of modern forest management practices (Martin, 1989).

Among ground beetles (Carabidae) the combined density has been found to correlate positively with the proportion of wet forest in a forest plot (Nilsson *et al.*, 1988). Thus, the present intensive forest drainage will probably reduce the insect density on the ground but other groups must also be studied before we can reach firm conclusions. We must, however, stop further drainage now because if we wait to act until we have complete knowledge, very few wet forests will remain (see also chapter 8).

Predatory carabid and staphylinid beetles seem to be very important predators on pupae of some folivorous insects (East, 1974; Kowalski, 1977). The latter group of insects has been studied most because of its supposed economic importance and conspicuous long-term cycles of abundance (e.g. Varley *et al.*, 1973). When the caterpillars reach peak densities, deciduous trees sometimes become totally defoliated in early summer. At such times the ground flora is exposed to much greater insolation compared with other years. Such defoliation years may be important for the reproductive success of forest herbs which are often limited by light availability.

10.7 CONCLUSIONS WITH CONSERVATION RECOMMENDATIONS

Most native organisms of Swedish hemiboreal forests are drastically affected by present forestry practices. These organisms have survived mainly because some landowners still do not use intensive, recommended methods. The few, small forest reserves have been of little significance except for a few, very rare species. However, new forest land is continuously taken into intensive use, much more quickly than new reserves are established. Therefore, the conservation strategy for hemiboreal forests must be changed *now* to save its diversity (species and genetic variation).

There are at least two possible strategies which may be successful. First, the rate or areal extent of forest reserve establishment must be increased at least to ten times the present level. New reserves should mainly be created on fertile soils since they are under-represented in present forest reserves (Nilsson, 1980). Furthermore, foresters must take conservation recommendations more seriously during forestry operations and voluntarily save patches of importance for 'demanding' organisms. Thus, the first conservation strategy is similar to the present one except for a higher ambition.

The second strategy stresses the interdependence of different forest patches in the landscape and aims at viable populations of native forest organisms on forestry land. I propose that all landowners must leave 5–20 (the number depending on forest type) big trees on each hectare of forest land to mature, die and decay naturally. Each forest owner should be allowed to retain old trees scattered over the area or in one patch. In this way there will be a variation between sun-exposed, harder snags and wetter, softer snags favouring different species (Palm, 1959). Nevertheless, there is a need for many undisturbed core areas in forest reserves to save the most demanding species, but the number of reserves needed is much lower for this second strategy. In spite of this, I estimate that both strategies will cost about the same if we are to save viable populations of native forest organisms. Probably more than 10% of the biomass of all large stems must be left to decay naturally to achieve this goal. There is an urgent research need to establish the actual population level and optimal dispersion of snags and big trees to save 'demanding' forest organisms, for example several woodpeckers and many lichens, mosses, fungi and insects dependent on mature trees.

We also need to know the importance of different mixes of forest trees for a range of organisms. What are the consequences for other organisms of reduced tree species richness in deciduous and mixed forests? How can we maintain tree species diversity and still produce valuable timber? Which tree species are most valuable from a conservation point of view and should therefore be left during forestry operations? These and similar other questions must be answered in the near future. However, forestry laws must be changed now to save both single, big trees and old stands generally. The same applies to riparian and swampy forests, especially those with a natural water regime (see also chapters 8 and 9). These measures stress the need to reduce recent man-made disturbance to a tolerable level. This is of course also true for air pollution. However, in addition we need to increase the level of some natural disturbances, those caused by forest fires and some large herbivores (beaver, wild boar and perhaps primitive cattle).

All these urgent conservation measures can be summarized in a single sentence. We must change disturbance frequencies and intensities to make them more similar to those in the virgin forests; that is, to provide organisms with the conditions to which they have become adapted over millions of years. If we change forestry methods with this in mind, I am sure conservation of organism diversity is compatible with the production of all the valuable raw material mankind needs from the forest.

REFERENCES

Ahlén, I. (1977). *Faunavård*. Liber, Stockholm.

Ahlén, I., Boström, U., Ehnström, B. & Pettersson, B. (1979). *Faunavård i Skogsbruket*. Skogsstyrelsen, Jönköping.

Ahlén, I. & Nilsson, S.G. (1982). Species richness and area requirements of forest bird species on islands with natural forests in Lake Mälaren and Hjälmaren.) *Vår Fågelvärld*, **41**, 161–84.

Ahti, T., Hämet-Ahti, L. & Jalas, J. (1968). Vegetation zones and their sections in northwestern Europe. *Ann. Bot. Fennici*, **5**, 169–211.

Anderbrant, O. & Schlyter, F. (1987). Ecology of the Dutch elm disease vectors *Scolytus laevis* and *S. scolytus* (Coleoptera: Scolytidae) in southern Sweden. *J. Appl. Ecol.*, **24**, 539–50.

Angelstam, P. (1983). Population dynamics of tetraonids, especially the black grouse *Tetrao tetrix* L. in boreal forests. PhD thesis, Uppsala University.

Aulén, G. (1988). Ecology and distribution history of the white-backed woodpecker *Dendrocopos leucotos* in Sweden *Rep. Depart. Wildl. Ecol., Swed. Univ. Agric. Sci., Uppsala*, No.14.

Bossema, I. (1979). Jays and oaks: an eco-ethological study of a symbiosis. *Behaviour*, **70**, 1–118.

Dhondt, A.A. & Eyckerman, R. (1980). Competition between the great tit and the blue tit outside the breeding season in field experiments. *Ecology*, **61**, 1291–6.

East, R. (1974). Predation on the soil dwelling stage of the winter moth at Wytham Wood. Berkshire. *J. Anim. Ecol.*, **43**, 611–26.

Ehnström, B. & Waldén, H.W. (1986). *Faunavård i Skogsbruket, Del. 2, Den lägre Faunan*. Skogsstyrelsen, Jönköping.

Elton, C.S. (1966). *The Pattern of Animal Communities*. Methuen, London.

Enemar, A. & Sjöstrand, B. (1972). Effects of the introduction of pied flycatchers *Ficedula hypoleuca* on the composition of a passerine bird community. *Ornis Scand.*, **3**, 79–89.

Enoksson, B. (1988). Prospective resource defence and its consequences in the nuthatch, *Sitta europaea* L. PhD thesis, Uppsala University.

Falinski, J.B. (1978). Uprooted trees, their distribution and influence in the primeval forest biotope. *Vegetatio*, **38**, 175–83.

Falinski, J.B. (1986). Vegetation dynamics in temperate lowland primeval forest. Ecological studies in Bialowieza forest. *Geobotany*, **8**, 1–537.

Falinski, J.B. (1988). Succession, regeneration and fluctuation in the Bialowieza Forest (NE Poland). *Vegetatio*, **77**, 115–28.

Falinski, J.B., Canullo, R. & Bialy, K. (1988). Changes in herb layer, litter fall and soil properties under primary and secondary tree stands in a deciduous forest ecosystem. *Phytocoenosis*, **1**, 1–49.

Falkengren-Grerup, U. (1987). Long-term changes in pH of forest soils in southern Sweden. *Environ. Pollut.*, **43**, 79–90.

Fox, J.F. (1977). Alternation and coexistence of tree species. *Amer. Nat.*, **111**, 69–89.

Gerell, R. (1988). Faunal diversity and vegetation structure of some deciduous forest in South Sweden. *Holarct. Ecol.*, **11**, 87–95.

Gustafsson, L. (1988). The biology of *Festuca altissima* in Sweden. *Svensk Bot. Tidskr.*, **81**, 337–49 (in Swedish with English summary).

Haapanen, A. (1965). Bird fauna of the Finnish forest in relation to forest succession, I. *Ann. Zool. Fennici*, **2**, 153–96.

Hallbäcken, L. & Tamm, C.O. (1986). Changes in soil acidity from 1927 to

1982–1984 in a forest area of south-west Sweden. *Scand. J. For. Res.*, **1**, 219–32.
Hallingbäck, T. (1986). Lunglavarna, *Lobaria*, på reträtt i Sverige. *Svensk Bot. Tidskr.*, **80**, 373–81.
Hallinbäck, T. (1989). Occurrence and ecology of the lichen *Lobaria scrobicularia* in southern Sweden. *Lichenologist*, **21**, 331–41.
Hallingbäck, T. & Olsson, K. (1987). The retreat of *Lobaria pulmonaria* in Scania. *Svensk Bot. Tidskr.*, **81**, 103–8 (in Swedish with English summary).
Hallingbäck, T. & Thor, G. (1988). The distribution and ecology of *Lobaria amplissima* in Sweden. *Svensk Bot. Tidskr.*, **82**, 125–39 (in Swedish with English summary).
Hansson, L. & Henttonen, H. (1985). Gradients in density variations of small rodents: the importance of latitude and snow cover. *Oecologia (Berl.)*, **67**, 394–402.
Jarvis, P.G. (1964). The adaptability of *Quercus petraea* (Matt.) Liebl. *J. Ecol.*, **52**, 545–71.
Jenni, L. (1987). Mass concentrations of bramblings *Fringilla montifringilla* in Europe 1900–1983: their dependence upon beech mast and the effect of snow-cover. *Ornis Scand.*, **18**, 84–94.
Jensen, T.S. (1982). Seed production and outbreaks of non-cyclic rodent populations in deciduous forests. *Oecologia (Berl.)*, **54**, 184–92.
Jensen, T.S. (1985). Seed–seed predator interactions of European beech *Fagus silvatica* and forest rodents, *Clethrionomys glareolus* and *Apodemus flavicollis*. *Oikos*, **44**, 149–56.
Kowalski, R. (1977). Further elaboration of the winter moth population models. *J. Anim. Ecol.*, **46**, 471–82.
Langley, P.J.W. & Yalden, D.W. (1977). The decline of the rarer carnivores in Great Britain during the 19th century. *Mammal Rev.*, **7**, 95–116.
Larsson, L.J. (1989). Svedjebruk i Värend och Sunnerbo. In *Skogen och Smålänningen. Historiska Föreningens i Kronobergs Län Skriftserie*, **6**, 61–99. Växjö.
Leiler, T.-E. (1973). Träd, skog och skalbaggar. *Sver. Nat., Årsbok* **64**, 107–12.
Lemée, G. (1989). Structure et dynamique de la hêtraie des réserves biologiques de la forêt de Fontainbleau: un cas de complexe climacique de forêt feuille monospecifique tempérée. *Acta Oecol., Oecol. Gener.*, **10**, 155–71.
Löfgren, O. & Moberg, R. (1984). Oceaniska lavar i Sverige och deras tillbakagång. *Swedish Environment Protection Board*, PM No. 1819 (in Swedish with English summary).
Löhrl, H. (1977). Nistökologische und ethologische Anpassungserscheinungen bei Höhlenbrütern. *Vogelwarte, Sonderheft*, pp. 92–100.
Malmström, C. (1939). Hallands skogar under de senaste 300 åren. *Medd Stat. skogsförs Anst.*, **31**, 171–300.
Malmer, N. (1973). Det föränderliga sydsvenska skogslanskapet. *Sver. Nat.*, **64**, 87–95.
Martin, O. (1989). Click beetles (Coleoptera; Elateridae) from old deciduous forests in Denmark. *Entomol. Medd.*, **57**, 1–107 (in Danish with English summary).
Nagel, J.-L. (1950). Changement d'essences. *J. Fort. Suisse*, **101**, 95–104.

Naiman, R.J., Johnston, C.A. & Kelly, J.C. (1988). Alteration of North American streams by beaver. *BioScience*, **38**, 753–62.

Nicolai, V. (1986). The bark of trees: thermal properties, microclimate and fauna. *Oecologia* (*Berl.*), **69**, 148–60.

Nilsson, S.G. (1979*a*). Effect of forest management on the breeding bird community in southern Sweden. *Biol. Conserv.*, **16**, 135–43.

Nilsson, S.G. (1979*b*). Density and species richness of some forest bird communities in South Sweden. *Oikos*, **33**, 392–401.

Nilsson, S.G. (1979*c*). Seed density, cover, predation and the distribution of birds in a beech forest in South Sweden. *Ibis*, **121**, 177–85.

Nilsson, S.G. (1980). Kriterier för värdering och urval av skogar lämpade som naturskogsreservat. *Swedish Environment Protection Board*, PM No. 1269.

Nilsson, S.G. (1984). The relation between the beech mast crop and the wintering of brambling, *Fringilla montifringilla*, and woodpigeon, *Columba palumbus*, in South Sweden. *Vår Fågelvärld*, **43**, 135–6 (in Swedish with English summary).

Nilsson, S.G. (1985). Ecological and evolutionary interactions between reproduction of beech *Fagus silvatica* and seed-eating animals. *Oikos*, **44**, 157–64.

Nilsson, S.G. (1986). Are bird communities in small biotope patches random samples from communities in large patches? *Biol. Conserv.*, **38**, 179–204.

Nilsson, S.G. (1987). Limitation and regulation of population density in the nuthatch *Sitta europea* (Aves) breeding in natural cavities. *J. Anim. Ecol.*, **56**, 921–37.

Nilsson, S.G., Björkman, C., Forslund, P. & Höglund, J. (1985). Nesting holes and food supply in relation to forest bird densities on islands and mainland. *Oecologia* (*Berl.*), **66**, 516–21.

Nilsson, S.G., Bengtsson, J. & Ås, S. (1988). Habitat diversity or area *per se*? Species richness of woody plants, carabid beetles and land snails on islands. *J. Anim. Ecol.*, **57**, 685–704.

Nilsson, S.G., Johnsson, K. & Tjernberg, M. (1991). Is avoidance by black woodpeckers of old nest holes due to predators? *Anim. Behav.*, **41**, 439–41.

Nitare, J. (1988). Changes in the fungal flora—research and conservation. *Svensk Bot. Tidskr.*, **82**, 485–9.

Palm, T. (1959). Die Holz- und Rinden-Käfer der süd- und mittelschwedischen Laubbäume. *Opusc. Entomol., Suppl.* **16**, 1–423.

Pettersson, B. (1985). Extinction of an isolated population of the middle spotted woodpecker *Dendrocopos medius* (L.) in Sweden and its relation to general theories on extinction. *Biol. Conserv.*, **32**, 335–53.

Shaw, M.W. (1974). The reproductive characteristics of oak. In *The British Oak. Its History and Natural History*, ed. M.G. Morris & F.H. Perring. Pendragon Press, Cambridge, pp. 162–81.

Skoglund, J. & Verwijst, T. (1989). Age structure of woody species populations in relation to seed rain, germination and establishment along the river Dalälven, Sweden. *Vegetatio*, **82**, 25–34.

Sonerud, G.A. (1985). Nest hole shift in Tengmalm's owl *Aegolius funereus* as defence against nest predation involving long-term memory in the predator. *J. Anim. Ecol.*, **54**, 179–92.

Southern, H.N. (1970). The natural control of a population of tawny owls *Strix aluco*. *J. Zool.* (*Lond.*), **162**, 197–285.

Swanberg, P.-O. (1981). Clutch size in the thick-billed nutcracker *Nucifraga c. caryocatactes* in Scandinavia in relation to the supply of hazel nuts in the individual winter stores. *Vår Fågelvärld*, **40**, 399–408 (in Swedish with English summary).
Tamm, C.O. (1956). Further observations on the survival and flowering of some perennial herbs, I. *Oikos*, **7**, 273–92.
Tamm, C.O. (1972). Survival and flowering of some perennial herbs, II. The behaviour of some orchids on permanent plots. *Oikos*, **23**, 23–8.
Tomialoj'c, L., Wesołowski, T. & Walankiewicz, W. (1984). Breeding bird community of primaeval temperate forest (Bialowieza National Park, Poland). *Acta Ornithol.*, **20**, 241–310.
Turcek, F.J. & Kelso, L. (1968). Ecological aspects of food transportation and storage in the Corvidae. *Commun. Behav. Biol.*, Part A, **1**, 277–97.
Van Balen, J.H., Booy, C.J.H., Franeker, J.A. & Osiek, E.R. (1982). Studies on hole-nesting birds in natural nest sites, 1. Availability and occupation of natural nest sites. *Ardea*, **70**, 1–24.
Varley, G.C., Gradwell, G.R. & Hassell, M.P. (1973). Insect Population Ecology: An *Analytical Approach*. Blackwell Scientific Publications, Oxford.
Weimarck, G. (1979). Svedjebruk i södra Sverige. *Bygd och Natur, Årsbok*, 1979.
Wibeck, E. (1909). Bokskogen inom Östbo och Västbo härad av Småland. *Sv. Skogsvårdsförenings Tidskr.*, **7**, 424–8.
Wilding, N., Collins, N. H., Hammond, P.M. & Webber, J.F. (ed.) (1989). *Insect–fungus Interactions*. Academic Press, London.
Wittig, R. (1986). Acidification phenomena in beech *Fagus silvatica* forests of Europe. *Water, Air Soil Pollut.*, **31**, 317–23.
Zackrisson, O. (1977). Influence of forest fires on the North Swedish boreal forest. *Oikos*, **29**, 22–32.

Chapter 11

Conservation Biology in Agricultural Habitat Islands

OLA JENNERSTEN[a], JON LOMAN[b],
ANDERS PAPE MØLLER[a], JEREMY ROBERTSON[c] & BJÖRN WIDÉN[d]

*[a]Department of Zoology,
Uppsala University,
Box 561,
S-751 22 Uppsala, Sweden*

*[b]Department of Ecology,
Lund University,
Ecology Building,
S-223 62 Lund, Sweden*

*[c]Department of Wildlife Ecology,
Swedish University of Agricultural Sciences,
Box 7002,
S-750 07 Uppsala, Sweden*

*[d]Department of Systematic Botany,
Lund University,
Ö. Vallgatan 18-22,
S-223 61 Lund, Sweden*

11.1 INTRODUCTION

Modern agriculture in temperate areas has a twofold impact on natural populations of plants and animals: it causes fragmentation and decreases the quality of habitats. Economy forces modern farmers to establish large fields that are easily worked. Ditches, habitat islands (piles of stone or shallow soils covered with vegetation) and irregularities in the fields are considered as obstacles and are therefore often removed. From a conservation point of view this development is often detrimental, because populations are fragmented, reducing migration between them.

Many species are well-adapted to habitats in the traditional agricultural landscape, for example, crop fields that are not artificially fertilized and habitats that have been used for fodder production, either as grazed pastures or as mown meadows. Agriculture, as well as reforestation of abandoned areas, has created habitats of low quality for many organisms, particularly during the last 50 years. This chapter deals with problems facing natural populations of animals and plants living in fragmented residual habitats of the agricultural landscape. We begin by describing the recent development of agriculture, then present five case studies from the research of each author. Finally, we discuss the results of these case studies, including the underlying theory and produce a synthesis relevant to conservation biology.

Man has practised agriculture in Sweden for roughly 6000 years (Berglund, 1969), but many large-scale effects on the flora and fauna are of recent origin. During the 19th century the amount of farmland increased markedly and was used more efficiently than earlier. The agricultural landscape was characterized by a small-scale mosaic of fields, meadows and grazed forests. A special feature of the Swedish agricultural landscape is the relatively high proportion of woodland. The mosaic has changed considerably during the last centry, in particular the last 50 years, since government grants, subsidies, and technological development have allowed greater specialization and efficiency. The area of cultivated land and the number of farms have decreased, while the size of fields has increased to follow the size of machinery (Table 11.1). There is less animal

Table 11.1
Changes in Swedish agriculture during the 1900s (from Svensson & Wigren, 1986)

Factor	*1917*	*1950*	*1980*
Farmers, % of population	45	23	5
Area of arable land (1000s ha)	3800	3600	3000
Area of meadows (1000s ha)	1000	650	350
Number of farms (1000s)	307	280	120
Number of dairy cows (1000s)	1800	1700	650
Artificial fertilizers (kg N/ha)	2	15	80
Pesticides (ton/year)	0	3000	8000

husbandry and so many meadows and pastures have either been converted to arable fields or, more commonly, been afforested with spruce trees. This period has also been characterized by an increasing reliance on pesticides (Odsjö, 1988) and artificial fertilizers (Table 11.1). Residual habitats (e.g. habitat islands, ditches and ponds) have been removed, and edges along the fields have been made as straight as possible to allow efficient farming. The prognosis is for an even more monotonous landscape since 'fewer will produce more on smaller areas' (Heurling, 1988). The flora and fauna of the residual habitats have been impoverished by many of these changes, and also because of edge effects from agricultural practices, especially drift of pesticides and fertilizers from adjacent fields. As much as three-quarters of the threatened vascular plants on the Swedish Red Data List belong to the traditional agricultural landscape (Ingelög, 1988), and the number of threatened species will probably increase in the near future. Re-establishment of traditional land use as a method of increasing habitat quality is unrealistic, except in small reserves. Instead, conservationists must identify the crucial stages in the life cycles of key species and any conservation management must be adjusted to satisfy the demands of these species.

The development in modern agriculture is thought to be harmful for animals and plants for several reasons:

(1) the disappearance of habitats essential for many animal and plant species, such as ponds (breeding sites for frogs), ditches (nest sites for bumble bees), and wet and flooded meadows (foraging sites for birds);
(2) a decrease in the quality of habitats has made them less suitable for certain animals and plants (e.g. change in grazing intensity);
(3) habitat islands have decreased in number, resulting in increased isolation which reduces disperal between islands (Merriam, 1984);
(4) small and isolated populations may face reduced gene flow and hence decreased heterozygosity (Franklin, 1980; Ledig, 1986). When a population becomes too small and/or isolated, it becomes vulnerable to extinction through genetic, demographic, and environmental factors that may interact (Soulé & Simbeloff, 1986);
(5) trophic relationships (e.g. predator–prey and pollinator–plant) are probably particularly sensitive to disturbances since effects on one trophic level can affect other levels (Janzen, 1974; Gilbert, 1980; Wilcove *et al.*, 1986).

We present and compare five case studies dealing with different taxa (plants, insects, and birds). We have adopted a biogeographical and an ecological approach. Presenting case studies is a way of reaching conclusions and making generalizations. If a pattern is found in one case it must be considered a possibility in other similar cases. Our aim is to illustrate realistic cases as a basis for a general discussion that will lead to an understanding of the conservation of residual habitats in the agricultural landscape and enable us to make practical suggestions for their preservation.

11.2 CASE STUDIES

11.2.1 The critical habitats for carabid beetles on farmland

From a conservation standpoint critical habitats are those habitats with the highest diversity of species, or habitats that are essential for the continued survival of some species on farmland. In this study carabid beetles, which are sensitive indicators of habitat quality (Thiele, 1977; Freitag, 1979), were surveyed on farmland around Sjö castle to the west of Uppsala in central Sweden, to measure directly the relative numbers of species and their diversity in the different habitat components. This enabled us to evaluate the different components of Swedish farmland and ascertain which habitats were essential to maintain the diversity of species, and make predictions on the likely effects of changes in agricultural management.

Carabids were sampled with a 450-m transect line of 13 pitfall traps set over a variety of farmland habitats. There were thus two traps in woodland, two on edges adjoining the fields, six on the fields and three on a large (*c.* 1 ha) wooded habitat island in the middle of the field. The pitfall traps consisted of two 850-ml plastic containers buried flush with the ground and 70 cm apart. A piece of wood, 70 cm long and 10 cm high, was placed along the ground between the two containers forming a barrier that deflected the beetles into either container (Wallin, 1985). There was much cannibalism among trapped beetles so a little detergent was added to a solution of 50% ethylene glycol in the bottom 4 cm of each container to kill and preserve the beetles. The traps were emptied each week between mid-May and the first week in October 1983.

Pitfall trapping has been criticized as an inappropriate method for comparing the relative numbers of carabids in different habitats (Greenslade, 1964; Duffey *et al.*, 1974; Thiele, 1977). The catches in pitfall traps depend not only on the densities of the different species but also on thier activity patterns which are related to temperature and seasonal differences in behaviour (reviewed in Thiele, 1977). This was not a problem in this study because the traps were established in the different habitats for most of the season and there was little difference in temperature among the trapping sites. Apparent differences in abundances of carabids in different habitas may be more a reflection of differences in the structure and density of vegetation rather than actual differences in population densities, because dense vegetation impedes their movements and reduces the probability of capture (Greenslade, 1964; Thiele, 1977). In the present study, this problem was alleviated by clearing the vegetation for about 50 cm around the traps. Clearing around traps makes them more efficient and reduces the differences in catchability between habitats that differ in vegetation structure and density (Greenslade, 1964).

Despite the problems, pitfall trapping is the only effective way of sampling a whole community of carabids over the whole season in a variety of habitats. Direct methods, such as quadrat sampling, are impractical in habitats like woodlands with low densities of carabids (Greenslade, 1964; Thiele, 1977). Furthermore, the direct methods can produce misleading data on the relative abundances of the different species in a community because they are a single snapshot of the community structure and the maximum numbers of different species peak at different times of the season (Thiele, 1977; Wallin, 1985). There are two indications that the result from the pitfall trapping in this study are reliable. First, most individuals were caught in arable fields, followed by edges, and fewest were caught in woodland. This follows the pattern found by quadrat sampling and mark-recapture estimates of densities. Second, Greenslade (1964) suggested that comparisons between different habitats are valid if the highest abundances are obtained from the densest, and more obstructed, habitat. We found the greatest abundances in the crop fields and edges, which had similar dense vegetation, and the lowest abundances in the sparsely vegetated woodlands and habitat islands.

Plotting the number of species and the number of individuals caught in the habitats on the transect gives an immediate impression of their distribution (Fig. 11.1), but there is no simple relationship between the numbers of species and the numbers of individuals. The highest numbers of individuals were found on arable fields where there were a few dominant species. The highest numbers of species seem to be associated with the bushy edges between fields and forest, but there were relatively few individuals at these high-diversity edges. Fewer species were found on

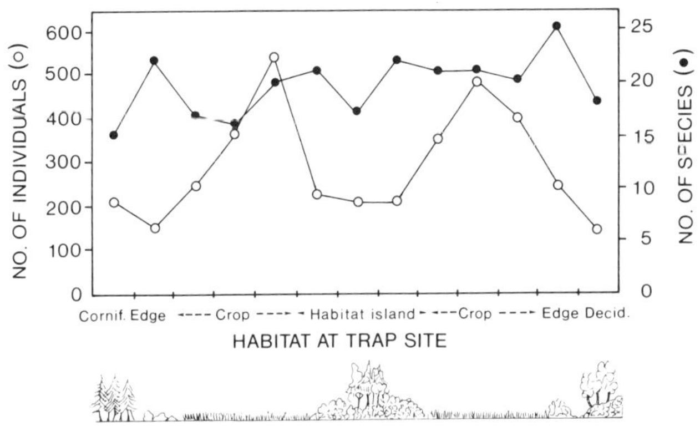

Fig. 11.1. The numbers of individuals and species of carabid beetles caught in a transect line of pitfall traps set over different habitats at a study site in central Sweden in 1983. The habitat at each trap is represented schematically under the graph. The numbers of individuals or species caught in the pairs of traps differed significantly between the habitats (individuals: $F_{3,76}=11{\cdot}73$, $p<0{\cdot}001$; species: $F_{3,76}=3{\cdot}77$, $p=0{\cdot}014$).

monocultures of arable field or woodland, and the lowest numbers of individuals were found in the forest where there were a few shade-tolerant species. This distribution gives an impression of the relative carabid fauna in the different habitats but it does not help us separate which habitats are critical in maintaining a high diversity of carabid beetles.

Measurements of diversity are strongly affected by sample size, because with more sampling effort there is more likelihood of collecting more species (Williams, 1964; Magurran, 1988). The sampling effort is uneven in the transect shown in Fig. 11.1 as most traps are in the crop fields. This problem can be overcome by rarefaction techniques that calculate the expected number of species in each sample if all the samples were of equal size (Hurlbert, 1971). However, all information on the relative numbers of the different species is lost in rarefaction and we are simply left with the number of species expected in each habitat (Williamson, 1973). To overcome this problem of unequal sampling we have used the data from two traps in each habitat, thus both traps in woodland, both traps on edges, two of the six traps on the fields, and two of the three traps on the habitat island. There was no significant difference in the numbers of individuals or species caught in the pairs of traps within each habitat, but there were significant differences between the habitats (Fig. 11.1).

A total of 2026 individuals of 49 species were caught in the eight traps

(Table 11.2). The species are ranked in decreasing total abundance and show the typical log normal pattern (Fisher *et al.*, 1943) of a few abundant species, some fairly abundant species and many relatively rare species. Three measures of species diversity are given in Table 11.2. First, the number of species, or species richness, is the simplest index but gives no indication of their relative abundance. Second, the inverse Berger-Parker index (May, 1975) gives a measure of the dominance of the most common species for each habitat, so that a low value indicates that the most common species is the major constituent in the sample and hence there is probably low diversity. Finally, diversity is measured with the log series, which is a robust parametric diversity index that is very useful in discriminating differences among habitats (May, 1975; Taylor *et al.*, 1976; Magurran, 1988). The values of the inverse Berger-Parker index and a jack-knife estimate overcome any effects of non-random sampling (Zahl, 1977; Magurran, 1988). It is clear from Table 11.2 that the most species were found in the edge habitats, which also had a low dominance and the highest diversity. The number of species was similar in the fields, woods and on the habitat island, but the lowest diversity was in the fields. Although there was one more species on the habitat island than in the woods its diversity was lower because of the greater dominance on the habitat island.

There are marked differences in the distribution of species among the four habitats. Table 11.2 lists the usual habitat preferences (Lindroth, 1986; Thiele, 1977) of the different species, and it is clear that most of the specimens are found in their expected habitats. There are several eurytopic or widespread species, such as *Pterostichus melanarius* and *P. niger*, that were caught in all four habitats, but there are also specialists such as *Amara brunnea* and *Calathus micropterus* that were only in the woods or on the wooded habitat island, or edge species such as *C. melanocephalus*, *A. lunicollis* and *A. communis*. The fewest species were found on the fields and many of them are eurytopic, although there are some that were confined to the fields, such as *Trechus quadristriatus* and *Bembidion quadrimaculatum*, while *A. apricaria* is a commensal species that is favoured by human activities such as farming. Although the habitat island was dominated by *P. melanarius* and *Carabus hortensis* it had a rich beetle fauna including many of the woodland species and some rarities such as *Cychrus caraboides* and *Licinus depressus*, both of which prey on land snails.

In order to summarize the relationships of species and their relative abundances in the different habitats, these data have been analysed with

Table 11.2
Carabid Beetles and their Relative Abundances in Four Habitats on Farmland at Sjö Castle between May and October 1983

	Species	*Dist*[a]	*Woods*	*Edge*	*Field*	*Island*	*Total*
1	*Pterostichus melanarius*	Et	9	99	169	147	424
2	*Bembidion lampros*	Et	3	39	241	4	287
3	*Carabus hortensis*	W	57	15		93	165
4	*Trechus secalis*	Et	9	69	58	7	143
5	*Pterostichus niger*	Et	9	41	44	34	128
6	*Calathus micropterus*	W	80	1		24	105
7	*Trechus quadristriatus*	F		2	98		100
8	*Bembidion quadrimaculatum*	F		8	89		97
9	*Amara brunnea*	W	81				81
10	*Pterostichus oblongopunctatus*	W/Et	33	1		26	70
11	*Agonum dorsale*	F		9	39	6	54
12	*Leistus ferrugineus*	Sh	16	12	1	23	52
13	*Harpalus rufipes*	Et	1	5	28	9	43
14	*Pterostichus strenuus*	Sh	17	15		1	33
15	*Pterostichus cupreus*	F		1	23	2	26
16	*Notiophilus biguttatus*	W/Et	19			7	26
17	*Calathus melanocephalus*	Op	5	14	3		22
18	*Amara lunicollis*	Et		20			20
19	*Trechus micros*	Sh			13	2	15
20	*Cychrus caraboides*	W		2		11	13
21	*Bembidion guttula*	Sh			11	1	12
22	*Amara apricaria*	Et			11		11
23	*Patrobus atrorufus*	W/Sh	3		1	6	10
24	*Amara communis*	Et/Op		10			10
25	*Carabus violaceus*	W/Op		5		4	9
26	*Synuchus vivalis*	Op		2	6		8
27	*Clivina fossor*	Op		1	6		7
28	*Badister bipustulatus*	Op		1		5	6
29	*Harpalus seladon*	Op/Et		4		1	5
30	*Loricera pilicornis*	W/Et	3	1	1		5
31	*Harpalus quadripunctatus*	W	2			3	5
32	*Amara aulica*	Sh		3	2		5
33	*Dromius sigma*	Sh		4			4
34	*Leistus rufescens*	W/Sh	1	2			3
35	*Trechus discus*	Sh		1	2		3
36	*Amara praetermissa*	W/Op	3				3
37	*Dromius agilis*	W	1			2	3
38	*Notiophilus palustris*	Sh		2			2
39	*Pterostichus nigrita*	Sh		1			1
40	*Amara familiaris*	Et/Op				1	1
41	*Bembidion bruxellense*	Et			1		1
42	*Harpalus winkleri*	W	1				1
43	*Amara eurynota*	Op	1				1
44	*Amara nitida*	Op		1			1
45	*Amara consularis*	Op/F			1		1
46	*Dromius quadrimaculatus*	W				1	1
47	*Bembidion doris*	Sh		1			1
48	*Agonum piceum*	Sh/W	1				1
49	*Licinus depressus*	Sh/W	1				1
	Number of species (species richness)		23	32	22	24	49
	Number of individuals		356	392	848	430	2026
	Dominance (inverse Berger-Parker)			5·75	4·09	4·21	2·93
	Diversity (log series α)			6·63	9·42	4·44	5·39

[a]Dist shows the typical habitats of each species, where Et = eurytopic or widespread, W = woods, F = fields, Sh = shady damp areas, and Op = open areas with sparse vegetation.

canonical correspondence analysis (CCA), a powerful multivariate ordination technique that readily identifies the habitat components important in determining the distribution of the species and the habitats contributing to species diversity (ter Braak, 1986, 1987; Jongman *et al.*, 1987). CCA combines regression and ordination into a direct multivariate gradient analysis that greatly improves the power to detect unimodal relationships between species and external variables, which cannot be detected with the usual parametric techniques such as principal components analysis. CCA reduces the complex multivariate data to two axes or eigenvectors (sometimes three eigenvectors in complex cases) that optimally show the maximum differences between the distribution of the species in relation to the weighted averages of the environmental factors and where the main species diversity lies (Jongman *et al.*, 1987). The statistical significance of the optimal distribution can also be tested with a series of Monte Carlo simulations and the output can be plotted graphically for visual interpretation. The relative association of each species with particular habitats is shown by their distance from the weighted average of each habitat.

Figure 11.2 shows the ordination diagram of the species scores, the

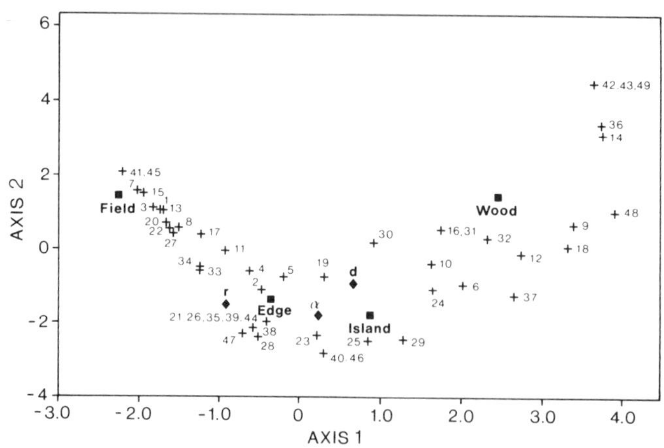

Fig. 11.2. Ordination diagram based on canonical correspondence analysis of the carabid beetles caught at Sjö in 1983 and the habitats around the pitfall traps (see Fig. 11.1) ($\lambda_1=0{\cdot}685$, $\lambda_2=0{\cdot}296$, scaling $a=1$, $p=0{\cdot}01$). +, the species; ■, weighted averages of the habitats; ◆, weighted averages of the diversity indices (**r** = species richness, **d** = dominance, **α** = log ratio diversity index). The species in the figure can be identified by referring their numbers to the list in Table 11.2.

habitat scores and the centroids of the diversity indices. The first two axes were significant (explaining 67% of the variance) and show that the species were associated with different habitats and that there was a simple gradient from those species in crop fields, through those on edges and habitat islands to those in woodland. Those species near the weighted average of the different habitats are associated with those habitats, thus the woodland species are on the right of the figure, the field species on the left and the edge and island species inbetween. The field species are distinct from the woodland species and those on the habitat island, and although there is overlap of the species on edges and habitat islands each has distinctive species. To test whether the observed differences could be attributed to chance, we ran Monte Carlo simulations with the first eigenvalue as a statistical test, and 99 random data sets all yielded a lower eigenvalue, showing that there are significant differences in species among the four habitats ($p<0{\cdot}01$). Note that the weighted averages for the diversity indices (richness *r*, dominance *d* and α) are placed near the edge and island habitats. This confirms that these habitats are the most diverse and suggests that their removal will disproportionally reduce the overall diversity of beetles on farmland. However, the carabids on these small habitat islands and narrow edges are probably vulnerable to the effects of agricultural practices on the fields, especially aerial drift of sprayed pesticides and fertilizers.

11.2.2 Bird diversity in the agricultural landscape

This case study (Loman & von Schantz, 1991) addresses the general question: Is there a relation between habitat island size and species diversity? Two aspects of island size are treated. First, is there a correlation between the size of a given habitat island and the number of species when all other factors are controlled for? It is also important to remove the effect of species numbers that merely comes from an increased number of individuals on larger islands. This is done with rarefaction, a technique that randomly reduces a sample to a standard size (Simberloff, 1979).

The second aspect is whether one large habitat island has more species than several small islands of the same total area? This much debated question has been analysed by grouping the investigated habitat islands (see below) into four groups of successively smaller islands but of the same total area.

Territory mapping (Anon., 1970) was used to determine the number of breeding pairs on 238 habitat islands in the agricultural landscape of Skåne, southernmost Sweden. The relative number of species was used as a measure of diversity and was calculated by dividing the actual number of species on a habitat island by the expected number of species obtained from the rarefaction analysis. For each habitat island there were also data on area, degree of isolation, shape (=relative amount of periphery) and six principal components of vegetation.

The relative number of species was negatively correlated with habitat island size after correcting for isolation, shape and vegetation (partial correlation analysis) (Fig. 11.3A). Thus, small habitat islands had relative-

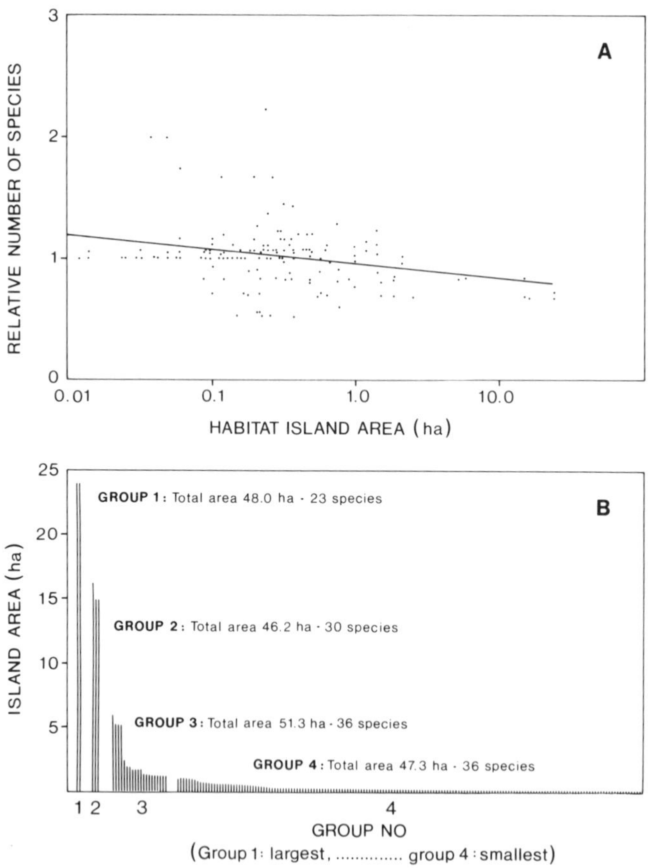

Fig. 11.3. (A) The relationship between the size of habitat island and relative number of bird species of 280 habitat islands in Skåne, southern Sweden ($r_p = -0{\cdot}23$, df=140, $p<0{\cdot}01$). The line denotes the linear regression. (B) The number of bird species per habitat island size group. The islands are grouped into four size classes (1, largest islands; 4, smallest islands) where each group has a similar total area.

ly more species than large islands. This may be because intraspecific competition precludes two pairs of the same species coexisting on a small habitat island but allows two pairs of different species to coexist.

Aggregating the habitat islands into four groups revealed that the small islands collectively accommodated more species than larger islands of the same total area (Fig. 11.3B). Two mechanisms may explain this pattern:

(1) edge species such as yellowhammers *Emberiza citrinella* may be more abundant on a collection of small islands than on a single large island because they have relatively more surrounding fields and/or forest periphery;
(2) the small islands may collectively have a higher variation of habitats and thus are able to accommodate more species than a single large habitat island.

The present study does not explain the mechanisms behind the observed pattern. However, from a conservation and species diversity point of view, it points to the relative merits of small habitat islands for birds. We do not say that large islands should be sacrificed in favour of small ones. However, mere size is not (despite the suggestion from some island theoreticians) a reason for discriminating against small habitat islands.

11.2.3 Bird nest predation and habitat island size

Studies of biogeography in habitat islands often make the tacit assumption that all individuals of a given species are similar. The simplest distribution of similar individuals of a particular species is the ideal free distribution (Fretwell, 1972). According to this distribution individuals occur in relation to their size provided islands exceed a minimum requirement. This will, however, only be the case if everything else is equal. This is usually not so because biological and stochastic processes vary in relation to habitat island size and because individuals differ in competitive ability (Fretwell, 1972). A number of different processes may affect the distribution and the abundance of species on habitat islands, including intraspecific and interspecific interactions and stochastic processes.

Nest predation is a major cause of reproductive failure in birds and may be affected by the size of habitat islands in a number of ways. First, small habitat islands have a relatively larger circumference than larger islands, and since predation often is an edge-effect (Gates & Gysel, 1978), individuals breeding on small islands will suffer from increased predation. Second, individuals nesting on small islands may differ from those nesting on larger ones. For example, risks of nest predation may be related to individual differences in the ability to build well-concealed nests that are relatively safe from predation. Third, smaller habitat islands may contain fewer and less suitable nest sites than larger ones. This study describes variation in the intensity of nest predation on blackbird *Turdus merula* nests in habitat islands of different size, and islands of different size may therefore differ in attractiveness. Data on the edge distribution and the reproductive success of blackbirds suggest that competitively superior individuals prefer large habitat islands to small ones (Møller, in press).

The frequency of nest predation by magpies *Pica pica* on blackbird nests was monitored in habitat islands covered with elm *Ulmus glabra* trees at Kraghede, Northern Jutland, a typical Danish farmland habitat (Møller, 1988). These species are suitable because they are common breeding birds, and breeding magpies are either present or absent on particular islands. The intensity of predation on blackbird nests was significantly negatively related to habitat island area, and the presence of a single pair of magpies increased the risk of nest predation (Fig. 11.4A).

Differential nest predation in relation to habitat island size could be caused by differences in the quality and thus the behaviour of individual blackbirds. By placing plasticine eggs mimicking blackbird eggs in old blackbird nests it was clearly demonstrated that the risk of nest predation still decreased with increasing habitat island area. Blackbirds on islands harbouring magpies more often suffered from nest predation than did blackbirds on other islands (Fig. 11.4B).

Differential nest predation in relation to habitat island area could also be due to differences in the availability of safe nest sites. This appeared not to be the case since blackbird nests on small islands, where nest predation was especially common, were more effectively hidden in safe sites such as buildings, coniferous trees and hedges (Møller, 1988). The risk of nest predation in small islands was therefore high despite the more effective concealment of nests. In conclusion, differential predation on blackbird nests in relation to the size of habitat islands appeared to be caused by area *per se* and not by other confounding variables.

As individuals are expected to maximize their life-time reproductive success, blackbirds should compete for access to sites that are relatively free from predation. Thus they are not predicted to follow an ideal free distribution relative to the size of habitat islands, but rather an ideal despotic distribution in which competitively superior individuals occupy the superior habitats (Fretwell, 1972). Is there any evidence for such a despotic distribution?

Habitat islands differed in their attractiveness to blackbirds. They

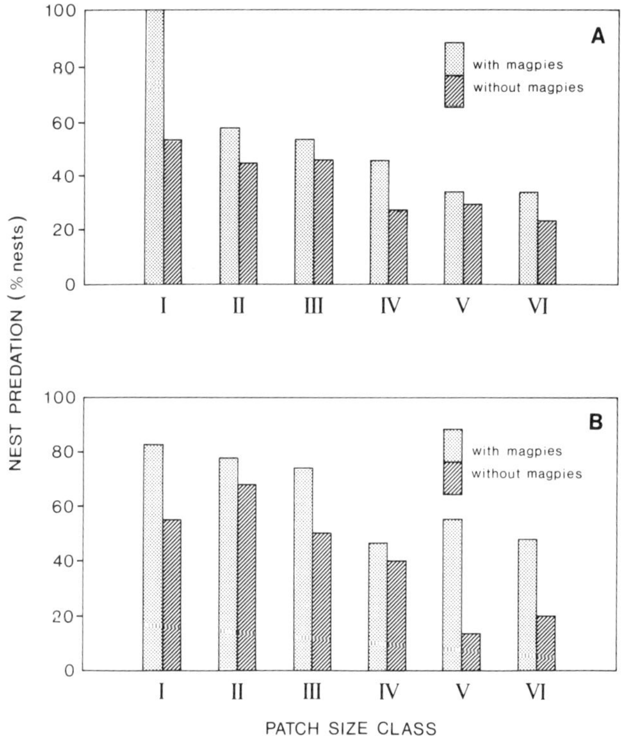

Fig. 11.4. (A) Predation rates on natural blackbird nests (% nests preyed upon) in relation to habitat island size in the Kraghede study area during 1971–1974 and 1977–1982. Sample sizes are 5 to 37 nests. (B) Predation rates on blackbird nest with plasticine eggs (% nests preyed upon) in relation to habitat island size during 1987. Sample size is 18 nests per bar. Patches either had (left bars) or did not have (right bars) breeding magpies. Patch size classes are I, 0·05–0·1 ha; II, 0·1–0·2 ha; III, 0·2–0·4 ha; IV, 0·4–0·8 ha; V, 0·8–1·6 ha; VI, 1·6–3·6 ha.

occurred on smaller islands less frequently than on large islands (Fig. 11.5A), but their population density on differently sized habitat islands showed on consistent pattern with habitat island area (Fig. 11.5B). The relatively high population density of blackbirds on small islands may be due to less intraspecific and interspecific competition for food (Møller, 1987).

The age distribution of blackbirds in relation to habitat island size provided evidence for an ideal despotic distribution, since young black-

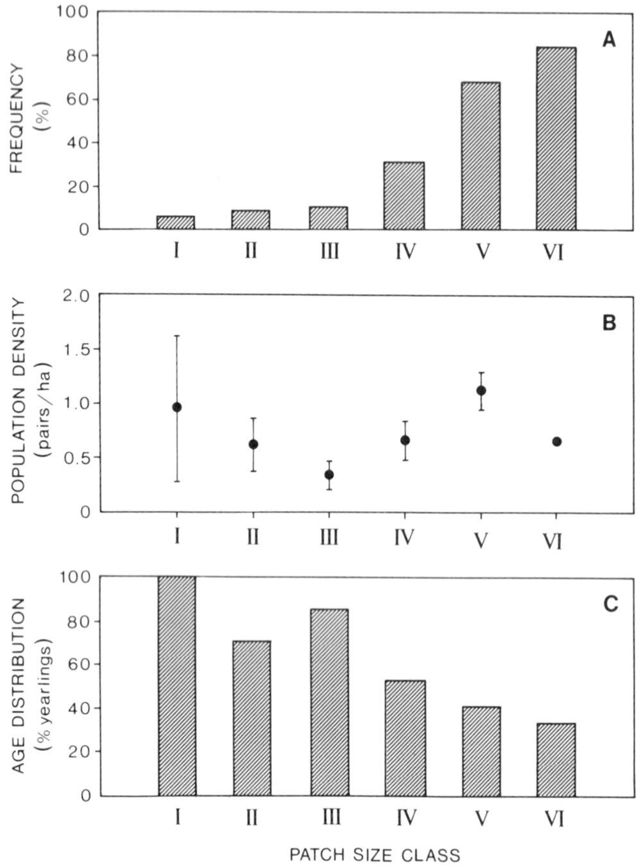

Fig. 11.5. (A) Frequency of breeding blackbirds (% patches with blackbirds) in relation to habitat island size during 1979–1982. (B) Population density of breeding blackbirds (pairs/ha) in relation to habitat island size during 1979–1982. (C) Age distribution (% yearlings) of breeding blackbirds in relation to habitat island size during 1971–1974 and 1977–1982. Blackbirds were caught in mist nets. For patch size classes see Fig. 11.4.

birds are competitively inferior (Bruun, 1969). The fraction of yearling blackbirds caught in mist nets during the reproductive season decreased from 100% in the smallest islands to less than 40% in the largest ones (Fig. 11.5C).

Reproductive success of blackbirds measured as clutch size was not significantly related to habitat island size (Fig. 11.6A). Mean clutch size

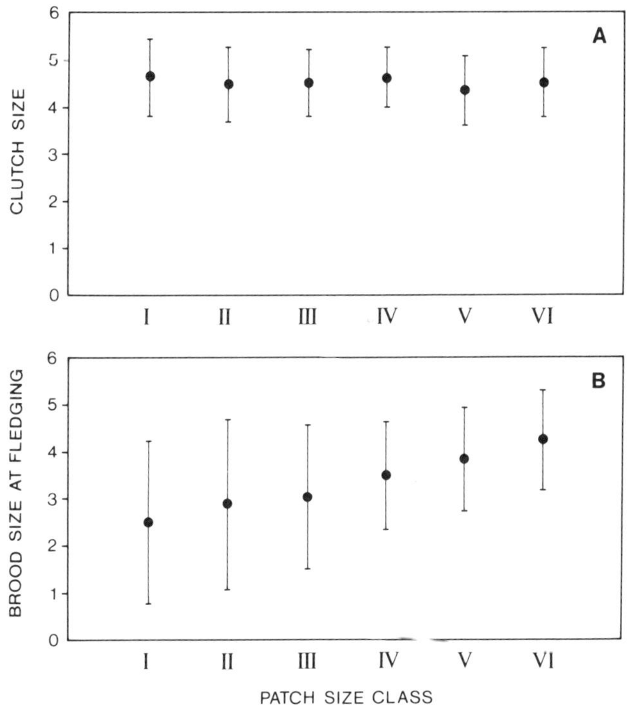

Fig. 11.6. (A) Clutch size of blackbirds in relation to habitat island size during 1971–1974 and 1977–1982. Sample sizes are 8 to 37 clutches. (B) Brood size of blackbirds at fledging in relation to habitat island size during 1971–1974 and 1977–1982. Sample sizes are 4 to 37 clutches. Values are mean ± SD (vertical lines). For patch size classes see Fig. 11.4.

was predicted to increase with increasing habitat island size, because older blackbirds, on average, lay larger clutches (Snow, 1958), and the blackbirds on large habitat islands were older. However, clutch size did not vary with island size (mean clutch size range: 4·4–4·6 eggs), but brood size increased from 2·5 fledglings per nest in the smallest islands to more than 4 in the largest ones (Fig. 11.6B). Differential reproductive success in relation to habitat island size was mainly caused by a lower nest predation risk in larger habitat islands.

11.2.4 Pollination ecology on habitat islands

Here we compare the abundance and diversity of flower-visiting insects of Lepidoptera and Hymenoptera and the seed set of perennial plants on habitat fragments in a modern agricultural area (island area) and in a traditional agriculture area consisting of a mosaic of different habitats (mainland area). We ask three questions:

(1) Are there differences in the abundance and diversity of flower-visiting insects between the island and the mainland areas?
(2) Are there differences in seed set between differently sized plant patches?
(3) Is there a difference in seed set between the two areas?

The study was conducted at two sites in the province of Dalsland, southwestern Sweden. The island site is the agricultural district of Dalbo consisting of eight habitat islands of roughly 0·1 to 0·5 ha. Each island is surrounded by arable fields, mainly of barley or oats. The mainland site is situated approximately 50 km north of the island site and consists of four meadows in a mosaic of forests, abandoned and cultivated fields, which are in the form of the traditional agricultural landscape. Thus the entire study area can be seen as a transect from modern to traditional agricultural land.

Insect and plant diversity were measured in transects on eight habitat islands at the island site and in four fields of the mainland site. Insect visits to specific plants were observed in 1 m × 2 m squares for 15-min periods. Seed set was analysed by marking individual plants prior to flowering (for details, see Jennersten *et al.*, 1988).

The island and the mainland sites were quite different in species composition (Table 11.3). A proportional similarity analysis (Wolda, 1981) showed that the occurrence of flowering plants and flower-visiting insects were grouped within areas (Fig. 11.7). The transects within the two areas were more similar in species composition than between the areas, although this is less clear-cut for plants than for insects. The encounter rate of insects per unit area and unit time differed significantly between the two areas. In July 1986 6·5 times as many flower-visiting insects were found at the mainland site as at the island site ($4{\cdot}77 \pm 0{\cdot}42$(SE), $n = 19$ and $0{\cdot}73 \pm 0{\cdot}10$, $n = 39$, insects/50 m^2/2 min, respectively). A nested ANOVA confirmed significant differences not only between sites, but also between transects within sites (Table 11.3). Thus, a spatial variation is evident that is higher at the island site than at the mainland site ($CV_I = 83{\cdot}73\%$, $CV_M = 38{\cdot}40\%$).

Data from both 1985 and 1988 demonstrated a significant temporal variation within the island site. In 1988, temporal variation in insect abundance was not significantly correlated with flower abundance nor

Table 11.3
Diversity (Shannon-Weaver H′) and the number of species of flowering plants and flower-visiting insects (Lepidoptera and Hymenoptera) at the mainland (M) and island (I) sites

Site	*Diversity*		*No. of species*	
	Plants	*Insects*	*Plants*	*Insects*
M1	2·750	2·866	18	24
M2	2·001	2·815	9	24
M3	2·532	2·740	16	22
M4	2·059	2·333	9	16
I1	2·085	1·538	10	7
I2	1·304	1·065	5	5
I3	2·047	1·748	8	7
I4	0·693	0·637	2	2
I5	1·277	0·956	4	3
I6	1·976	2·146	8	9
I7	1·273	1·128	4	5
I8	1·288	1·295	4	4

Differences between sites are significant (ANOVA: $F=373{\cdot}2$, df$=1$, $P<0{\cdot}001$), as are differences between transects within sites (ANOVA: $F=11{\cdot}2$, df$=11$, $P<0{\cdot}001$).

with number of flowering species of plants but with the occurrence of bountiful plant species. For example, when *Sedum telephium* began to flower, bumble bees, and in particular honeybees, appeared in large numbers on the islands. The number of visiting insects was positively correlated with the abundance of *S. telephium* flowers on islands ($\tau=0{\cdot}74$, $p<0{\cdot}05$, $n=8$), and accounted for 42·3% of all flower visits ($n=710$). Other 'popular' plants were *Linaria vulgaris* (14·8%), and *Hypericum maculatum* and *H. perforatum* (12·0%), which when common attracted numerous bumble bees. The two *Hypericum* species offer no nectar but only pollen.

Possible explanations for the differences between the two sites are (1) effects of isolation: the islands are separated by large areas of arable land with low availability of food source plants for insects (except irregular occurrences of crops like clover and rape) as well as nest sites for bumble bees; (2) environmental effects: the area occupied by the islands are not cultivated because the soils are too shallow or are bedrock outcrops, and thus are sensitive to periods of drought. For example, at the end of June 1988 following two months of very dry weather almost no plants flowered.

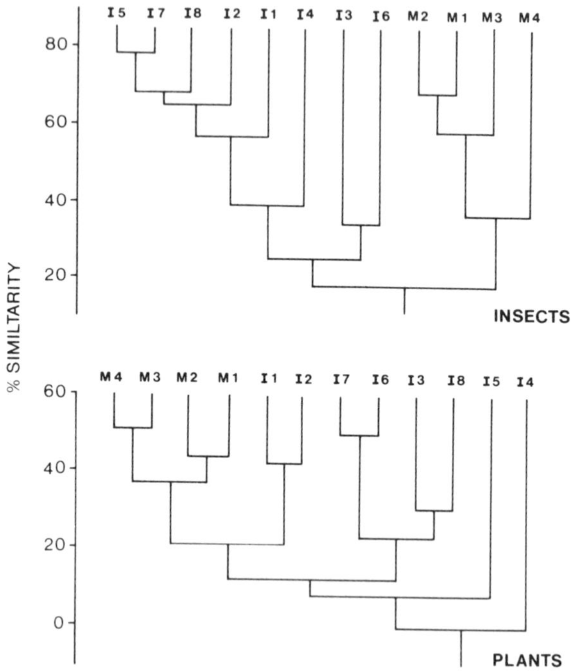

Fig. 11.7. Proportional similarity (Sörensen–Renkonen–Czekanowski's index; $PS=\Sigma_i \min(p_i, q_i)$, Wolda, 1981) on eight habitat island at the island site and four meadows at the mainland site. Top graph describes flower-visiting insects (Lepidoptera + Hymenoptera) and bottom graph flowering plants found in transects.

A patch of a plant species can be regarded as a model for a fragmented and an isolated population. Geometrical features of an island do affect the seed set of a plant species (e.g. Handel, 1983), and for example, in the sticky catchfly *Viscaria vulgaris* (Caryophyllaceae) individuals in large patches produce more seeds than individuals in small patches (Fig. 11.8). Curiously, the visitation rate per flower of the main pollinator, long-tongued bumble bees (e.g. *Bombus hortorum*), is independent of patch size, so individual flowers in small patches have a similar probability of a visit compared to flowers in large patches. One hypothesis to explain this phenomenon might be the foraging behaviour of bumble bees. Individual bumble bees are known to concentrate their foraging to one food source plant (their 'major' food source, *sensu* Heinrich, 1976, 1979). They also visit the flowers of a number of other plants, but to a

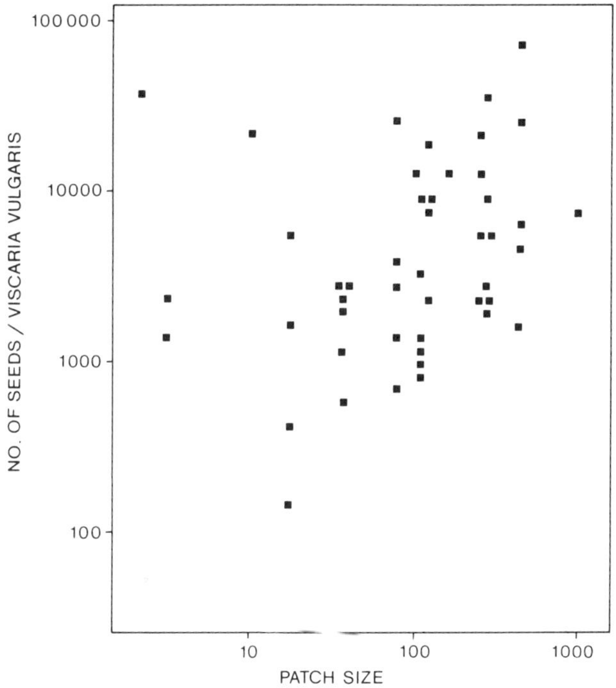

Fig. 11.8. Number of seeds per *Viscaria vulgaris* plant in relation to plant patch size (=number of plants/patch) in Dalsland, southwestern Sweden, 1982. Kendall's $\tau = 0{\cdot}27$, $n = 47$, $p < 0{\cdot}01$. Note the log-log scale. From Jennersten (1985).

lesser extent (their 'minor' food sources). This behaviour might help the bumble bees keep track of changes in their environment. It is possible that *V. vulgaris* acts as a 'major' food source when growing in large patches and as a 'minor' when in small patches. Other studies have found positive correlations between patch size and number of flowers per foraging bout of bumble bees (Sie & Baltus, 1987). The result from the plants' point of view may be a much more efficient transfer of conspecific pollen grains in large patches and a higher probability of heterospecific pollen transfer in small ones. Thus there would be a higher deposition of *V. vulgaris* pollen on stigmas of flowers in large patches despite more visits per flower in small patches.

Intriguingly, other factors may prevent an increase in patch size. In *Viscaria vulgaris* predispersal seed predation was also positively correlated with patch size; a higher proportion of seed capsules are attacked by mainly a geometrid moth (*Perizoma hydrata*) and a curculionid beetle (*Sibinia viscariae*) in large compared to small patches ($\tau = 0{\cdot}22$, $n = 813$, $p < 0{\cdot}001$). Additionally, the presence of a sterilizing fungal disease (*Ustilago violacea*) was positively related to patch size (Jennersten, 1985), possibly for the same reason as for high seed set—the more efficient foraging of bumble bees, which is the main disease vector.

The maiden pink *Dianthus deltoides*, a perennial, protandrous, caryophyllaceous herb of sunny habitats on sandy soils, grows commonly at the mainland site and on two of the island sites. Its flowers close at night—hence it can only be visited diurnally by dipteran flies and in particular by butterflies, its primary pollinators.

D. deltoides flowers received fewer visits at the island site compared to the mainland site due to the lower abundance of anthophilous insects (Jennersten, 1988). The seed set also differed between the sites, being higher at the mainland site in both 1986 and 1987, but not in 1988 when precipitation was unusually high during the flowering period of *D. deltoides.* A hand pollination experiment in 1987 revealed pollen availability as the limiting factor since hand pollination increased seed set 4·1 times at the island site but not at all at the mainland site (Fig. 11.9). The seet set of hand-pollinated flowers was similar at the two sites. Resource limitation is not a probable explanation for these differences because the addition of water and fertilizer had no effect.

Bagged flowers also produced similar numbers of seeds at the two sites. In addition, the seed set of bagged flowers did not differ from the seed set of naturally pollinated flowers at the island site, indicating that a large proportion of the seeds produced at the island site are self-fertilized. The flowers are clearly self-compatible and self-fertilized seeds germinate. The island site has probably been under influence of agriculture for at least one hundred years and it is therefore likely that the two islands have been isolated for a similarly long time. One may therefore speculate over the genetic status of the island plants. Self pollination leads to a lower heterozygosity and hence a higher probability of the expression of deleterious alleles that may interfere with reproductive capacity or make the plants more vulnerable to environmental stress.

A seed germination test revealed that seeds from the island population had a narrower amplitude in terms of germination time: most seeds from the island site germinated within 10 days (90·0%) while only 37·7% of the mainland seeds germinated within the same period ($G = 80{\cdot}1$, df$=1$, $p < 0{\cdot}001$), thus indicating that the island population may be genetically less heterogeneous than the mainland site.

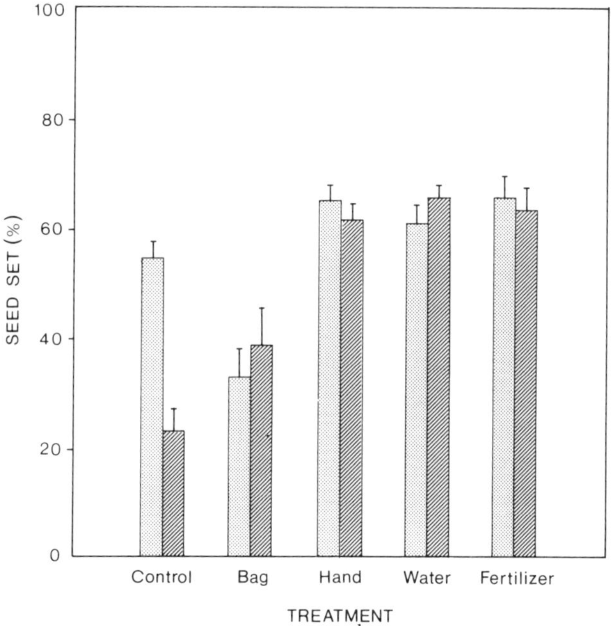

Fig. 11.9. Seed set (% ovules that developed into seeds) in *Dianthus deltoides* following different experimental treatments: flowers were bagged to prevent insect visits (Bag), flowers were artificially hand-pollinated (Hand), hand-pollinated and given water (Water), hand-pollinated and given artificial fertilizer (Fertilizer), and not manipulated to provide a natural control (Control) at the mainland site (▥) and the island site (▨). Vertical bars denote standard error of mean. Modified from Jennersten (1988).

11.2.5 Reproductive success of the field fleawort *Senecio integrifolius* in relation to habitat fragmentation and habitat quality

This study estimates fitness parameters in relation to habitat fragmentation (patch size) and habitat use (grazing) of natural populations of a threatened plant, the field fleawort *Senecio integrifolius* (Asteraceae). Three questions have been asked:

(1) Are there differences in seed set among populations with different population sizes?
(2) Are there quality differences among habitats due to grazing intensity that are reflected in differences in reproductive success among populations?

(3) What is the relative contribution of population size and habitat quality to the reproductive success?

Senecio integrifolius is a perennial herb of grazed pastures usually producing one leaf rosette bearing a single stem. Individual plants flower iteroparously with an interval of several years. Flowering frequency and generation time differ among populations depending on grazing intensity; plants at intensively grazed sites flower more frequently and become older than plants at ungrazed sites (Widén, 1987). The plant is self-incompatible (Widén, in prep.) and is visited by a variety of insects, mainly dipteran flies (Widén, 1986). Germination is light-dependent and recruitment occurs mainly during the first autumn. The species lacks a soil seed bank. (For details of the dynamics of natural populations, see Widén, 1986, 1987, 1988.)

The study was carried out in six of the eight known Swedish populations of *S. integrifolius*. The populations vary in size and structure from less than 10 flowering stems in one or several small patches of a few square metres each, to over 8000 flowering stems in one continuous population of *c.* 1 ha (Widén, 1988). The species had a wider distribution in Skåne, the southernmost province of Sweden, during the last century (Nilsson & Gustafsson, 1977). Any migration among these populations is very unlikely, mainly because they are distributed in well-separated island habitats in the modern agricultural landscape with few adjacent sites of sufficient quality. The main reason for the retreat of the species is the use of artificial fertilizers and lack of grazing by cattle (Widén, 1987). The extant populations are all situated in habitats that are not artificially fertilized due to their topography, and at sites that are still grazed or were grazed until recently by cattle. In the following we will define the quality of the habitats of the natural populations in terms of how the site is grazed.

Seed set was measured in natural populations, 1980–1982. Seed set (% seeds per head) was positively correlated with population size (Fig. 11.10(A)). This is probably an effect of density-dependent seed set, an hypothesis that was experimentally tested in cultivation. Plants derived from seeds sampled in the largest natural population (Tosteberga) were planted in four patches in a semi-natural habitat in the Fredriksdal Botanical Garden, Helsingborg. Twenty-five plants were evenly distributed in each of 1, 4, 25 and 100 m^2. From Fig. 11.10(B) it is evident that there is a density effect on seed set in the experimental population. Hand pollination in one natural population and in cultivation revealed that

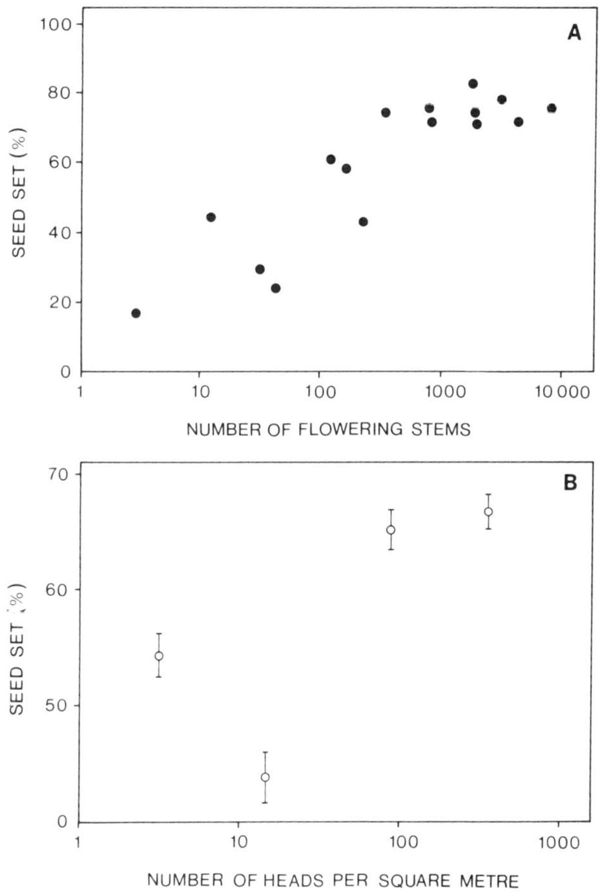

Fig. 11.10. Seed set in *Senecio integrifolius*. (A) Average seed set and population size in six natural populations, 1980–1982. Each population is represented by one dot each year (for two populations only 1980–1981). (B) Seed set (mean ±SE) in cultivated plants derived from Tosteberga. Twenty-five plants at each plot were planted at four densities.

pollen limitation explains most of the variation in seed set (Widén, in prep.).

Individual plants were mapped in permanent plots, 1980–1982, in four populations with different levels of grazing intensity. The number of stems produced as well as the number of successful stems (not grazed before seed dispersal) were recorded. The probability of a stem being

Table 11.4
The probability of reproductive success at different life-cycle stages in four natural populations of *Senecio integrifolius* (1980–1982)

Life cycle stage	*Kverrestad*	*Grödby*	*Tosteberga*	*Benestad*
Seed set	0·79	0·75	0·74	0·42
Successful seed dispersal	0·90	0·74	0·38	0·05
Germination	0·01	0·08	0·11	0·75

Individual plants were recorded in five permanent plots (1 m^2) at each site every second week during the vegetative season. The sites are arranged from low to high intensity of grazing (left to right). The estimates are based on data from Widén (1987).

grazed before seed dispersal varied from 0·1 at the site with the lowest grazing intensity to 0·95 at the site most intensively grazed (Table 11.4). Thus, very few plants dispersed their seeds at the most heavily grazed site (Widén, 1987). Based on the number of successful stems and the average number of seeds per head, seed production in permanent plots was estimated for 1980–1982. Although the number of stems differed fourfold among populations, the number of seeds produced in permanent plots differed 220 times (Widén, 1987, 1988).

The number of seedlings recorded from July 1980 to October 1982 in the permanent plots varied 25 times among populations. These seedlings represent from 1% germination of the produced seeds at the lightly grazed site to 75% germination at the heavily grazed site (Table 11.4). Thus, intensive grazing has a detrimental effect on one life cycle stage (seed production), but a very favourable effect on another (seed germination).

In populations with less than 100 flowering stems seed set is less than half of that in large populations (Fig. 11.10). However, the influence on reproductive success from population size is much less than the influence from habitat quality, e.g. grazing intensity (Table 11.4).

Grazing is an essential component of a high-quality habitat for *S. integrifolius*. Intensive grazing favours seed germination (Table 11.4) and survival of seedlings and older plants (Widén, 1987). The sites of two of the eight populations of *S. integrifolius* have not been grazed, or only lightly grazed, for many years (Widén, 1987), and these populations are now the smallest with less than 10 flowering stems annually. Reintroduction of grazing would increase population size at these sites. However, intensive grazing can decrease the number of sccessful stems to a small fraction of the potential of the population and thus decrease the population growth rate. The third smallest of the extant populations (Benestad) is overgrazed, much reducing the potential population size. It would be easy to improve the quality of the habitat also under intensive grazing management by changing the timing of grazing. Grazing usually starts in early May and seed dispersal of *S. integrifolius* is usually over in late June. If grazing starts at the time of seed dispersal most stems would be successful, and this would rapidly increase the population size. Thus, the quality of the habitat could be improved markedly by changing the grazing practice.

11.3 GENERAL CONCLUSIONS

An important aim of nature conservation is the preservation of a diverse fauna and flora. One way of achieving this is to preserve large areas of pristine habitats, but such reserves are usually in remote areas and relatively inaccessible. Most people's experience of nature is of residual habitats in managed environments such as the agricultural landscape, which although not pristine are worth preserving. Most of the fauna and flora of farmland survives on these suboptimal remnants, but the main stronghold of the species that originated in natural open habitats (e.g. steppe species) is on farmland. Nature conservation can be seen from different perspectives with two extremes on a continuous gradient, the global and the local perspective. From a global point of view, the preservation of all species is a worthy task. From a local perspective, it is important to maintain a high species diversity. It is essential to stress these differences because different goals may call for an emphasis on different conservation measures. The present paper takes the local perspective, and we ask: how do we preserve species diversity and a mosaic of remnant habitats in farmland?

The species-rich traditional habitats of farmland are being converted to arable fields or coniferous plantations that support few species, and modern management removes many of the small islands of scrub and other remnant habitats that might form refuges for some species. A mixture of habitat islands, meadows, scrub and deciduous woodland is necessary to maintain a high species diversity. The increasing dominance of large featureless arable fields and coniferous plantations will seriously impoverish the fauna and flora of much of south and central Sweden, and the only remaining diversity will be in nature reserves where the traditional habitats are protected. The quality of remnant habitats in farmland can, however, be improved by paying attention to crucial life stages of indicator species when planning the management regime. Grazing has been one of the most important factors for the present distribution of many plants and animals adapted to the traditional agricultural landscape. Even in the near future, grazing in habitats where no artificial fertilizers are used will be essential for preserving high species-rich pastures with minor adjustments of the time during the season when grazing is allowed. This, however, requires more knowledge of life-cycle parameters of sensitive species than presently available.

What are the implications of interspecific interactions for nature conservation? The intensity of intraspecific and interspecific interactions is likely to vary with the size of habitat islands, in the same way as predation on blackbird nests and the probable intraspecific competition for territories. Such differences in the suitability of islands lead to an ideal despotic distribution because differently sized islands differ in profitability. It is possible that animals other than birds moving among islands will experience similar interactions and as a consequence will show an ideal despotic distribution.

The reproductive success of blackbirds differed markedly among habitat islands of different size. Smaller islands are likely to be population sinks with a negative population growth, and they are only maintained by a steady flow of individuals from more productive islands, i.e. sources. This probably also applies to other species. Sources and sinks can only be identified if essential demographic parameters such as natality and mortality are known (Pulliam, 1988). The true sources may be far outside small study areas and hence cannot easily be identified. The major lesson for nature conservation is that vast areas of suboptimal habitats, even those having high population densities, cannot be assumed to be sources (van Horne, 1983; Pulliam, 1988). Intensively managed agricultural areas, because they have few habitats and are structurally homogeneous, are likely to be sinks for most species. However, they might also be source areas for a few species that are preadapted to this type of environment, for example steppe species.

Habitat islands were characterized by a low abundance of flower-visiting insects, but with high spatial and temporal variation. This variation was to a large extent governed by the abundance of highly rewarding nectar/pollen plants that attracted insects from afar (e.g. honeybees). Thus availability of flowering plants on habitat islands is important for a rich pollinator guild. Too few pollinators can result in a reduced seed set, increased self-fertilization and, eventually, a lower degree of heterozygosity and extinction. However, in plants reduced heterozygosity is not as serious as in animals since plants can deal with, for example, environmental heterogeneity through phenotypic plasticity. Isolated, animal-pollinated plants can adapt by changing their breeding system. For example, *Dianthus deltoides* may be able to change from protandry to homogamy with a high degree of self-pollination.

A problem facing conservationists is whether to preserve a single large or several small reserves (SLOSS). A large reserve supports larger populations that are genetically and demographically more stable than populations on a small reserve. However, the single large population is vulnerable to catastrophy such as disease, which would be less likely to affect all parts of a subdivided population on several small reserves. In the context of an agricultural landscape relatively more species may occur on small habitat islands (e.g. Fig. 11.3B). However, it is important to identify the population sources and sinks before being able to determine effectively whether to preserve a number of small islands or a single large one of the same area.

ACKNOWLEDGEMENTS

We acknowledge the support from the Swedish Environmental Protection Agency (O.J., J.L., J.R., B.W.), the Swedish Natural Science Research Council (O.J., A.P.M.) and the World-Wide Fund For Nature (J.R.).

REFERENCES

Anon. (1970). Recommendations for an international standard for a mapping method in bird census work. *Bull. Ecol. Res. Commun.*, **9**, 49–52.

Berglund, B. (1969). Vegetation and human influence in South Scandinavia during prehistoric time. *Oikos*, Suppl., **12**.

Bruun, B. (1969). Territorial behaviour and dominance at a feeding station. *Dansk Ornithol. For. Tidsskr.*, **63**, 170–3.

Duffey, E., Morris, M.G., Sheail, J., Ward, L.K., Wells, D.A. & Wells, T.C.E. (1974). *Grassland Ecology and Wildlife Management*. Chapman & Hall, London.

Fisher, R.A., Corbet, A.S. & Williams, C.B. (1943). The relation between the number of species and the number of individuals in a random sample of an animal population. *J. Anim. Ecol.*, **12**, 42–58.

Franklin, I.R. (1980). Evolutionary changes in small populations. In *Conservation*

Biology. An Evolutionary–Ecological Perspective, ed. M.E. Soulé & B.A. Wilcox, Sinauer Associates, Sunderland, Massachusetts, pp. 135–49.

Fretwell, S.D. (1972). *Populations in a Seasonal Environment*. Princeton University Press, Princeton, New Jersey.

Frietag, R. (1979). Carabid beetles and pollution. In *Carabid Beetles, Their Evolution, Natural History and Classification*, ed. T.L. Erwin, G.E. Ball & D.R. Whitehead. Junk, The Hague, pp. 507–21.

Gates, J.E. & Gysel, L.W. (1978). Avian nest dispersion and fledgling success in field–forest ecotones. *Ecology*, **59**, 871–83.

Gilbert, L.E. (1980). Food web organization and the conservation of neotropical diversity. In *Conservation Biology*, ed. M.E. Soulé & B.A. Wilcox. Sinauer, Sunderland, Massachusetts, pp. 11–33.

Greenslade, P.J.M. (1964). Pitfall trapping as a method for studying populations of Carabidae (Coleoptera). *J. Anim. Ecol.*, **33**, 301–10.

Handel, S.N. (1983). Pollination ecology, plant population structure and gene flow. In *Pollination Biology*, ed. L. Real. Academic Press, London, pp. 163–211.

Heinrich, B. (1976). The foraging specialization of individual bumble bees. *Ecol. Monogr.*, **46**, 105–28.

Heinrich, B. (1979). *Bumble bee Economics*. Harvard University Press, Cambridge, Massachusetts.

Heurling, B. (ed.) (1988). Mark & vatten år 2010—Framtidsbedömningar om kulturlandskapets utveckling. Bostadsdepartementet Ds 1988:35, Stockholm.

Hurlbert, S.H. (1971). The non-concept of species diversity: a critique and alternative parameters. *Ecology*, **52**, 577–86.

Ingelög, T. (1988). Conservation and the status of the swedish flora. *Svensk. Bot. Tidskr.*, **82**, 376–8 (in Swedish).

Janzen, D.H. (1974). The deflowering of Central America. *Nat. Hist.*, **83**, 48–53.

Jennersten, O. (1985). Pollination and fungal disease transmission: interactions between *Viscaria vulgaris*, *Ustilago* and insects. PhD thesis, University of Uppsala.

Jennersten, O. (1988). Pollination in *Dianthus deltoides* (Caryophyllaceae): effects of habitat fragmentation on visitation and seed set. *Conserv. Biol.*, **2**, 359–66.

Jennersten, O., Berg, L. & Lehman, C. (1988). Phenological differences in pollinator visitation, pollen deposition and seed set in the sticky catchfly, *Viscaria vulgaris*. *J. Ecol.*, **76**, 1111–32.

Jongman, R.H.G., ter Braak, C.J.F. & van Tongeren, O.F.R. (1987). *Data Analysis in Community and Landscape Ecology*. Pudoc, Wageningen.

Ledig, T.F. (1986). Heterozygosity, heterosis, and fitness in outbreeding plants. In *Conservation Biology*, ed. M.E. Soulé. Sinauer Associates, Sunderland, Massachusetts, pp. 77–104.

Lindroth, C.H. (1986). *The Carabidae (Coleoptera) of Fennoscandia and Denmark*. Brill, Copenhagen.

Loman, J. & Von Schantz, T. (1991). Birds in a farmland—more species in smaller than in larger habitat islands. *Conserv. Biol.*, **5**, 176–88.

Magurran, A.E. (1988). *Ecological Diversity and its Measurement*. Croom Helm, London.

May, R.M. (1975). Patterns of species abundance and diversity. In *Ecology and*

Evolution of Communities, ed. M.L. Cody & J.L. Diamond. Harvard University Press, Cambridge, Massachusetts, pp. 81–120.

Merriam, G. (1984). Connectivity: a fundamental ecological characteristic of landscape pattern. In *Proceedings of the First International Seminar of the International Association of Landscape Ecology* (*IALE*), ed. I. Brandt & P. Agger. Universitetsforlag GeoRuc, Roskilde, pp. 5–15.

Møller, A.P. (1987). Breeding birds in habitat islands: random distribution of species and individuals? *J. Biogeogr.*, **14**, 225–36.

Møller, A.P. (1988). Nest predation and nest site choice in passerine birds in habitat islands of different size: a study of magpies and blackbirds. *Oikos*, **53**, 215–21.

Møller, A.P. (in press). Clutch size, nest predation and distribution of avian unequal competitors in a patchy environment. *Ecology*.

Nilsson, Ö. & Gustafsson, L-Å. (1977). Project Linné reports 49–63. *Svensk Bot. Tidskr.*, **71**, 205–24 (in Swedish with English summary).

Odsjö, T. (1988). Use of pesticides and other chemicals in Swedish agriculture. *Vår Fågelvärld Suppl.*, **12**, 63–84 (in Swedish with English summary).

Pulliam, H.R. (1988). Sources, sinks, and population regulation. *Amer. Nat.*, **132**, 652–61.

Sie, A. & Baltus, M-S. (1987). Island size, pollinator behaviour, and pollinator limitation in catnip. *Ecology*, **68**, 1679–90.

Simberloff, D. (1979). Rarefaction as a distribution free method of expressing and estimating diversity. In *Ecological Diversity in Theory and Practice*, ed. J.F. Grassle, G.P. Patil, W.K. Smith & C. Taillie. International Cooperative Publishing House, Fairland, Maryland, pp. 159–76.

Simberloff, D. & Abele, L.G. (1982). Refuge design and island biogeographic theory: effects of fragmentation. *Amer. Nat.*, **120**, 41–50.

Snow, D.W. (1958). *A Study of Blackbirds.* Allen & Unwin, London.

Soulé, M.E. & Simberloff D. (1986). What do genetics and ecology tell us about the design of nature reserves? *Biol. Conserv.*, **35**, 19–40.

Svensson, R. & Wigren, M. (1986). A survey of the history, biology and preservation of some retreating synanthropic plants. *Sym. Bot. Ups.*, **25**(4), 1–74.

Taylor, L.R., Kempton, R.A. & Woiwod, I.P. (1976). Diversity statistics and the log-series model. *J. Anim. Ecol.*, **45**, 255–71.

ter Braak, C.J.F. (1986). Canonical correspondence analysis: a new eigenvector technique for multivariate direct gradient analysis. *Ecology*, **67**, 1167–79.

ter Braak, C.J.F. (1987). Ordination. In *Data Analysis in Community and Landscape Ecology*, ed. R.H.G. Jongman, C.J.F. ter Braak & O.F.R. van Tongeren. Pudoc, Wageningen, pp. 91–173.

Thiele, H.V. (1977). *Carabid Beetles in their Environment.* Springer, New York.

van Horne, B. (1983). Density as a misleading indicator of habitat quality. *J. Wildl. Manage.*, **47**, 893–901.

Wallin, H. (1985). Spatial and temporal distribution of some abundant carabid beetles (Coleoptera: Carabidae) in cereal fields and adjacent habitats. *Pedabiologia*, **28**, 19–34.

Widén, B. (1986). *Senecio integrifolius*—a threatened species in Sweden. Swedish Environment Protection Board, PM No. 3055 (in Swedish with English summary).

Widén, B. (1987). Population biology of *Senecio integrifolius* (Compositae), a rare plant in Sweden. *Nord. J. Bot.*, **7**, 687–704.
Widén, B. (1988). Population studies and conservation management of plants. *Svensk. Bot. Tidskr.*, **82**, 449–57 (in Swedish with English summary).
Wilcove, D.S., McLellan, C.H. & Dobson, A.P. (1986). Habitat fragmentation in the temperate zone. In *Conservation Biology*, ed. M.E. Soulé. Sinauer Associates, Sunderland, Massachusetts, pp. 77–104.
Williams, C.B. (1964). *Patterns in the Balance of Nature and Related Problems in Quantitative Ecology*. Academic Press, London.
Williamson, M.H. (1973). Species diversity in ecological communities. In *The Mathematical Theory of the Dynamics of Biological Populations*, ed. M.S. Bartlett & R.W. Horns. Academic Press, London, pp. 325–35.
Wolda, H. (1981). Similarity indices, sample size and diversity. *Oecologia (Berl.)*, **50**, 296–302.
Zahl, S. (1977). Jack-knifing an index of diversity. *Ecology*, **58**, 907–13.

Epilogue: Towards an Applied Ecological Theory?

LENNART HANSSON

Department of Wildlife Ecology, Swedish University of Agricultural Sciences, S-750 07 Uppsala, Sweden

This book opened with a survey of ecological principles for nature conservation derived partly from general ecological theory. Conservation problems were then surveyed for a large number of boreal and temperate environments, with an analysis of the applicability and utility of present ecological theory in relation to such problems. The authors were well aware of modern developments in ecological theory and had also prepared themselves in several seminars together with more theoretically inclined ecologists. Their experience suggests that present theory is fairly limited in its ability to solve pressing conservation problems.

More general theories, such as those relating to population genetics, island biogeography and minimum population sizes, have been given less emphasis than more recent theories such as that on landscape ecology and disturbance, or on demographic and dispersal strategies, which have partly been developed in close cooperation with conservationists. The use of basic ecological theory may be limited either because it is too abstract or too little understood by empiricists, or because it has little relevance to practical problems.

Conservation problems as encountered in natural environments are specific; theory will hardly ever directly solve practical ecological questions (Hall, 1988). In the present context, theory was used for its heuristic value, to supply ideas on new ways of approaching the problems. Some ecologists even believe that each ecological problem has a unique solution and that only specific descriptive and experimental research is warranted (e.g. Simberloff, 1986).

However, in other scientific fields there are parallel developments of pure and applied theory. In the physical sciences we find theoretical physics and technology and in biology theoretical physiology and medicine. So why should there not be a similar development in ecology? There may of course be some particular problems with applied ecology. In physics the units examined (elementary particles, molecules, crystals) are globally identical, in medicine each individual is genetically specific but there are global adaptive mechanisms tending to preserve a stable functioning of individual organisms. On the other hand, units in ecology are diverse and each has a peculiar history and beneficient adaptations are absent or dubious (to say the least) for the populations and communities of most interest in conservation. Basic ecological theory mainly considers adaptations of individuals towards physical and biotic environments. An applied ecological theory for the management of populations or communities can thus only weakly rely on evolution as a theoretical basis. Nevertheless, useful ideas may be developed from purely mechanistic reasoning (an example being the original island biogeographic theory) and as generalizations from empirical research. Such theories will certainly need both deductive and inductive elements (cf. Mentis, 1988).

In order to obtain a good heuristic background, and if possible also a predictive science, attempts should be made to develop general ecological theory as well as experience from successful management towards an integrated science of applied ecology. The alternative is not encouraging: each conservation problem will have to be attacked from scratch and there can be little doubt that we would eventually drown in an ocean of detail.

REFERENCES

Hall, C.A.S. (1988). An assessment of several of the historically most influential theoretical models in ecology and of the data provided in their support. *Ecol. Modelling*, **43**, 5–31.

Mentis, (1988). Hypothetico-deductive and inductive approaches in ecology. *Funct. Ecol.*, **2**, 5–14.

Simberloff, D. (1986). The contribution of population and community biology to conservation science. *Ann. Rev. Ecol., Syst.*, **19**, 473–511.

Index

DURHAM UNIVERSITY LIBRARY
3 0104 01231609 1